Wave Scattering Theory for Small Bodies of Arbitrary Shapes

Wave Scattering Theory for Small Bodies of Arbitrary Shapes

Alexander G. Ramm

Kansas State University, USA

World Scientific

NEW JERSEY • LONDON • SINGAPORE • BEIJING • SHANGHAI • HONG KONG • TAIPEI • CHENNAI • TOKYO

Published by

World Scientific Publishing Co. Pte. Ltd.

5 Toh Tuck Link, Singapore 596224

USA office: 27 Warren Street, Suite 401-402, Hackensack, NJ 07601

UK office: 57 Shelton Street, Covent Garden, London WC2H 9HE

British Library Cataloguing-in-Publication Data
A catalogue record for this book is available from the British Library.

WAVE SCATTERING THEORY FOR SMALL BODIES OF ARBITRARY SHAPES

ISBN 978-981-98-2348-2 (hardcover)
ISBN 978-981-98-2349-9 (ebook for institutions)
ISBN 978-981-98-2350-5 (ebook for individuals)

For any available supplementary material, please visit
https://www.worldscientific.com/worldscibooks/10.1142/14589#t=suppl

Desk Editor: Rhaimie B Wahap

Typeset by Stallion Press
Email: enquiries@stallionpress.com

To Luba and Olga

Preface

Wave scattering by bodies, small in comparison with the wavelength, is of interest in many applications: light scattering by cosmic and other dust, light scattering in colloidal solutions, scattering of waves in media with small inhomogeneities, such as holes in metal, for example, ultrasound mammography, ocean acoustics, etc. In 1871 Rayleigh started his classical work on wave scattering by small bodies. He understood that the main input in the far-zone field, scattered by a dielectric body, small in comparison with the wavelenghth λ, is made by the dipole radiation. However, he did not give methods for calculating this radiation for bodies of arbitrary shapes. The body is small if $ka < 0.1$, where $k = \frac{2\pi}{\lambda}$ is the wavenumber, and a is the characteristic dimension of the body. Practically in some cases one may consider the body small if $ka < 0.2$. Thomson (1893) understood that the main part of the far-zone field, scattered by a small perfectly conducting body, consists not only of the electric dipole radiation, but also of the magnetic dipole radiation which is of the same order of magnitude. Many papers and books dealing with the wave scattering from small bodies and its applications have been published since then.

However only in the author's works ([85], [112], [113]) have analytic formulas for calculation with arbitrary accuracy of the electric and magnetic polarizability tensors for bodies of arbitrary shapes been derived. These formulas allow one to calculate with the desired accuracy the dipole radiation from bodies of arbitrary shapes, and the electric and magnetic polarizability tensors for these bodies in terms of their geometries and material properties (dielectric permeability ϵ, magnetic permittivity μ, and conductivity σ) of the bodies. Using these formulas the author has derived analytic formulas for the S-matrix for acoustic and electromagnetic wave scattering by small bodies of arbitrary shapes. The author has obtained two-sided estimates for various functionals of practical interest in scattering theory, such as electrical capacitances of the conductors of arbitrary shapes and elements of the polarizability tensors of dielectric bodies of arbitrary shapes. These results allow the author to solve the inverse radiation problem.

Iterative methods for calculating static fields play an important role in the theory developed in this monograph. These methods are presented for interior and exterior boundary-value problems and for various boundary conditions. Boundary-value problems are reduced to boundary integral equations, and these equations are solved by means of iterative processes. There is a common feature of the static problems we study. Namely, these problems are reduced to solving Fredholm integral equations at the largest eigenvalue (smallest characteristic value, which is reciprocal to the eigenvalue) of the corresponding compact integral operator. The right-hand side of the equation is such that this equation is solvable. The largest eigenvalue is semisimple, that is, it is a simple pole of the resolvent of the corresponding compact operator. For this class of solvable operator equations at their largest eigenvalues the author had developed convergent iterative processes which allow one to solve the correponding equation stably with respect to small perturbations of the data. The above material is presented in Chapters 1–7, which are based on monograph [113].

The Fredholm alternative and a characterization of bounded and unbounded Fredholm operators of zero index are given in Chapter 8. The dependence on a parameter of the resolvents of analytic and meromorphic families of Fredholm operators is studied. Our presentation is based on works ([88], [81], [136]). This presentation is simple, short, and can be used in courses for graduate students.

Boundary-value problems for elliptic second-order equations are studied in rough domains, i.e., in domains with non-smooth boundaries, far less smooth than the Lipschitz boundaries ([103], [104], [30], [31], [107]). These results are presented in Chapter 9.

Low frequency asymptotics for solutions of exterior boundary-value problems are obtained (see [121], [133], [127], [142], [87], [90], [101]). These results are presented in Chapter 10.

The inverse problem of finding small subsurface inhomogeneities from the scattering data measured on the surface is discussed in ([105], [36]). These results are presented in Chapter 11.

The Modified Rayleigh Conjecture (MRC) is formulated and proved ([116]). An efficient numerical method for solving obstacle scattering problems is proposed and justified mathematically on the basis of MRC ([34], [118], [119], [125], [37]). Part of these results is presented in Chapter 12.

Methods, optimal with respect to accuracy, for calculating multiple integrals with weakly singular integrands are developed ([10]). These results are presented in the Appendix.

Most of the problems treated in the book are three-dimensional, because for two-dimensional problems the specific and often powerful tool of conformal mapping is available. The iterative methods have some advantages over grid methods and, to a certain extent, over variational methods:

(1) they give analytic approximate formulas for the field and for some functionals of the field of practical importance (such as capacitance and polarizability tensor),
(2) the formulas for the functionals can be used in a computer program for calculating these functionals for bodies of arbitrary shape,
(3) iterative methods are convenient to use on computers.

From a practical point of view, the above methods reduce solving the boundary-value problems to calculating some multiple integrals. Of special interest is the case of integrands with weak singularities. One of the main results of the book are analytical approximate formulas for scattering matrices for small bodies of arbitrary shapes. These formulas answer many practical questions, for example, how the scattering depends on the shape of the body or on the boundary conditions, how one calculates the effective field in a medium consisting of many small particles, and many other questions. In particular, these formulas allow one to solve the inverse radiation problem, which can be formulated as follows: If (E, H) is the field scattered by a small probe placed at the point x in an electromagnetic field (E_0, H_0), how does one calculate $(E_0(x), H_0(x))$ from knowledge of the scattered field (E, H)? This is an inverse problem of radiation theory or inverse radiomeasurements problem.

We also present two-sided variational estimates of capacitances and polarizability tensors. This book is based mostly on the author's papers and results. But the subject is classical and there have been many papers and books written on this subject. Some of them are cited in the bibliography, but the bibliography is incomplete.

Chapters 6, 8–10, 12, and Appendix A can be read independently of other chapters. Other chapters build on each other: in Chapter 7 results from Chapter 5 are used, in Chapters 1–3 the results from Chapter 6 are used, in Chapter 5 the results from Chapters 1–3 are used, and in Chapter 11 the results from Chapter 7 are used. The basic topic discussed in the Appendix A is the many-body wave scattering problem for small bodies. Based on the solution to this problem a method for creating materials with a desired refraction coefficient is formulated.

The many-body wave scattering problem the author solved asymptotically exactly under the assumption $a \ll d \ll \lambda$. Here a is the characteristic size of a small body, d is the minimal distance between neighboring bodies, and λ is the wavelength. Under this assumption the multiple scattering is essential.

The practical interest of the author's results is very high: the materials with a desired refraction coefficient (for example, materials with negative refraction and wave-focusing materials) are of obvious interest in practice. The author has published his results on creating materials with a desired refraction coefficient in several monographs and many papers cited in these monographs. Author's CV, list of publications and a brief description of scientific results of the author are included. The author has tried to make the presentation in this book essentially self-contained. The sign $\Box$ denotes the end of a proof.

Contents

Introduction

This book addresses largely three-dimensional problems. Scattering problems for bodies, small in comparison with the wavelength, are reduced to static problems. Complex variable methods (conformal mappings) for solving static two-dimensional problems have been widely discussed in the literature. The problems solvable in closed form are collected in [13], [33], [43], [58], [143], [71], [73]. The method of separation of variables has been used to solve the static problems for ellipsoids and its limiting forms (disks, needles), for a half-plane, wedge, plane with an elliptical aperture, hyperboloid of revolution, parabaloid of revolution, cone, thin spherical shell, spherical segment, two conducting spheres, and some other problems. Electrostatic fields in a flaky (layered) medium with parallel and sectorial boundaries have been studied [33], [143]. Some of the problems were solved in closed form using integral equations, e.g., the problems for a disk, spherical shell, plane with a circular hole, etc. Wiener-Hopf, dual, and singular integral equations were used [33], [143], [76], [164]. Electrostatic problems for a finite circular hollow cylinder (tube) were studied in [158] by numerical methods. The capacitance per unit length of the tube and the polarizability of the tube were calculated. The authors reduced the integral equation for the surface charge to an infinite system of linear algebraic equations and solved the truncated system on a computer. Their method depends heavily on the particular geometry of the problem and does not allow one to handle any local perturbations of the shape of the tube. In [68] the variational methods of Ritz, Trefftz, the Galerkin method, and the grid method are discussed in connection with the static problems. However, no specific properties of these problems are used. These methods are presented in a more general setting in [53], [66]. In practice, these methods are time-consuming, and variational methods in three-dimensional static problems probably have some advantages over the grid method. A vast literature exists on the calculation of the capacitances of perfect conductors [43], [79]. In [43] there is a reference section which gives the capacitances of conductors of certain shapes. In [78], [79] a systematic exposition of variational methods for estimation of the capacitances and other functionals of practical interest is given. In [153] there are some programs for calculating the two-dimensional static fields using integral equations method.

In [148] some geometrical properties of the lines of electrical field strength are used for approximate calculations of the field. This approach is empirical.

One of the objectives of this book is to present systematically the usage of integral equations for calculating static fields and some practically useful functionals of these fields, in particular, capacitances and polarizability tensors of bodies of arbitrary shape. The method gives approximate analytical formulas for calculations of these functionals with the desired accuracy. These formulas can be used to construct a computer program for calculating capacitances and polarizability tensors. The many-body problems are also discussed as well as the problems for flaky-homogeneous bodies, e.g., coated particles. Two-sided variational estimates of capacitances and polarizability tensors are given. The problems for open thin metallic screens are considered as well as those for perfect magnetic films. Calculating the magnetic polarizability of perfect magnetic films is important because such films are used as memory elements of computers. The above-mentioned formulas for capacitances and polarizhbility tensors allow one to give approximate analytical formulas for the scattering matrix in the problem of wave scattering by small bodies of arbitrary shape. This is done for scalar and electromagnetic waves. The dependence of the scattering matrix on the boundary conditions on the surface of the scatterer is investigated. The wave scattering in a medium consisting of many small particles is studied and equations for the effective (self-consistent) field in such a medium are derived. This makes it possible to discuss the inverse problem of determining the properties of such a medium from knowledge of the waves scattered by this medium.

The theory of wave scattering by small bodies was originated by Rayleigh (1871), who studied various aspects of this theory until his death in 1919. During the last century many papers were published in this field, but the first analytical approximate formulas for polarizability tensors and the scattering matrix were derived in [110], [84], [95], [132] and summarized in the monograph [143].

Here these and other results are presented systematically. The author hopes that these results can be used by engineers, physicists, and persons interested in atmospheric and ocean sciences, radiophysics, and colloidal chemistry. Radiowave scattering by rain and hail; light scattering by cosmic dust, muddy water, and colloidal solutions; methods of nondestructive control; ultrasound mammography; detection of mines in the water or ground; finding small holes in metals; and radiomeasurement techniques are just a few examples of many possible applications of the theory of wave scattering by small bodies of arbitrary shapes.

In addition to the theory of wave scattering by small bodies, the following topics are discussed:

(a) the Modified Rayleigh Conjecture (MRC) and its applications to solving obstacle scattering problems [116], [34], [118], [119], [125],
(b) a characterization of Fredholm operators with index zero and the singularities of the resolvent of analytic Fredholm operator-functions, [136], [88], [81],
(c) boundary-value and scattering problems in rough domains (less smooth than Lipschitz domains), [104], [31], [30],
(d) low-frequency behavior of solutions to operator equations and solutions to boundary-value problems, [87], [90], [108], [133],
(e) finding small subsurface inhomogeneities from scattering data, [105], [36], [107],
(f) wave scattering by many small bodies, [146], [113],
(g) optimal methods for calculation of weakly singular multidimensional integrals, [10].

The structure of the book is explained in the table of contents. A modest background in analysis is required from the reader. The book is essentially self-contained. There are new mathematical results in the book, but the book is addressed not only to mathematicians, but to a wide audience that applies mathematics. This audience includes numerical analysts, physicists, engineers, and graduate students. The book can be used in graduate courses for students in several areas of science, including integral equations and their applications, numerical mathematics, wave scattering, electrodynamics, and PDE.

This monograph is based mainly on the author's papers and some material from his earlier monographs [113], [133], [143], [120]. Chapter 9 is based on [144], [104], [31], and the presentation follows closely that in [31]. The Appendix is based on [10].

The author thanks Springer Verlag, Kluwer Academic/Springer and other publishers for permissions to use the material from his published papers and books.

Chapter 1

Basic Problems

1.1 Statement of Electrostatic Problems for Perfect Conductors

1. The basic equations of electrostatics are well known [58]:

$$\operatorname{curl} E = 0, \quad \operatorname{div} D = \rho, \quad D = \varepsilon E, \tag{1.1}$$

where E is the electric field, D is the induction, $\rho(x)$ is the charge distribution, and ε is the dielectric constant of the medium. If the medium is homogeneous and isotropic, then ε is constant; if it is isotropic but unhomogeneous, then $\varepsilon = \varepsilon(x)$, $x = (x_1, x_2, x_3)$. In the general case $\varepsilon = \varepsilon_{ij}(x)$, $1 \leq i,\ j \leq 3$, is a tensor. The boundary condition on the surface Γ of a conductor is of the form

$$E_t|_\Gamma = N \times E|_\Gamma = 0, \tag{1.2}$$

where N is the unit outer normal to Γ. If σ is the surface charge distribution then

$$D_N = (D, N) = \sigma. \tag{1.3}$$

The vectors E and D are to be finite and can have discontinuities only on the surfaces of discontinuity of $\varepsilon(x)$, i.e., on the surfaces which are the boundaries of domains with different electrical properties (interface surfaces). The boundary conditions on such surfaces are

$$E_{1t} = E_{2t}, \quad D_{1N} = D_{2N}, \tag{1.4}$$

where 1 and 2 stand for the first and second medium, respectively. A perfect conductor in electrostatics is a body with $\varepsilon = +\infty$. Let us define an insulator in electrostatics as a body with $\varepsilon = 0$, i.e., on its surface

$$D_N|_\Gamma = 0. \tag{1.5}$$

This definition is useful because a superconductor behaves in a magnetic field H like an insulator in the electric field $E = H$. Indeed, on the surface of the superconductor the boundary condition

$$B_N|_\Gamma = 0 \tag{1.6}$$

holds, where B is the magnetic induction [58].

2. Many problems of practical interest in quasistatic electrodynamics can be reduced to static problems.

For example, let a conductor Ω be placed in a harmonic electromagnetic field. Let the wave length λ of the field be much larger than the characteristic dimension a of Ω, $\lambda \gg a$. In practice $\lambda > 0.2a$ is often enough. If the depth δ of the skin layer is small, $\delta \ll a$, then the calculation of the field scattered by this body can be reduced to the static problem

$$\operatorname{div} B = 0, \quad \operatorname{curl} B = 0 \text{ in } \Omega_e, \tag{1.7}$$

$$B_N|_\Gamma = -B_{0N}|_\Gamma, \quad B(\infty) = 0. \tag{1.8}$$

Here Ω_e is the exterior of the domain Ω, B_0 is the magnetic induction at the location of Ω. One can assume that B_0 is constant since $a \ll \lambda$, i.e., the exterior field does not change significantly within the distance a. It is clear that the problem of (1.7)-(1.8) is equivalent (formally) to the problem of the insulator in the exterior electrostatic field $E_0 = B_0$.

It is worthwhile to mention that many problems of thermostatics, hydrodynamics, and elastostatics can be reduced to static problems similar to the above.

3. Let us formulate three basic problems of electrostatics.

Problem 1.1 A conductor is placed in a given electrostatic field. Find the charge distribution σ induced on its surface.

Problem 1.2 A conductor has total charge Q. Find the surface charge distribution σ.

Problem 1.3 A conductor is charged to a potential V. Find σ.

In these problems the conductor may be a single body or a system of bodies.

4. In most books on electrostatics the third boundary condition is not discussed. Nevertheless some practical problems (such as the calculation of

the resistance of linearly polarizable electrodes, and the calculation of the skin effect) can be reduced to the static boundary value problem with the third boundary condition.

5. Let us formulate the basic problems of electrostatics as problems of the potential theory. Let ε be a constant. Then from (1.1) it follows that

$$E = -\nabla\phi, \quad \triangle\,\phi = -\rho. \tag{1.9}$$

In the domain free of charge one has

$$\triangle\,\phi = 0. \tag{1.10}$$

If the given exterior field is $E_0 = -\nabla\phi_0$, then

$$\phi = \phi_0 + v, \tag{1.11}$$

and v satisfies (1.10). The boundary condition (1.2) takes the form

$$\phi|_\Gamma = \text{const}, \tag{1.12}$$

while (1.3) takes the form

$$-\epsilon\frac{\partial\phi}{\partial N}\bigg|_\Gamma = \sigma. \tag{1.13}$$

Problem 1.1 can be formulated as follows:

Find the solution ϕ of (1.10) of the form (1.11), subject to condition (1.12), such that

$$v(\infty) = 0 \text{ and } \int_\Gamma \frac{\partial\phi}{\partial N}ds = 0. \tag{1.14}$$

Condition (1.14) means that the total surface charge on the conductor is zero (the electroneutrality condition). Since $\int_\Gamma(\partial\phi_0/\partial N)ds = 0$, condition (1.14) implies:

$$\int_\Gamma \frac{\partial v}{\partial N}ds = 0. \tag{1.15}$$

Problem 1.2 can be formulated as follows:

Find the solution ϕ of (1.10) subject to (1.12) and such that

$$-\varepsilon\int_\Gamma \frac{\partial\phi}{\partial N}ds = Q, \quad \phi(\infty) = 0. \tag{1.16}$$

The constant in condition (1.12) should be found in the process of solving **Problem 1.1** and **Problem 1.2**. This constant is the potential of the conductor. It is known and easy to prove that **Problem 1.1** and **Problem 1.2** have unique solutions. Indeed, the corresponding homogeneous problems are $\triangle\ \phi = 0$ in Ω_e, $\phi|_\Gamma = 0$, $\phi(\infty) = 0$, and $\triangle\ \phi = 0$ in Ω_e, $-\varepsilon\frac{\partial\phi}{\partial N}|_\Gamma = 0$, $\phi(\infty) = 0$. The only solution to the first problem is $\phi = 0$ by the maximum principle, and the only solution to the second problem is $\phi = 0$ by the strong maximum principle ([29]).

6. If the conductor is a thin unclosed metallic screen, then the edge condition must be satisfied. Let F denote the screen and L denote its edge. Then the edge condition can be written as

$$|\phi(x)| \sim \{g(x)\}^{1/2}, \quad g(x) \equiv \min_{t\in L} |x - t|. \tag{1.17}$$

The function $g(x)$ is the distance from the point x to the edge. From (1.17) it follows that

$$|E| = |-\nabla\phi| \sim \{g(x)\}^{-1/2}, \quad \sigma(s) \sim \{g(s)\}^{-1/2}, \tag{1.18}$$

where $s \in F$. Condition (1.17) is easy to understand if one notes that the potential near the edge of the wedge behaves like $r^\nu \sin(\nu\theta)$, where (r, θ) are polar coordinates, $\nu = (2 - \theta_0\pi^{-1})^{-1}$, and θ_0 is the angle of the wedge. If $\theta_0 = 0$ (this is the case of the screen) then $\nu = 0.5$ and one obtains (1.17).

1.2 Statement of the Basic Problem for Dielectric Bodies

1. Let a dielectric body Ω with the dielectric constant ε_i be placed in a medium with the dielectric constant ε_e. A basic electrostatic problem is to find the electric field which occurs if one places the body in the given electrostatic field $E_0 = -\nabla\phi_0$. This problem can be formulated as

$$\triangle\ \phi = 0 \text{ in } \Omega \text{ and } \Omega_e, \tag{1.19}$$

$$\varepsilon_i\left(\frac{\partial\phi}{\partial N}\right)_i = \varepsilon_e\left(\frac{\partial\phi}{\partial N}\right)_e \text{ on } \Gamma, \tag{1.20}$$

$$\phi = \phi_0 + v, \quad v(\infty) = 0. \tag{1.21}$$

Here and below $(\partial\phi/\partial N)_{i(e)}$ are the limiting values on Γ of the normal derivatives in the interior (exterior) domains.

For v one has the problem

$$\triangle\ v = 0 \text{ in } \Omega \text{ and in} \Omega_e, \tag{1.22}$$

$$\varepsilon_i\left(\frac{\partial v}{\partial N}\right)_i = \varepsilon_e\left(\frac{\partial v}{\partial N}\right)_e + (\varepsilon_e - \varepsilon_i)\left(\frac{\partial \phi_0}{\partial N}\right) \text{ on } \Gamma, \quad v(\infty) = 0. \tag{1.23}$$

If the body Ω is inhomogeneous, then

$$\operatorname{div}\big(\varepsilon(x)\nabla\phi\big) = 0 \text{ in } \Omega. \tag{1.24}$$

2. Let us give an example of a practical problem which leads to a boundary value problem with the third boundary condition

$$\left(\frac{\partial \phi}{\partial N} + h\phi\right)\bigg|_\Gamma = f, \quad h = \text{const}. \tag{1.25}$$

Suppose that on the surface of a perfect conductor there is a thin film, e.g., an oxide film. Let ψ be the potential of the conductor and let ϕ be the potential of the exterior surface of the film. In electrochemistry it is assumed that $\phi - \psi$ is proportional to the current $j = -\gamma\nabla\phi$, where γ is the specific conductivity of the film. Therefore $\phi - \psi = -b\gamma(\partial\phi/\partial N)$, where the constant b is the coefficient of proportionality. This condition is clearly of the form (1.25) with $h = (b\gamma)^{-1}, f = -h\psi$. The same condition will appear in the problem with an impedance surface or with a surface covered by a thin dielectric film.

In electrochemistry the surfaces of the metallic electrodes are not equipotential because of the electrochemical polarizations. The potential of the electrodes depends on the normal component of the electric current. If this dependence is linear one gets condition (1.25).

1.3 Reduction of the Basic Problems to Fredholm's Integral Equations of the Second Kind

1. Let us state several formulas from potential theory which will be used below. Let

$$v(x) = \int_\Gamma \frac{\sigma(t)dt}{4\pi r_{xt}}, \quad w(x) = \int_\Gamma \frac{\partial}{\partial N_t}\frac{1}{4\pi r_{xt}}\mu(t)dt, \tag{1.26}$$

where $r_{xt} = |x - t|$ and N_t is the exterior (outer) unit normal to Γ at the point t. Then

$$\left(\frac{\partial v}{\partial N}\right)_{ie} = \frac{A\sigma \pm \sigma}{2}, \quad w_{ie} = \frac{A^*\mu \mp \mu}{2}, \tag{1.27}$$

where the upper (lower) sign corresponds to i (e), and

$$A\sigma = \int_\Gamma \sigma(t) \frac{\partial}{\partial N_s} \frac{1}{2\pi r_{st}} dt, \quad A^* \mu = \int_\Gamma \mu(t) \frac{\partial}{\partial N_t} \frac{1}{2\pi r_{st}} dt, \tag{1.28}$$

where Γ is the surface of Ω. Unless otherwise specified we assume that Γ is smooth.

In Chapter 9 a wide class of non-smooth (rough)boundaries is considered. This class includes Lioschitz boundaries as a proper subclass.

Note that

$$\triangle\, v = 0, \quad \triangle\, w = 0 \text{ in } \Omega \text{ and } \Omega_e. \tag{1.29}$$

Formulas (1.26)–(1.29) are well known [38]. For smooth surfaces the following formula ([38]) holds:

$$\left(\frac{\partial w}{\partial N}\right)_i = \left(\frac{\partial w}{\partial N}\right)_e. \tag{1.30}$$

The above properties of the potential hold if the densities σ and μ are continuous. If the densities are Hölder continuous, the derivatives of the potentials have additional smoothness properties, which we do not state because they will not be used. A function f is called Hölder continuous if for some constants $c > 0$ and α, $0 < \alpha \leq 1$,

$$\big|f(t) - f(s)\big| \leq c|t - s|^\alpha.$$

The potential theory is developed for Lipschitz surfaces([20], [157]).

2. In order to reduce **Problem 1.1** from Section 1.1 to Fredholm's integral equation, let us look for a solution of this problem of the form

$$\phi = \phi_0 + \int_\Gamma \frac{\sigma(t)dt}{4\pi\varepsilon_e r_{xt}}. \tag{1.31}$$

The unknown function $\sigma(t)$ has physical interpretation as the surface charge distribution. The function ϕ in (1.31) satisfies equation (1.10), condition (1.11), and the first condition in (1.14). Substitution of (1.31) into (1.13) with $\varepsilon = \varepsilon_e$, yields

$$\sigma = -A\sigma - 2\varepsilon_e \frac{\partial \phi_0}{\partial N}, \quad \int_\Gamma \sigma\, dt = 0. \tag{1.32}$$

The second equation is condition (1.15). If $\varepsilon \neq \varepsilon_e$ and the medium has dielectric constant ε, then

$$\sigma_\varepsilon = \frac{\varepsilon_e}{\varepsilon}\sigma \tag{1.33}$$

where σ_ε is the surface charge distribution in this new problem and σ is the solution of (1.32), i.e., the surface charge distribution in the original problem.

Exercise 1.1 Prove this statement.

It is known [38] that every solution of the equation $\sigma = -A\sigma$ is of the form $\sigma = \text{const}\,\omega(t)$, where $\omega(t) \geq 0$, $\int_\Gamma \omega(t)dt > 0$. The function $\omega(t)$ is the equilibrium charge distribution on the surface Γ of the conductor. Every solution of the adjoint equation $\mu = -A^*\mu$ is of the form $\mu = \text{const}$. From this and from Fredholm's alternative it follows that problem (1.32) has a unique solution. Existence is guaranteed since $\int_\Gamma (\partial\phi_0/\partial N)ds = 0$, while uniqueness follows from the second condition in (1.32).

3. Let us look for a solution of **Problem 1.2** of the form

$$\phi = \int_\Gamma \frac{\sigma(t)dt}{4\pi\varepsilon_e r_{st}}, \tag{1.34}$$

where

$$\sigma = -A\sigma, \quad \int_\Gamma \sigma\, dt = Q, \tag{1.35}$$

Problem (1.35) has the unique solution

$$\sigma = Q\omega(s), \tag{1.36}$$

where $\omega(s)$ is the solution of (1.35) corresponding to $Q = 1$ i.e., an equilibrium charge distribution of the total charge $Q = 1$ on the surface Γ of the conductor. It is easy to prove that every solution of the first equation (1.35) is a constant multiple of $\omega(t)$. Indeed, if ω_1 and ω_2 are two solutions to equation (1.35), then $\omega_1 - \lambda\omega_2$ solves this equation for any $\lambda = \text{const}$. Choose λ so that $\int_\Gamma (\omega_1 - \lambda\omega_2)ds = 0$. Then $\omega_1 = \lambda\omega_2$. Indeed, if σ solves (1.35), then $v(\sigma) = \text{const}$ in D, $v = const$ on Γ, and $\sigma = -\frac{\partial v}{\partial N_e} > 0$. Thus, if $\int_\Gamma \sigma dt = 0$, then $\sigma = 0$. Our argument proves that $\dim N(I + A) = 1$, where $N(B) := \{u : Bu = 0\}$ is the null-space of an operator B.

4. Let us now consider the interior and exterior problems

$$\triangle\,\phi = 0 \text{ in } \Omega, \quad \frac{\partial\phi}{\partial N} + h\phi|_\Gamma = f, \tag{1.37}$$

$$\triangle\,\phi = 0 \text{ in } \Omega_e, \quad \frac{\partial\phi}{\partial N} - hu|_\Gamma = f, \tag{1.38}$$

where $h = h_1 + ih_2$, $h_1 \geq 0, h_2 \leq 0$, $|h| > 0, h = \text{const}$. It is easy to prove that problems (1.37) and (1.38) have unique solutions. If one looks for a solution of the form $\phi = v$, where v is defined in (1.26), then the density σ of the potential v satisfies the equation

$$\sigma + T\sigma = -A\sigma + 2f \tag{1.39}$$

for problem (1.37), and

$$\sigma + T\sigma = A\sigma - 2f \tag{1.40}$$

for problem (1.38). Here A is defined in (1.28) and

$$T\sigma \equiv h \int_\Gamma \frac{\sigma\, dt}{4\pi r_{st}}. \tag{1.41}$$

For the Dirichlet problems

$$\triangle\, u = 0 \text{ in } \Omega, \quad u|_\Gamma = f, \tag{1.42}$$

$$\triangle\, u = 0 \text{ in } \Omega_e, \quad u|_\Gamma = f, \tag{1.43}$$

one looks for the solution of the form $u = w$, where w is defined in (1.26), and for μ obtains the equations

$$\mu = A^*\mu - 2f, \tag{1.44}$$

$$\mu = -A^*\mu + 2f, \tag{1.45}$$

respectively.

5. In order to reduce the basic problem of electrostatics for dielectric bodies to integral equations let us look for a solution of the form (1.31). Using (1.27) and the boundary condition (1.23) one obtains the equation

$$\sigma = -\gamma A\sigma - 2\gamma\varepsilon_e \frac{\partial \phi_0}{\partial N}, \quad \gamma = \frac{\varepsilon_i - \varepsilon_e}{\varepsilon_i + \varepsilon_e}, \tag{1.46}$$

where ε_i is the dielectric constant of the body. If $\varepsilon_i = \infty$ then $\gamma = 1$. This is the case of a perfect conductor and in this case (1.46) is identical to (1.32). If $\varepsilon_i = 0$ then $\gamma = -1$. This is the case of an insulator and in this case (1.46) becomes

$$\sigma = A\sigma + 2\varepsilon_e \frac{\partial \phi_0}{\partial N}. \tag{1.47}$$

6. If p conductors are placed in the exterior field $E_0 = -\nabla\phi$, then one looks for a potential of the form

$$\phi = \phi_0 + \sum_{j=1}^{p} \int_{\Gamma} \frac{\sigma_j(t)}{4\pi\varepsilon_e r_{xt}} dt. \tag{1.48}$$

From (1.48) and the boundary conditions

$$-\varepsilon_e \frac{\partial \phi}{\partial N}\bigg|_{\Gamma_m} = \sigma_m, \quad 1 \le m \le p, \tag{1.49}$$

one obtains the system of integral equations

$$\sigma_j(t_j) = \sum_{m=1, m\neq j}^{p} T_{jm}\sigma_m - A_j\sigma_j - 2\varepsilon_e \frac{\partial \phi_0}{\partial N}, \quad 1 \le j \le p, \tag{1.50}$$

where

$$T_{jm}\sigma_m = \int_{\Gamma_m} \frac{\partial}{\partial N_{t_j}} \frac{1}{2\pi r_{t_j t_m}} \sigma_m(t_m) dt_m, \tag{1.51}$$

$$A_j\sigma_j = \int_{\Gamma_j} \frac{\partial}{\partial N_{t_j}} \frac{1}{2\pi r_{t_j s_j}} \sigma_j(s_j) ds_j, \tag{1.52}$$

and the electroneutrality conditions should be satisfied

$$\int_{\Gamma_j} \sigma_j(t)dt = 0, \quad 1 \le j \le p. \tag{1.53}$$

7. If p dielectric bodies are placed in the exterior field $E_0 = -\nabla\phi_0$, then the potential is of the form (1.48) and from the boundary conditions

$$\varepsilon_j \left(\frac{\partial \phi}{\partial N}\right)_i = \varepsilon_e \left(\frac{\partial \phi}{\partial N}\right)_e \text{ on } \Gamma_j, \quad 1 \le j \le p, \tag{1.54}$$

one obtains the system of integral equations

$$\sigma_j(t_j) = -k_j \sum_{m=1, m\neq j}^{p} T_{jm}\sigma_m - k_j A_j\sigma_j - 2k_j\varepsilon_e \frac{\partial \phi_0}{\partial N_{I_j}}, \tag{1.55}$$

where

$$k_j := \frac{\varepsilon_j - \varepsilon_e}{\varepsilon_j + \varepsilon_e}, \quad 1 \le j \le p, \tag{1.56}$$

ε_j is the dielectric constant of the jth body, T_{jm} and A_j are defined in (1.51) and (1.52), and, unless for some j_0 one has $\varepsilon_{j_0} = \infty$, there are no additional conditions on σ_j. Otherwise one should impose the electroneutrality condition (1.53) on σ_{j_0}.

8. Let us consider a flaky-homogeneous (layered) body placed in the exterior field $E_0 = -\nabla\phi_0$. Taking again the potential of the form (1.48) and using the boundary conditions

$$\varepsilon_j \left(\frac{\partial \phi}{\partial N}\right)_i = \varepsilon_{j-1} \left(\frac{\partial \phi}{\partial N}\right)_e \quad \text{on } \Gamma_j, \tag{1.57}$$

one finds the system of integral equations

$$\sigma_j(t_j) = -\gamma_j \sum_{m=1, m\neq j}^{p} T_{jm}\sigma_m - \gamma_j A_j \sigma_j - 2\gamma_j \varepsilon_e \frac{\partial \phi_0}{\partial N_{t_j}}, \tag{1.58}$$

where

$$\gamma_j = \frac{\varepsilon_j - \varepsilon_{j-1}}{\varepsilon_j + \varepsilon_{j-1}} \tag{1.59}$$

and T_{jm}, A_j are defined in (1.51), (1.52).

1.4 Reduction of the Static Problems to Fredholm's Integral Equations of the First Kind

If the body is an open thin metallic screen it is not easy to reduce the static problems to a convenient Fredholm equation of the second kind, see e.g. [24]. Nevertheless it is easy to obtain Fredholm's integral equations of the first kind for the problem and to solve these equations by an iterative process.

Let us consider **Problem 1.1** from Section 1.1. Looking for a potential of the form (1.31), using boundary condition (1.12) and denoting the constant potential on the surface of the conductor by V one gets

$$\int_\Gamma \frac{\sigma(t)dt}{4\pi\varepsilon_e r_{st}} = V - \phi_0, \quad s \in \Gamma. \tag{1.60}$$

The constant V is to be found from the condition

$$\int_\Gamma \sigma(t)dt = 0. \tag{1.61}$$

Problem 1.2 from Section 1.1 leads in a similar way to the equation

$$\int_{\Gamma} \frac{\sigma(t)dt}{4\pi\varepsilon_e r_{st}} = V, \quad s \in \Gamma, \tag{1.62}$$

which is uniquely solvable if V is given. If the constant V is not given, but the total charge Q is given:

$$\int_{\Gamma} \sigma \, dt = Q, \tag{1.63}$$

and if $\eta(t)$ solves (1.62) with $V = 1$, then problem (1.62)-(1.63) has the solution

$$\sigma(t) = \frac{Q}{Q_1}\eta(t), \tag{1.64}$$

where

$$Q_1 = \int_{\Gamma} \eta(t)dt. \tag{1.65}$$

Problem 1.3 from Section 1.1 is equivalent to equation (1.62) without additional conditions.

Chapter 2

Iterative Processes for Solving Fredholm's Integral Equations for Static Problems

2.1 An Iterative Process for Solving the Problem of Equilibrium Charge Distribution and Charge Distribution on a Conductor Placed in an Exterior Static Field

1. In Section 1.3, **Problem 1.1** from Section 1.1 was reduced to problem (1.32). It is known [38] that the operator A in (1.32) is compact in $L^2(\Gamma)$ and in $C(\Gamma)$ provided that Γ is smooth (it is sufficient to assume that the equation of the surface in the local coordinates is $x_3 = f(x_1, x_2)$ and ∇f is Hölder continuous). It is also known [38] that $\lambda = -1$ is the smallest characteristic value of A which is simple. This means that $\lambda = -1$ is a simple pole of the resolvent $(A - \lambda I)^{-1}$ and the corresponding null space is one-dimensional, i.e., every solution of the equation $\sigma = -A\sigma$ is of the form $\sigma = \text{const}\, \omega(t)$, where $\omega(t)$ is the solution normalized by the condition $\int_\Gamma \omega\, dt = 1$. Let G_1 denote the null space of the operator $I + A^*$, where A^* is defined in (1.28). It is known [38] and can be verified directly that $\mu = 1$ is a solution of the equation $\mu = -A^*\mu$. By the Fredholm alternative G_1 is one dimensional. Let $G_1^\perp$ be the orthogonal complement to G_1 in $H = L^2(\Gamma)$. Then $G_1^\perp$ is the set of functions satisfying the condition $\int_\Gamma \sigma\, dt = 0$. If ϕ_0 is the electrostatic potential then $\int_\Gamma (\partial\phi_0/\partial N) dt = 0$. The theoretical basis for the iterative processes of this chapter is given in Chapter 6. In order to apply Theorem 6.1 from Section 6.1 one has to check that equation $\sigma = -A\sigma$ has only the trivial solution in $G_1^\perp$. Every solution to this equation is of the form $\sigma = c\omega(t), c = \text{const}, \int_\Gamma \omega(t) dt > 0$. Therefore $\int_\Gamma \sigma\, dt = 0$ implies that $c = 0$ and $\sigma = 0$. Theorem 6.1 and the above argument show that the following theorem holds.

Theorem 2.1 *Problem 1.1 in Section 1.1 has a unique solution σ, given* *by the iterative process*

$$\sigma_{n+1} = -A\sigma_n - 2\varepsilon_e \frac{\partial \phi_0}{\partial N}, \quad \sigma_0 = -2\varepsilon_e \frac{\partial \phi_0}{\partial N}, \quad \sigma = \lim_{n\to\infty} \sigma_n. \tag{2.1}$$

This process converges as a geometrical series with ratio q, $0 < q < 1$, where q depends only on the shape of Γ.

Remark 2.1 *If Γ is a sphere then $q = 1/3$. The number $q = |\lambda_1 \lambda_2^{-1}|$, according to Theorem 6.1. Here $\lambda_1, \lambda_2, \lambda_3, \dots$ are the characteristic values of A (i.e., $\phi_j = \lambda_j A \phi_j$ for some $\phi_j \neq 0$) numbered so that $|\lambda_1| < |\lambda_2| \leq |\lambda_3| \leq \cdots$. One can calculate λ_1 and λ_2, numerically using the methods given in [44], [53], and find q.*

2. Let us solve **Problem 1.2** by the iterative process given in Theorem 6.2. Problem 1.2 was reduced to problem (1.35). Its solution is of the form (1.36) and $\int_\Gamma \omega\, dt = 1$. Since $f = 1$ satisfies the condition $f_{G_1} \neq 0$ where f_{G_1} is the projection of f onto G_1 (note that G_1 is spanned by the function ω) one can use Theorem 6.2 from Section 6.1. This yields

Theorem 2.2 *Problem 1.2 has a unique solution σ, given by the iterative process*

$$\sigma_{n+1} = -A\sigma_n, \quad \sigma_0 = Q/S, \quad \sigma = \lim_{n\to\infty} \sigma_n, \tag{2.2}$$

where $s = \operatorname{meas}\Gamma$. The process converges at the rate given in Theorem 2.1.

Remark 2.2 *It is easily seen that*

$$\int_\Gamma \sigma_n dt = \int_\Gamma \sigma_{n-1} dt = \cdots = \int_\Gamma \sigma_0 dt = Q. \tag{2.3}$$

Indeed

$$-\int_\Gamma A\sigma\, dt = -\int_\Gamma \int_\Gamma \frac{\partial}{\partial N_t} \frac{1}{2\pi r_{st}} \sigma(s) ds\, dt = \int_\Gamma \sigma(s) \times$$
$$\left\{ \int_\Gamma - \frac{\partial}{\partial N_t} \frac{1}{2\pi r_{st}} dt \right\} ds = \int_\Gamma \sigma(s) ds$$

Here we have used the known [38] formula

$$-\int_\Gamma \frac{\partial}{\partial N_t} \frac{1}{2\pi r_{st}} dt = 1.$$

Relation (2.3) means that the iterative process (2.2) redistributes the fixed total charge on the surface, thus causing the surface change to approach the equilibrium distribution.

3. Suppose **Problem 1.2** is solved and $\omega(t)$ has been found. Then it is easy to solve **Problem 1.3**. Indeed, let V_0 be the potential of the conductor with $Q = 1$, i.e.,

$$\int_\Gamma \frac{\omega(t)}{4\pi\varepsilon_e r_{st}} = V_0.$$

Then the solution of **Problem 1.3** is

$$\sigma(t) = VV_0^{-1}\omega(t).$$

One can verify this directly.

Exercise 2.1 Do it.

2.2 An Iterative Process for Solving the Problem for Dielectric Bodies in an Exterior Static Field

1. The above problem is reduced in Section 1.3 to equation (1.46), where $-1 < \gamma < 1$, provided that $\varepsilon_i > 0$, $\varepsilon_e > 0$, $\varepsilon_i \neq 0$, and $\varepsilon_i \neq \infty$. It was already stated that all the characteristic values of A lie in the domain $|\lambda| \geq 1$. Therefore one can use Theorem 6.4 from Section 6.1. This Theorem implies the existence and uniqueness of solution of (1.46) and the convergence of the iterative process

$$\sigma_{n+1} = -\gamma A\sigma_n - 2\gamma\varepsilon_e\frac{\partial\phi_0}{\partial N}, \quad \sigma_0 = \sigma_0; \quad \sigma = \lim_{n\to\infty}\sigma_n, \tag{2.4}$$

where $\sigma_0 \in L^2(\Gamma)$ is arbitrary, to the solution of (1.46). The rate of convergence is that of the geometrical series with ratio q, $0 < q < |\gamma|^{-1}$. If $\sigma_0 = -2\gamma\varepsilon_e(\partial\phi_0/\partial N)$ then process (2.4) converges for $-1 \leq \gamma \leq 1$ and $q \leq |\lambda_2|^{-1}$, where λ_2 is the second characteristic value of A.

2. Suppose a flaky-homogeneous body described in Section 1.3 is placed in the exterior static field with the potential ϕ_0. The system of integral equations for this problem is (1.58).

Theorem 2.3 *The system* (1.58) *has a unique solution given by the iterative process*

$$\begin{aligned}\sigma_j &= \lim_{n\to\infty}\sigma_j^{(n)},\\ \sigma_j^{(n+1)}(t_j) &= -\gamma_j\Sigma_{m=1,m\neq j}^{p}T_{jm}\sigma_m^{(n)} - \gamma_j A_j\sigma_j^{(n)} - 2\gamma_j\varepsilon_e\frac{\partial\phi_0}{\partial N_{t_j}},\end{aligned} \tag{2.5}$$

$$\sigma_j^{(0)} = -2\gamma_j \varepsilon_e \frac{\partial \phi_0}{\partial N_{t_j}}, \quad 1 \le j \le p, \tag{2.6}$$

which converges as a geometrical series with ratio q, $0 < q < 1$, where q depends only on the shapes of Γ_j.

Proof. Let us write (1.58) as

$$\sigma = -B\sigma + f, \tag{2.7}$$

where

$$\sigma = \big(\sigma_1, \ldots, \sigma_p\big), \quad f = \left(-2\varepsilon_e \gamma_1 \frac{\partial \phi_0}{\partial N_{t_1}}, \ldots, -2\varepsilon_e \gamma_p \frac{\partial \phi_0}{\partial N_{t_p}} \right),$$

and B is the matrix operator of the form

$$B = \begin{pmatrix} \gamma_1 A_1 \ \gamma_1 T_{12} \cdots \gamma_1 T_{1p} \\ \cdots\cdots\cdots\cdots\cdots\cdots \\ \gamma_p T_{p1} \ \gamma_p T_{p2} \cdots \gamma_p A_p \end{pmatrix}. \tag{2.8}$$

This operator acts in the space $H = L^2(\Gamma)$ of vector-valued functions with inner product

$$(\sigma, \omega) = \sum_{j=1}^{p} \int_{\Gamma_j} \sigma_j(t)\omega_j(t)dt. \tag{2.9}$$

In order to prove Theorem 2.1 it is sufficient to show that the equation

$$\sigma = -\lambda B\sigma \tag{2.10}$$

has only trivial solution for $|\lambda| \le 1$ (see Theorem 6.4 from Section 6.1). Suppose $|\lambda| \le 1$ and σ is a nontrivial solution of (2.10). Let us rewrite (2.10) as

$$\sigma_j = -\lambda\gamma_j \left(A_j \sigma_j + \sum_{m=1, m\neq j}^{p} T_{jm}\sigma_m \right). \tag{2.11}$$

If

$$v = \sum_{j=1}^{p} \int_{\Gamma_j} \frac{\sigma_j dt}{4\pi\varepsilon_e r_{xt}},$$

then

$$\left(\frac{\partial v}{\partial N_i} - \frac{\partial v}{\partial N_e} \right)\bigg|_{\Gamma_j} = -\lambda\gamma_j \left(\frac{\partial v}{\partial N_i} + \frac{\partial v}{\partial N_e} \right)\bigg|_{\Gamma_j}, \quad 1 \le j \le p,$$

and

$$(1+\lambda\gamma_j)\frac{\partial v}{\partial N_i} = (1-\lambda\gamma_j)\frac{\partial v}{\partial N_e} \text{ on } \Gamma_j, \quad 1 \le j \le p, \tag{2.12}$$

Let D_0 be the exterior domain with boundary Γ_1, D_p be the interior domain with boundary Γ_p, and D_j be the domain with boundary $\Gamma_j \cup \Gamma_{j+1}$. Let a_j, $1 \le j \le p$, be arbitrary constants. Consider the identity

$$\sum_{j=0}^{p} a_j \int_{D_j} |\nabla v|^2 dx = \sum_{j=1}^{p} \int_{\Gamma_j} \bar{v}_j \left(a_j \frac{\partial v}{\partial N_i} - a_{j-1}\frac{\partial v}{\partial N_e}\right) ds. \tag{2.13}$$

From (2.12) and (2.13) it follows that

$$\sum_{j=0}^{p} a_j \int_{D_j} |\nabla v|^2 dx = \sum_{j=1}^{p} \int_{\Gamma_j} \bar{v} \left(a_j - a_{j-1}\frac{1+\lambda\gamma_j}{1-\lambda\gamma_j}\right) \frac{\partial v}{\partial N_i} ds. \tag{2.14}$$

If $|\gamma_j| < 1$ and $|\lambda| \le 1$ then $|\lambda\gamma_j| < 1$. Let us set

$$a_0 = \varepsilon_e, \quad a_j = a_{j-1}\frac{1+\lambda\gamma_j}{1-\lambda\gamma_j}, \quad 1 \le j \le p.$$

Then (2.14) shows that $v \equiv 0$ and therefore $\sigma = 0$, i.e., $\sigma_j = 0$, $1 \le j \le p$. If $|\lambda| = 1$, $\lambda \ne 1$ then $\lambda\gamma_j \ne 1$ and the same argument shows that $\sigma = 0$. If $\lambda = 1$ and $\lambda_{j_0} = 1$ then $\varepsilon_{j_0} = \infty$ and $v|_{\Gamma_j} = \text{const}$. In this case one is interested in the potential in the domain exterior to Γ_{j_0} and derives an identity similar to (2.14), namely

$$\begin{aligned}
&\sum_{j=0}^{j_0-1} a_j \int_{D_j} |\nabla v|^2 dx \\
&= \sum_{j=1}^{j_0-1} \int_{\Gamma_j} \bar{v}\left(a_j \frac{\partial v}{\partial N_i} - a_{j-1}\frac{\partial v}{\partial N_e}\right) ds + \int_{\Gamma_{j_0}} \bar{v} a_{j_0-1}\frac{\partial v}{\partial N_e} ds \\
&= \sum_{j=1}^{j_0-1} \int_{\Gamma_j} \bar{v}\left(a_j - a_{j-1}\frac{1+\lambda\gamma_j}{1-\lambda\gamma_j}\frac{\partial v}{\partial N_i}\right) ds - a_{j_0-1}\int_{\Gamma_{j_0}} \bar{v}\frac{\partial v}{\partial N_e} ds.
\end{aligned} \tag{2.15}$$

Because of the electroneutrality condition

$$\int_{\Gamma_{j_0}} \frac{\partial v}{\partial N_e} ds = 0, \tag{2.16}$$

and the boundary condition on the surface of the perfect conductor $v|_{\Gamma_{j_0}} =$ const, the last integral in (2.15) vanishes. Therefore it follows from (2.15) that $\sigma \equiv 0$ provided that (2.16) holds. Note that (2.16) is equivalent to the equality

$$\int_{\Gamma_{j_0}} \sigma_{j_0} ds = 0. \tag{2.17}$$

Let us prove that $\lambda = -1$ is a semisimple characteristic value of the operator B. This will be important for construction of iterative methods of solution of equation (2.7) (cf. Theorem 6.1 in Chapter 6). A characteristic number λ is called semisimple if the equation $\sigma = \lambda B\sigma$ has nontrivial solutions and the equation $u = \lambda Bu + \sigma$ has no solution for any nonzero σ which is a solution of $\sigma = \lambda B\sigma$.

In Chapter 6 it is proved that if B is compact then λ is semisimple if and only if it is a simple pole of the resolvent $(I - zB)^{-1}$.

Suppose that

$$\sigma = -B\sigma, \quad \sigma \neq 0, \quad u = -Bu + \sigma. \tag{2.18}$$

Let $\int_{\Gamma_j} \sigma_j dt = Q_j$, $\int_{\Gamma_j} u_j dt = q_j$. Note that [38]

$$\int_\Gamma \frac{\partial}{\partial N_t} \frac{1}{2\pi r_{xt}} dt = \begin{cases} 0, & x \notin D, \\ -1, & x \in \Gamma, \\ -2, & x \in D, \end{cases} \tag{2.19}$$

where D is a bounded domain with a smooth boundary Γ. Integrating (2.18) over Γ yields

$$q_j = \gamma_j q_j + 2\gamma_j \sum_{m>j} q_m + Q_j, \quad j = 1, 2, \ldots, j_0, \tag{2.20}$$

because

$$\begin{aligned} \int_{\Gamma_j} dt\, T_{jm}\sigma_m &= \int_{\Gamma_j} dt \int_{\Gamma_m} \frac{\partial}{\partial N_t} \frac{1}{2\pi r_{ts}} \sigma_m(s) ds \\ &= \int_{\Gamma_m} ds\, \sigma_m(s) \int_{\Gamma_j} dt \frac{\partial}{\partial N_t} \frac{1}{2\pi r_{ts}} = q_m \begin{cases} 0, & m < j \\ -2, & m > j \end{cases}. \end{aligned} \tag{2.21}$$

Therefore (2.20) is a linear system with an upper triangular matrix. We have already showed that if $\int_\Gamma \sigma_{j_0} dt = Q_{j_0} = 0$ then $\sigma \equiv 0$. Since we assume that $\sigma \not\equiv 0$ we have $Q_{j_0} \neq 0$. Since $\gamma_{j_0} = 1$ the last equation in (2.20) reads $q_{j_0} = q_{j_0} + Q_{j_0}$. Thus $Q_{j_0} = 0$ and $\sigma \equiv 0$. This contradicts the assumption that $\sigma \not\equiv 0$. Therefore $\lambda = -1$ is a semisimple characteristic value of B.

The statement of Theorem 2.3 follows now from Theorem 6.1. Note that we need this theorem only in the case in which $\varepsilon_{j_0} = \infty$ because in this case -1 is the characteristic value of B. If each ε_j is finite then the operator B has no characteristic values in the unit disk $|\lambda| \le 1$ and the iterative process (2.5) converges for any initial approximation, not necessarily satisfying the condition

$$\int_\Gamma f\,dt = 0. \tag{2.22}$$

This condition is satisfied by the initial approximation (2.6). □

3. Let us consider an iterative process for solving the many-body problem in the exterior static field.

In Section 1.3 this problem was reduced to system (1.55) in the case of dielectric bodies and to system (1.50) and conditions (1.53) for the case of perfect conductors. Since the case of perfect conductors can be treated as an instance of dielectric bodies with $\varepsilon_j = \infty$, let us consider system (1.55) and rewrite it as an operator equation

$$\sigma = -\tilde{B}\sigma + f, \tag{2.23}$$

where

$$\tilde{B}_{jm} = k_j T_{jm}(1 - \delta_{jm}) + k_j \delta_{jm} A_j, \quad f_j = -2k_j \varepsilon_e \frac{\partial \phi_0}{\partial N_{t_j}}, \tag{2.24}$$

and k_j is defined in (1.56).

Theorem 2.4 *If $|k_j| < 1$, $1 \le j \le p$, then equation (2.23) has a unique solution σ for any $f \in H = L^2(\Gamma)$, given by the iterative process*

$$\sigma_{n+1} = -\widetilde{B}\sigma_n + f, \quad \sigma = \lim_{n\to\infty} \sigma_n, \tag{2.25}$$

where $\sigma_0 \in H$ is arbitrary. Process (2.25) converges no more slowly than a convergent geometrical series. If $k_j = 1$ for some j then equation (2.23) has a solution for any $f \in H$ such that

$$\int_\Gamma f\,ds = 0. \tag{2.26}$$

This solution satisfies the condition

$$\int_\Gamma \sigma\,dt = 0. \tag{2.27}$$

There is only one solution of equation (2.23) *with* f *satisfying* (2.26) *in the class of functions* $\sigma \in H$ *satisfying* (2.27)*. This solution can be found by the iterative process* (2.25) *where* σ_0 *satisfies condition* (2.27)*, e.g.,* $\sigma_0 = f$*. The process converges at least as fast as a convergent geometrical series.*

A proof of Theorem 2.4 is similar to the proof of Theorem 2.3 and can be left to the reader as an exercise.

2.3 A Stable Iterative Process for Finding the Equilibrium Charge Distribution

The iterative process for solution of this problem is given in Theorem 2.2. However this process is unstable in the following sense. Consider the process with perturbations

$$\sigma_{n+1} = -A\sigma_n + \varepsilon_n, \quad \|\varepsilon_n\| \le \varepsilon. \tag{2.28}$$

Since -1 is a characteristic value of A the operator $(I+A)^{-1}$ is not defined everywhere in H and the process (2.28) can diverge. For example if $\varepsilon_n = f$, $\|f\| < \varepsilon$, $\int_\Gamma f\,ds > 0$, and $\sigma_0 = f$, then process (2.28) diverges. Indeed, in this case $\sigma_n = \sum_{m=0}^{n}(-1)^m A^m f$. The Neumann series $\sum_{m=0}^{\infty}(-1)^m A^m f$ does not converge for the elements $f \in N(I+A)$, where $N(I+A)$ is the null space of the operator $I+A$.

We have already seen that $\sigma \in N(I+A)$ has the property $\int_\Gamma \sigma\,dt \neq 0$. Therefore every f such that $\int_\Gamma f\,dt \neq 0$ can be represented as $f = c\sigma + f_1$, where $c = \text{const} \neq 0$ and $\int_\Gamma f_1 dt = 0$. Since -1 is a semisimple characteristic value the operator $(I+A)^{-1}$ is defined at f_1 and is not defined at σ. Hence $(I+A)^{-1}$ is not defined at f and σ_n does not converge as $n \to \infty$. One can verify this by a direct calculation using the identity

$$-\int_\Gamma A\sigma\,dt = -\int_\Gamma \sigma\,dt \tag{2.29}$$

which is valid for any $\sigma \in H$ (see Remark 2.2 in Section 2.1). Integrating σ_n over Γ yields $\int_\Gamma \sigma_n dt = q(n+1)$, where $q = \int_\Gamma f\,dt \neq 0$. Therefore $\int_\Gamma \sigma_n dt \to \infty$ and σ_n does not converge in H. This simple argument gives the rate of divergence of the process (2.28).

This motivates the problem of constructing a stable iterative process for solving the problem (1.35). Let $Q = 1$ in (1.35), $S = \text{meas}\,\Gamma$, $\phi = S^{-1}$,

$\omega = \phi + h$. Then from the equation $\omega = -A\omega$ it follows that

$$h = -Ah + F, \quad F = -\phi - A\phi, \quad \int_\Gamma \phi\, dt = 1. \tag{2.30}$$

Note that from (2.29) it follows that

$$\int_\Gamma F\, dt = 0. \tag{2.31}$$

The following theorem gives a stable iterative process for solution of (2.30). This theorem is a particular case of the abstract Theorem 6.2.

Theorem 2.5 *The iterative process*

$$h_{n+1} = -Ah_n - \frac{1}{S}\int_\Gamma h_n dt + F, \quad h_0 = F, \tag{2.32}$$

where F *is defined in* (2.30), *converges in* $H = L^2(\Gamma)$ *no more slowly than a convergent geometrical series to an element* h, *and* $\omega = h + S^{-1}$ *is the unique solution of the problem* (1.35) *for* $Q = 1$. *Furthermore, the process* (2.32) *is stable: i.e., if*

$$g_{n+1} = -Ag_n - \frac{1}{S}\int_\Gamma g_n dt + F + \varepsilon_n, \quad h_0 = F, \quad |\varepsilon_n| \le \varepsilon, \tag{2.33}$$

then

$$\limsup_{n\to\infty} \|g_n - h\| = O(\varepsilon). \tag{2.34}$$

Remark 2.3 *The process* (2.32) *converges in* $C(\Gamma)$ *if* Γ *is smooth.*

2.4 An Iterative Process for Calculating the Equilibrium Charge Distribution on the Surface of a Screen

1. The basic equation (see Section 1.4) is

$$\int_\Gamma \frac{\eta(t)dt}{4\pi\varepsilon_e r_{st}} = 1. \tag{2.35}$$

Here Γ can be the surface of a metallic body or the surface of a metallic screen (an infinitely thin body). First consider the case of the solid body. Let

$$a(t) = \left\{\int_\Gamma \left(4\pi\varepsilon_e r_{st}\right)^{-1} ds\right\}^{-1}. \tag{2.36}$$

From the abstract results given in Section 6.4 one gets the following theorem:

Theorem 2.6 *Let $\eta_n = a(t)\psi_n$, where $a(t)$ is defined in* (2.36),

$$\psi_{n+1} = \big(I + A_1\big)\psi_n + 1, \quad \psi_0 = 1 \tag{2.37}$$

$$A_1\psi = \int_\Gamma \big(4\pi\varepsilon_e r_{st}\big)^{-1} a(t)\psi(t)dt. \tag{2.38}$$

Then ψ_n converges in $H = L^2(\Gamma)$, and $\lim_{n\to\infty}\eta_n = \eta$ is the solution of equation (2.35).

Consider now the case in which Γ is the surface of a metallic screen. Let G be the edge of Γ,

$$h(t) = g^{-1/2}(t), \tag{2.39}$$

where $g(x)$ is defined in (1.17), and let

$$a_1(t) = h(t)\left\{\int_\Gamma \frac{h(s)ds}{4\pi\varepsilon_e r_{st}}\right\}^{-1} \tag{2.40}$$

Let $H_- = L^2(\Gamma; a_1^{-1}(t))$, where $L^2(\Gamma; p)$ is the L^2 space with the norm $\|f\|^2 = \int_\Gamma |f|^2 p\,dt$.

Theorem 2.7 *If $a(t)$ is replaced by $a_1(t)$ in Theorem 2.6, then the sequence η_n constructed in Theorem 2.6 converges in H_- to the solution of equation* (2.35).

2. Consider problem (1.60)-(1.61). If η solves (2.35), then $V\eta$ solves equation (1.60) with $\phi_0 = 0$. Let τ solve the equation

$$\int_\Gamma \frac{\tau(s)ds}{4\pi\varepsilon_e r_{st}} = \phi_0. \tag{2.41}$$

This equation can be solved by the iterative processes given in Theorems 2.6 and 2.7. The constant V can be found from condition (1.61):

$$V = \int_\Gamma \tau(t)dt\bigg(\int_\Gamma \eta(t)dt\bigg)^{-1}. \tag{2.42}$$

Let us summarize the above as a theorem.

Theorem 2.8 *The solution of problem* (1.60)-(1.61) *can be obtained by the formulas:*

$$\sigma = \lim_{n\to\infty} \sigma_n, \quad \sigma_n = V_n \eta_n = \tau_n \tag{2.43}$$

where η_n *is defined in Theorem 2.6 for the case of the volume conductor and in Theorem 2.7 for the case of the metallic screen,* τ_n *is defined by means of the iterative processes given in Theorems 2.6 and 2.7:*

$$\tau_{n+1} = (I + A_1)\tau_n + \phi_0, \quad \tau_0 = \phi_0, \tag{2.44}$$

$$V_n = \int_\Gamma \tau_n(t)dt \left(\int_\Gamma \eta_n dt \right)^{-1}. \tag{2.45}$$

Remark 2.4 *It can be proved (see, e.g.,* [143, *Appendix 10*] *or* [133]*) that the operator* $Tf = \int_\Gamma \frac{f(t)dt}{4\pi\varepsilon_e r_{st}}$ *maps* $H_q(\Gamma)$ *onto* $H_{q+1}(\Gamma)$*, where* $H_q = W_2^q(\Gamma)$, $-\infty < q < \infty$*, is the Hilbert scale of Sobolev spaces and* $\Gamma \in C^\infty$ *is a compact closed surface. The operator* T *is a pseudodifferential elliptic operator of order* -1 *(see* [56], [1]*).*

Chapter 3

Calculating Electric Capacitance

3.1 Capacitance of Solid Conductors and Screens

1. Suppose that the total charge of a conductor is Q and its potential is V. Then

$$Q = CV \tag{3.1}$$

and the coefficient C is called the capacitance of the conductor. If $\sigma(t)$ is the surface charge distribution, then

$$\int_\Gamma \frac{\sigma(t)dt}{4\pi\varepsilon_e r_{st}} = V, \quad s \in \Gamma, \tag{3.2}$$

and

$$\int_\Gamma \sigma \, dt = Q. \tag{3.3}$$

Thus

$$C = \int_\Gamma \sigma \, dt \left(\int_\Gamma \frac{\sigma \, dt}{4\pi\varepsilon_e r_{st}} \right)^{-1}. \tag{3.4}$$

The function $\sigma(t)$ can be calculated by the iterative processes given in Section 2.3 and Section 2.4. If σ_n is an approximation to σ then the potential

$$\int_\Gamma \frac{\sigma_n dt}{4\pi\varepsilon_e r_{st}} = V_n(s) \tag{3.5}$$

is not constant on Γ. In this case we introduce the averaged potential

$$V_n = S^{-1} \int_\Gamma V_n(s)ds, \quad S := \operatorname{meas}\Gamma. \tag{3.6}$$

If $\sigma_n \to \sigma$ in $H = L^2(\Gamma)$ then $V_n \to V$ and

$$C^{(n)} = Q_n/V_n = \int_\Gamma \sigma_n dt \left(\frac{1}{S} \int_\Gamma ds \int_\Gamma \frac{\sigma_n(t)dt}{4\pi\varepsilon_e r_{st}} \right)^{-1}. \tag{3.7}$$

is an approximation to C. The iterative process (2.2) satisfies condition (2.3),

$$\int_\Gamma \sigma_n dt = Q, \quad n = 1, 2, \ldots. \tag{3.8}$$

In this case (3.7) can be written as

$$C^{(n)} = 4\pi\,\varepsilon_e S^2 \left(\int_\Gamma \int_\Gamma r_{st}^{-1} \delta_n(t) dt\, ds \right)^{-1}, \tag{3.9}$$

where δ_n is the n-th approximation to the solution of the problem

$$\delta = -A\delta, \quad \int_\Gamma \delta(s) ds = S, \tag{3.10}$$

and A is defined as usual (see (1.28)). One can construct δ_n by means of the iterative process

$$\delta_{n+1} = -A\delta_n, \quad \delta_0 = 1. \tag{3.11}$$

Theorem 2.2 and formula (3.9) imply the following theorem.

Theorem 3.1 *Let*

$$C^{(n)} = 4\pi\,\varepsilon_e S^2 \Bigg\{ \left(-\frac{1}{2\pi} \right)^n \int_\Gamma \int_\Gamma \frac{dt\, ds}{r_{st}} \underbrace{\int_\Gamma \cdots \int_\Gamma}_{n} \psi(t, t_1) \cdots \psi(t_{n-1}, t_n) dt_1 \cdots dt_n \Bigg\}^{-1}, \tag{3.12}$$

where $S = \operatorname{meas}\Gamma$ *and*

$$\psi(t, s) = \frac{\partial}{\partial N_t} \frac{1}{r_{ts}}. \tag{3.13}$$

Then

$$\left| C - C^{(n)} \right| \le cq^n, \tag{3.14}$$

where $c > 0$ *and* $0 < q < 1$ *depend on the shape of the conductor but do not depend on* n. *The following inequality holds:*

$$4\pi\,\varepsilon_e S^2 J^{-1} = C^{(0)} \le C, \tag{3.15}$$

where

$$J = \int_\Gamma \int_\Gamma r_{st}^{-1} dt\, ds. \tag{3.16}$$

Proof. The first statement of Theorem 3.1 follows from Theorem 2.2, and the second statement will be proved in Section 3.3. □

Remark 3.1 *The following empirical method is used for calculating capacitances: they assumed that the surface charge distribution of the total charge Q is uniform, i.e., $\sigma = QS^{-1}$, calculated the averaged potential*

$$V = S^{-1} \int_\Gamma ds \int_\Gamma \frac{QS^{-1}dt}{4\pi\varepsilon_e r_{st}}$$

and found an approximation to C by the formula:

$$C \approx QV^{-1} = 4\pi\varepsilon_e S^2 J^{-1}. \tag{3.17}$$

This is the zeroth approximation (3.12). *Theorem 3.1 gives additional information: first, the inequality* (3.15), *which says that the zeroth approximation is a lower bound for C, and second, the way to compute C with any desired accuracy by using the n-th approximation. Therefore Theorem 3.1 gives a justification of the empirical rule described above.*

Remark 3.2 *One can use the iterative process given in Section 2.4 to calculate electrical capacitances of conductors. Let η be the solution of equation* (3.2) *with $V = 1$ and η_n be the approximation of the $n-$th order to η. Then $V_n \approx 1$ for large n and formula* (3.7) *takes the form*

$$C_n \approx Q_n = \int_\Gamma \eta_n dt. \tag{3.18}$$

The subscript n in (3.18) *indicates that C_n in* (3.18) *differs from $C^{(n)}$ in* (3.12).

2. If the conductor is a thin metallic screen one can use formula (3.18). The empirical method described in Remark 3.1, i.e., formula (3.17), is not very accurate for screens. For example if the screen is a circular disk the error in calculating the capacitance by formula (3.17) is 7.5%.

3.2 Variational Principles and Two-Sided Estimates of Capacitance

1. Variational principles for capacitances have been widely discussed in the literature. The well-known book [79] should be mentioned first. A reference book [43] on electrical capacitances is a collection of numerical results and formulas for calculating of capacitance. Our purpose is to give some methods for deriving two-sided estimates for capacitance. Some of the results seem to be new, e.g., a necessary and sufficient condition for the Schwinger stationary principle to be extremal and estimates of the capacitance of a conductor placed in inhomogeneous dielectric medium.

2. We start with the following theorem.

Theorem 3.2 *Let A be a symmetric linear operator in a Hilbert space H with domain of definition $D(A)$. The equality*

$$(Af, f) = \max_{\phi \in D(A)} \frac{|(Af, \phi)|^2}{(A\phi, \phi)} \tag{3.19}$$

holds if and only if $A \geq 0$, i.e., $(A\phi, \phi) \geq 0$ for all $\phi \in D(A)$. By definition, $|(Af, \phi)|^2/(A\phi, \phi) = 0$ if $(A\phi, \phi) = 0$.

Remark 3.3 *Let $Af = g$. In many physical problems (some examples will be given later) the quantity (f, g) has physical significance. J. Schwinger (see, e.g., [39]) used the stationary representation of this quantity*

$$(f, g) = st_{\phi \in D(A)} \frac{|(g, \phi)|^2}{(A\phi, \phi)}, \tag{3.20}$$

where st is the sign of the stationary value. In practice it is important to know when this representation is extremal. Theorem 3.2 answers this question and provides a tool for deriving lower bounds for (Af, f).

Remark 3.4 *For the equality*

$$(Af, f) = \min_{\phi \in D(A)} \frac{|(Af, \phi)|^2}{(A\phi, \phi)} \tag{3.21}$$

to hold it is necessary and sufficient that $A \leq 0$.

Proof of Theorem 3.2. If $A \geq 0$ then $|(Af, \phi)|^2 \leq (Af, f)(A\phi, \phi)$ for all $f, \phi \in D(A)$. This is just the Cauchy inequality for the nonnegative bilinear form $[f, \phi] = (Af, \phi)$. Hence $(Af, f) \geq |(Af, \phi)|^2/(A\phi, \phi)$ and equality holds for $\phi = \lambda f$, $\lambda =$ const. Thus (3.19) follows and *the sufficiency part is proved.*

If $A \le 0$ then $-A \ge 0$ and

$$(-Af, f) = \max_{\phi \in D(A)} \frac{|(-Af, \phi)|^2}{(-A\phi, \phi)}. \tag{3.22}$$

Since $\max(-x) = -\min x$, where x is a real variable, one can see that (3.22) is equivalent to (3.21).

Let us prove the necessity of the condition $A \ge 0$. Suppose that $(A\psi, \psi) < 0$ and $(A\omega, \omega) > 0$. Let $\phi = \omega + \lambda\psi$, where λ is a real number, and (3.19) holds. Then

$$(Af, f) \ge \frac{|(Af, \omega)|^2 + 2\lambda Re(Af, \omega)(\psi, Af) + \lambda^2 |(Af, \psi)|^2}{(A\omega, \omega) + 2\lambda Re(A\psi, \omega) + \lambda^2 (A\psi, \psi)}. \tag{3.23}$$

Since $(A\omega, \omega)(A\psi, \psi) < 0$, the denominator of this fraction has two real zeros. Because the fraction is bounded from above, the numerator has the same roots as the denominator. The product of these roots for the denominator in (3.23) is equal to $\frac{(A\omega,\omega)}{(A\psi,\psi)} < 0$. Therefore, the product of these roots for the numerator is negative, and one gets the inequality:

$$\frac{|(Af, \omega)|^2}{|(Af, \psi)|^2} < 0, \tag{3.24}$$

which is a contradiction. Therefore $A \ge 0$ or $A \le 0$. The case $A \le 0$ is impossible. Indeed, (3.19) implies that $(Af, f) \ge |(Af, \phi)|^2/(A\phi, \phi)$, and, if $(A\phi, \phi) < 0$, one gets $(Af, f)(A\phi, \phi) \le |(Af, \phi)|^2$. Thus

$$(-Af, f)(-A\phi, \phi) \le |(-Af, \phi)|^2, \tag{3.25}$$

which contradicts the Cauchy inequality for the *nonnegative* operator $-A$. Therefore $A \ge 0$. *The necessity part is proved.* □

Remark 3.5 *Let $A = A^*$. Then*

$$(Af_i, f_j) = st \frac{(Af_i, \phi_j)(\phi_i, Af_j)}{(A\phi_i, \phi_j)}. \tag{3.26}$$

If $A \ge 0$ then for $i = j$ one can replace st by max in (3.26).

3. It is now easy to derive some lower bounds for capacitance. Let Γ be the surface of a perfect conductor which is charged to the potential $V = 1$. If σ is the surface charge distribution, then

$$A\sigma \equiv \int_\Gamma \frac{\sigma(t)dt}{4\pi\varepsilon_e r_{st}} = 1, \tag{3.27}$$

and

$$C = \int_\Gamma \sigma \, dt. \tag{3.28}$$

Since the integral operator A in (3.27) is selfadjoint and positive on $H = L^2(\Gamma)$, Theorem 3.2 says that

$$C = \max\left\{ \left(\int_\Gamma \sigma(t)dt\right)^2 \left(\int_\Gamma\int_\Gamma \frac{\sigma(t)\sigma(s)ds\,dt}{4\pi\varepsilon_e r_{st}}\right)^{-1} \right\}, \tag{3.29}$$

where the maximum is taken over all $\sigma \in C(\Gamma)$ if Γ is a smooth closed surface. From (3.29) the well-known principle of Gauss [79] follows immediately:

$$C^{-1} = \min_\sigma \left(Q^{-2} \int_\Gamma \sigma(t)u(t)dt \right). \tag{3.30}$$

This principle says that if the total charge Q is distributed on the surface Γ with the density $\sigma(t)$ and $u(t)$ is the potential of this charge distribution on Γ, then the minimal value of the right-hand side of (3.30) is C^{-1} and this minimal value is attained by the equilibrium charge distribution (i.e., by the solution of (3.27)).

From (3.29) it is easy to obtain some lower bounds for C. For example, if $\sigma = 1$ then (compare with (3.15))

$$C \geq C^{(0)} \equiv \frac{4\pi\varepsilon_e S^2}{J}, \quad S = \operatorname{meas}\Gamma, \quad J = \int_\Gamma\int_\Gamma \frac{ds\,dt}{r_{st}}. \tag{3.31}$$

One can take

$$\sigma_m = \sum_{j=1}^{m} c_j \phi_j, \tag{3.32}$$

where $\{\phi_j\}$ is a linearly independent system of functions in H and c_j are constants which are to be determined from the condition that the right-hand side of (3.29) is maximal. Then σ_m is an approximation to the equilibrium charge distribution and the value of the right-hand side of (3.29) is an approximation to C.

4. Let us formulate two classical variational principles for capacitances: the Dirichlet and Thomson principles [79]. The Dirichlet principle gives an upper bound for C. The Thomson principle is equivalent to the Gauss principle. Therefore combining the Dirichlet principle and (3.29) one can obtain two-sided estimates for C.

The Thomson principle is:

$$C^{-1} = \min \int_{D_e} \varepsilon_e |E|^2 dx, \tag{3.33}$$

where D_e is the exterior of the domain with boundary Γ, and the minimum is taken over the set of vector fields satisfying the conditions

$$\operatorname{div} E = 0, \quad \int_\Gamma (N, \varepsilon_e E) dt = 1, \tag{3.34}$$

where N is the outer unit normal to Γ at the point t. The minimum in (3.33) is attained at the vector $E = -\nabla u$, where

$$\triangle\, u = 0 \text{ in } D_e, \quad u|_\Gamma = \text{const}, \quad u(\infty) = 0, \quad -\varepsilon_e \int_\Gamma \frac{\partial u}{\partial N} dt = 1. \tag{3.35}$$

The Dirichlet principle is:

$$C = \min \int_{D_e} \varepsilon_e |\nabla u|^2 dx, \tag{3.36}$$

where the minimum is taken over the set of functions $u \in C^1(D_e)$ such that

$$u|_\Gamma = 1, \quad u(\infty) = 0. \tag{3.37}$$

This minimum is attained at the function u which is the solution to the problem

$$\triangle\, u = 0 \text{ in } D_e, \quad u|_\Gamma = 1, \quad u(\infty) = 0. \tag{3.38}$$

Both principles are particular cases of the principles formulated and proved in the next section.

5. If Γ is the surface of a screen the admissible functions in the variational principles should satisfy the edge condition: if L is the edge of Γ then

$$u \sim \{g(x)\}^{1/2}, \quad \sigma \sim \{g^{-1/2}(x)\}, \quad g(x) \equiv \min_{t \in L} |x - t|. \tag{3.39}$$

3.3 Capacitance of Conductors in an Anisotropic and Nonhomogeneous Medium

Let $\varepsilon = \varepsilon_{ij}(x)$ be the tensor (a positive definite matrix) of dielectric permettivity of the medium and let D be a perfect conductor with a smooth boundary Γ. The problem of finding the capacitance of this conductor placed in an inhomogeneous anistropic medium is of interest in many practical cases. For example, suppose a metallic body is placed partially in water. If the characteristic dimension of the conductor is small in comparison with the wavelength in the medium with large dielectric constant the capacitance determines the scattering amplitude. We assume for simplicity that $\varepsilon_{ij}(x) \in C^1(D_e)$ and $\varepsilon_{ij}(x) = \varepsilon_{ij}$ does not depend on x for sufficiently large x. This assumption guarantees that the basic results about existence of solutions to static problems are the same as for the Laplace operator corresponding to homogeneous medium. The variational principles 1 and 2, formulated below, are analogous to the classical Dirichlet and Thomson principles:

Principle 1

$$C = \min \int_{D_e} (\varepsilon \nabla u, \nabla u) dx, \tag{3.40}$$

where the minimum is taken over the C^1–functions $u(x)$ such that

$$u|_\Gamma = 1, \quad u(\infty) = 0. \tag{3.41}$$

In the statement of these principles the usual notations

$$(a, b) = \sum_{j=1}^{3} a_j b_j, \quad (\varepsilon a)_i = \sum_{j=1}^{3} \varepsilon_{ij}(x) a_j \tag{3.42}$$

are used.

Principle 2

$$C^{-1} = \min \int_{D_e} (\varepsilon E, E) dx, \tag{3.43}$$

where the minimum is taken over the set of vector fields satisfying the conditions

$$\operatorname{div}(\varepsilon E) = 0 \text{ in } D_e, \quad \int_\Gamma (N, \varepsilon E) dt = 1. \tag{3.44}$$

Proof of Principle 1. Assume that

$$\operatorname{div}(\varepsilon \nabla u) = 0 \text{ in } D_e \tag{3.45}$$

and that (3.41) is valid. The Euler equation for the functional in (3.40) is (3.45). Therefore (3.45) and (3.41) are necessary conditions for the function which solves (3.40), (3.41). The solution of (3.45) and (3.41) exists and is unique. Let us show that the functional in (3.40) attains its minimum at this solution and this minimum is equal to the capacitance C. Let $\eta \in C^1(D_e)$ satisfy the conditions

$$\eta|_\Gamma = 0, \quad \eta(\infty) = 0. \tag{3.46}$$

Then

$$\begin{aligned}
&\int_{D_e} (\varepsilon\nabla u + \varepsilon\nabla\eta, \nabla u + \nabla\eta)dx \\
&= \int_{D_e} (\varepsilon\nabla u, \nabla u)dx + \int_{E_e} (\varepsilon\nabla\eta, \nabla\eta)dx + 2Re\int_{D_e} (\varepsilon\nabla u, \nabla\eta)dx \\
&\geq \int_{D_e} (\varepsilon\nabla u, \nabla u)dx.
\end{aligned} \tag{3.47}$$

One can assume that u and η are real-valued functions, and then the sign Re can be dropped in the above equation.

We took into consideration that the matrix ε is positive definite and

$$\int_{D_e} (\varepsilon\nabla u, \nabla\eta)dx = -\int_\Gamma (N, \eta\varepsilon\nabla u)ds - \int_{D_e} \eta \,\mathrm{div}(\varepsilon\nabla u)dx = 0. \tag{3.48}$$

Furthermore,

$$\int_{D_e} (\varepsilon\nabla u, \nabla u)dx = -\int_\Gamma (N, \varepsilon\nabla u)u\, dt = \int_\Gamma (D, N)dt = Q, \tag{3.49}$$

where D is the electrical induction. Therefore the minimum in (3.40) is equal to the capacitance C if u is the solution to problem (3.41), (3.45). □

Proof of Principle 2. From (3.49) it lollows that the right-hand side of (3.43) ss equal to C^{-1} if $E = -A\nabla u$, where u is the solution to (3.41), (3.45) and the constant A is defined as

$$A = \left\{ -\int_\Gamma (N, \varepsilon\nabla u)dt \right\}^{-1} = Q^{-1}. \tag{3.50}$$

In (3.50) one has $Q = C$, because $Q = Cu$, and $u = 1$ on Γ, see (3.41).

Let us show that any other E satisfying (3.49) gives a larger value to functional (3.46). Indeed, let E be as above, and let $E + h$ solve (2.44).

Then

$$\begin{aligned}
&\int_{D_e} (\varepsilon E + \varepsilon h, E + h)dx \\
&\quad = \int_{D_e} (\varepsilon E, E)dx + \int_{D_e} (\varepsilon h, h)dx + 2Re\int_{D_e} (\varepsilon E, h)dx \qquad (3.51)\\
&\quad \geq \int_{D_e} (\varepsilon E, E)dx.
\end{aligned}$$

Here the following identity was used:

$$\begin{aligned}
\int_{D_e} (\varepsilon E, h)dx &= -A\int_{D_e} (\nabla u, \varepsilon h)dx \\
&= A\int_{D_e} u \operatorname{div}(\varepsilon h)dx + A\int_{\Gamma} u(N, \varepsilon h)dt = 0.
\end{aligned} \qquad (3.52)$$

We have proved *Principle 2.* □

Remark 3.6 *If*

$$\varepsilon_{ij} = \delta_{ij} = \begin{cases} 1, & i = j \\ 0, & i \neq j \end{cases},$$

then principles 1 and 2 are the Dirichlet and Thomson principles.

Remark 3.7 *Principles 1 and 2 give estimates of the capacitance from above and from below.*

Example 3.1 Let us take

$$E = -A\varepsilon^{-1}\nabla u, \qquad (3.53)$$

where ε^{-1} is the inverse matrix of ε, u is an arbitrary harmonic function in D_e (i.e., $\triangle\, u = 0$ in D_e), and

$$A^{-1} = -\int_{\Gamma} \frac{\partial u}{\partial N} dt. \qquad (3.54)$$

Then condition (3.44) is satisfied. Let

$$u(x) = \frac{1}{S}\int_{\Gamma} \frac{dt}{2\pi r_{xt}}, \qquad S = \operatorname{meas}\Gamma. \qquad (3.55)$$

Then the constant A, defined in (3.54), is equal to 1. Therefore it follows from (3.43) that

$$C \geq 4\pi^2 S^2 \left\{ \int_{D_e} (\varepsilon^{-1}\nabla v, \nabla v) dx \right\}^{-1}, \tag{3.56}$$

where

$$v(x) \equiv \int_\Gamma r_{xt}^{-1} dt. \tag{3.57}$$

If $\varepsilon_{ij}(x) = \varepsilon_e \delta_{ij}$, where $\varepsilon_e = const$, i.e., the medium is isotropic and homogeneous, then (3.56) and Green's formula imply that

$$C \geq 4\pi^2 S^2 \varepsilon_e \left\{ -\int_\Gamma ds \left(\int_\Gamma \frac{dt}{r_{st}} \frac{\partial}{\partial N_s} \int_\Gamma \frac{dt}{r_{st}} \right) \right\}^{-1}. \tag{3.58}$$

Example 3.2 Let $\varepsilon_{ij}(x) = \varepsilon(x)\delta_{ij}$,

$$u(x) = |x|^{-1}, \quad E = \frac{Ax}{|x|^3 \varepsilon(x)}, \tag{3.59}$$

where

$$A = \left\{ \int_\Gamma \frac{(t, N)}{|t|^3} dt \right\}^{-1} = \frac{1}{4\pi}. \tag{3.60}$$

From (3.43) it follows that

$$C \geq 16\pi^2 \left\{ \int_{D_e} \frac{dx}{|x|^4 \varepsilon(x)} \right\}^{-1}. \tag{3.61}$$

In particular if $D_e = \{x : |x| \geq a\}$ and $\varepsilon(x) = \varepsilon(|x|) = \varepsilon(r)$, then

$$C \geq 16\pi^2 \left\{ 4\pi \int_a^\infty dr \, \frac{1}{r^2 \varepsilon(r)} \right\}^{-1} = 4\pi \left\{ \int_a^\infty \frac{dr}{r^2 \varepsilon(r)} \right\}. \tag{3.62}$$

Actually, in this case C is equal to the right-hand side of (3.62) because (3.59) is the electrostatic field corresponding to the equilibrium charge distribution on the sphere $r = a$ if $\varepsilon(x) = \varepsilon(r)$.

Example 3.3 Let all of the space be divided into n parts bounded by conical surfaces. Suppose that the jth cone cuts the solid angle ω_j on the unit sphere and the vertices of the cones are in the center of a metallic ball with radius a. Let the dielectric constant of the jth cone be $\varepsilon_0\varepsilon_j(r)$. Then (3.61) says that

$$C \geq 16\pi^2\varepsilon_0 \left\{ \sum_{j=1}^{n} \omega_j \int_a^\infty \frac{dr}{r^2\varepsilon_j(r)} \right\}^{-1} . \tag{3.63}$$

In particular, if $\omega_1 = \omega_2 = 2\pi$ then

$$C \geq 8\pi\varepsilon_0 \left\{ \int_a^\infty \frac{dr}{r^2\varepsilon_1(r)} + \int_a^\infty \frac{dr}{r^2\varepsilon_2(r)} \right\}^{-1} . \tag{3.64}$$

This example covers the case of the ball halfway immersed in the water.

It is clear from the above examples that *Principle 2* is easy to use in practice: the only difficulty is of computational nature. In application of principle 1 there is an additional difficulty of finding a set of functions which satisfy condition (3.41). If the surface Γ is a coordinate surface in some known coordinate system it is easy to find such functions and Principle 1 gives upper bounds on C. A more general situation is discussed in Example 3.5 below.

Example 3.4 Let us take Example 3.3 and substitute $u = a/r$ in (3.40). This yields

$$C \leq \varepsilon_0 \sum_{j=1}^{n} \omega_j a^2 \int_a^\infty r^{-2}\varepsilon_j(r) dr . \tag{3.65}$$

In particular, if $\omega_1 = \omega_2 = 2\pi$ one obtains

$$\begin{aligned} &8\pi\varepsilon_0 \left\{ \int_a^\infty r^{-2}\big[\varepsilon_1(r) + \varepsilon_2(r)\big] dr \right\}^{-1} \\ &\quad \leq C \leq 2\pi\varepsilon_0 a^2 \int_a^\infty r^{-2} \cdot \big[\varepsilon_1(r) + \varepsilon_2(r)\big] dr, \end{aligned} \tag{3.66}$$

from (3.64) and (3.65). For $\varepsilon_1(r) = \varepsilon_2(r) = 1$, estimate (3.66) gives the exact value of C. One can improve the estimates taking more complicated admissible functions.

Example 3.5 Suppose that $r = F(\theta, \phi)$ is the equation of the surface of the conductor. Set $u = F(\theta, \phi)/|x|$ in (3.40). Then condition (3.41) holds and (3.40) yields the following upper bound on C:

$$\begin{aligned} C &\le \varepsilon_e \int_{S^2} d\phi d\theta \sin\theta \int_{F(\theta,\phi)}^{\infty} dr r^2 \left(\frac{|F|^2}{r^4} + \frac{|F_\theta|^2}{r^4} + \frac{|F_\phi|^2}{r^4 \sin^2\theta} \right) \\ &= \varepsilon_e \int_{S^2} \frac{d\phi d\theta \sin\theta}{F(\theta,\phi)} \left(|F|^2 + |F_\theta|^2 + \sin^{-2}\theta |F_\phi|^2 \right), \quad \varepsilon_e = \text{const}\,. \end{aligned} \tag{3.67}$$

This formula is useful if the integral on its right-hand side converges.

3.4 Physical Analogues of Capacitance

In heat transfer theory, in quasistatic electrodynamics, and in other areas of applied science, mathematical formulation of the problems can be reduced to the solution of the Laplace equation. Therefore in these areas of physics there exist some quantities analogous to the capacitance.

For example heat conductance in a homogeneous medium can be defined as

$$G_T = \frac{k}{\varepsilon} C, \tag{3.68}$$

where k is the coefficient of thermal conductivity, ε is the dielectric constant, C is the electrical capacitance of the conductor, and G_T is the heat conductance of the body with the same shape as the shape of the conductor.

If G_M is the magnetic conductance and μ is the magnetic constant then

$$G_M = \frac{\mu}{\varepsilon} C. \tag{3.69}$$

If G is the electric conductance and γ is the coefficient of electrical conductivity, then

$$G = \frac{\gamma C}{\varepsilon}. \tag{3.70}$$

3.5 Calculating the Potential Coefficients

1. Let n conductors be placed in a homogeneous medium with the dielectric constant $\varepsilon = 1$. Let Γ_j be the surface of the jth conductor. Because the equations of electrostatics are linear there is a linear dependence between

the potentials V_j of the conductors and their total charges Q_j,

$$Q_j = \sum_{i=1}^{n} C_{ij} V_j, \quad 1 \le i \le n. \tag{3.71}$$

The coefficients C_{ij}, $i \neq j$ are called the electrical inductance coefficients and the coefficients C_{jj} are called the capacitance coefficients.

The quadratic form

$$U = \frac{1}{2} \sum_{i,j=1}^{n} C_{ij} V_j V_i \tag{3.72}$$

is the energy of the electrostatic field. Therefore this form is positive definite. It is well known that this is the case if and only if all the principal minors of the matrix C_{ij} are positive (Sylvester's criterion). In particular

$$C_{jj} > 0, \quad C_{jj} C_{ii} > C_{ij}^2, \quad \det\left(C_{ij}\right) > 0, \tag{3.73}$$

and

$$C_{ij} = C_{ji}, \quad 1 \le i, j \le n \tag{3.74}$$

since the matrix C_{ij} is real-valued. We can rewrite (3.71) as

$$V_i = \sum_{j=1}^{n} C_{ij}^{(-1)} Q_j, \quad 1 \le i \le n. \tag{3.75}$$

The coefficients $C_{ij}^{(-1)}$ are called *the potential coefficients*. The following inequalities hold

$$C_{jj}^{(-1)} > 0, \quad C_{ij}^{(-1)} > 0; \quad C_{ij} < 0. \tag{3.76}$$

The first inequality in (3.76) holds because $C_{ij}^{(-1)}$ is a positive definite matrix if C_{ij} is. In order to prove the last inequality in (3.76) let us take $V_m = 0$ if $m \neq j$ and $V_j = 1$, then formula (3.71) shows that $Q_i = C_{ij}$. Therefore we must show that $Q_i < 0$. But $Q_i = -\varepsilon_e \int_{\Gamma_i} (\partial u / \partial N) ds$. Thus it is sufficient to prove that $(\partial u/\partial N)|_{\Gamma_i} \ge 0$. Here u is the electrostatic potential generated by the jth conductor, provided that the other conductors have zero potentials. The function u is a harmonic function (i.e., $\Delta\, u = 0$) and $u(\infty) = 0$, $u|_{\Gamma_j} = 1$. Since u is harmonic it cannot have extremal points inside the domain of definition. Therefore $0 < u < 1$ between the conductors.

Since $u|_{\Gamma_i} = 0$ according to our assumption, it is clear that $(\partial u/\partial N)|_{\Gamma_i} \geq 0$, and the last inequality in (3.76) is proved. The second inequality in (3.76) can be proved similarly.

2. The problem of determining the equilibrium charge distribution on the surfaces of a system of conductors can be reduced to the following system of integral equations (see (2.23), where k_j should be replaced by 1 and $f = 0$):

$$\sigma = -\widetilde{B}\sigma,$$
$$(\tilde{B}\sigma)_j = \sum_{m\neq j, m=1}^{n} T_{jm}\sigma_m + A_j\sigma_j, \quad 1 \leq j \leq n, \quad \sigma = \big(\sigma_1, \ldots, \sigma_n\big), \tag{3.77}$$

$$\int_{\Gamma_j} \sigma_j dt = Q_j, \quad 1 \leq j \leq n. \tag{3.78}$$

Here Q_j is the total charge of the jth conductor. (See Section 2.2 and Section 2.3.)

Theorem 3.3 *The solution to problem* (3.77)-(3.78) *exists, is unique, and can be found by the iterative process*

$$\sigma^{(k+1)} = -\tilde{B}\sigma^{(k)}, \quad \sigma_j^{(0)} = Q_j S_j^{-1}, \quad 1 \leq j \leq n, \quad S_j = \operatorname{meas}\Gamma_j. \tag{3.79}$$

This theorem follows from Theorem 6.2.

Let us derive some approximate formulas for the potential coefficients. Taking $Q_j = \delta_{jm}$ in (3.75) yields

$$C_{im}^{(-1)} = V_i. \tag{3.80}$$

Let us substitute in the system of integral equations

$$\sum_{j=1}^{n} \int_{\Gamma_j} \frac{\sigma_j(t)dt}{4\pi\varepsilon_e r_{tt_i}} = V_i, \quad 1 \leq i \leq n, \tag{3.81}$$

$\sigma_j^{(0)} = Q_j S_j^{-1}\delta_{jm}$ instead of $\sigma_j(t)$. Taking into account (3.75) one obtains

$$C_{im}^{(-1)} \approx \frac{1}{4\pi\varepsilon_e S_m} \int_{\Gamma_m} \frac{dt}{r_{tt_i}}, \quad 1 \leq i \leq n. \tag{3.82}$$

The right-hand side of this formula is not constant on Γ_i because $\sigma_j^{(0)}$ is not the exact solution to (3.81). Therefore we take as an approximation to

$C_{im}^{(-1)}$ the average of the right-hand side of (3.82). This yields

$$C_{im}^{(-1)} \approx \tilde{C}_{im}^{(-1)} \equiv \frac{1}{4\pi\varepsilon_e S_m S_i} \int_{\Gamma_i} \int_{\Gamma_m} \frac{ds\,dt}{r_{st}}, \quad 1 \le i, m \le n. \tag{3.83}$$

One can improve formula (3.83) by using the higher order approximations to σ_i say $\sigma^{(k)}$ defined in (3.79). In order to find some approximation to C_{ij} one can invert the matrix $C_{ij}^{(-1)}$, using the approximate values of $C_{ij}^{(-1)}$ given above.

3. Let us derive variational principles for the potential coefficients. To do so we take the potential energy of the electrostatic field

$$U = \frac{1}{2} \sum_{i,j=1}^{n} C_{ij}^{(-1)} Q_i Q_j \tag{3.84}$$

and set $Q_i = \delta_{im}$. This yields

$$2U = C_{mm}^{(-1)}. \tag{3.85}$$

Among various surface charge distributions such that

$$\int_{\Gamma_i} \sigma_i(t) dt = \delta_{im}, \quad 1 \le i \le n, \tag{3.86}$$

the distribution, corresponding to the actual electrostatic field, minimizes U. Thus

$$C_{mm}^{(-1)} = \min \sum_{i,j=1}^{n} \int_{\Gamma_i} \int_{\Gamma_j} \frac{\sigma_i(t)\sigma_j(s) ds\,dt}{4\pi\varepsilon_e r_{st}}, \tag{3.87}$$

where the minimum is taken over the set of σ_j satisfying condition (3.86).

In order to derive a variational principle for C_{mm} we take $V_i = \delta_{im}$ in (3.72). This yields

$$2U = C_{mm}. \tag{3.88}$$

The potential energy U of the electrostatic field with the potential $u(x)$ can be written as

$$U = \frac{1}{2} \int_{D_e} \varepsilon_e |\nabla u|^2 dx, \tag{3.89}$$

where D_e is the domain outside of the conductors. Let u satisfy the conditions

$$u|_{\Gamma_m} = 1, \quad u|_{\Gamma_i} = 0, \quad i \ne m, \quad u(\infty) = 0, \quad u \in C^1(D_e). \tag{3.90}$$

Then

$$C_{mm} = \min \int_{D_e} \varepsilon_e |\nabla u|^2 dx, \tag{3.91}$$

where the minimum is taken over the set of functions u satisfying condition (3.90).

Let $m \neq j$ and assume

$$\int_{\Gamma_i} \sigma_i dt = \delta_{ij} + \delta_{im}, \quad 1 \leq i \leq n. \tag{3.92}$$

From (3.84) and (3.92) it follows that

$$2U = C_{mm}^{(-1)} + 2C_{mj}^{(-1)} + C_{jj}^{(-1)}. \tag{3.93}$$

Therefore

$$C_{mm}^{(-1)} + 2C_{mj}^{(-1)} + C_{jj}^{(-1)} = \min \sum_{i,k=1}^{n} \int_{\Gamma_i} \int_{\Gamma_i} \frac{\sigma_i(t)\sigma_k(s) ds\, dt}{4\pi\varepsilon_e r_{st}}, \tag{3.94}$$

where the minimum is taken over the set of functions σ_i satisfying condition (3.92).

If $C_{jj}^{(-1)}$, $1 \leq i \leq n$, are already calculated, then one can calculate $C_{mj}^{(-1)}$ from (3.94).

Chapter 4

Numerical Examples

4.1 Introduction

Given in Sections 2 and 3 algorithms for calculating electrostatic fields and linear functionals of these fields, such as electrical capacitances, were reduced in these sections to calculating certain multiple integrals. From the point of view of numerical analysis one should integrate functions with at worst weak singularities. The numerical integration of such functions is a problem of independent interest. It has been discussed in detail for functions of one variable [21], [55], [54], but less is known about calculating multidimensional integrals of functions with weak singularities. The basic idea in the one-dimensional case is to integrate explicitly the singular part of the integer and thus to reduce the problem to the integration of a smooth function. This problem is well understood.

In the multidimensional case the first step in the above program was not discussed sufficiently.

In [10] optimal methods for calculating multidimensional integrals with weakly singular integrands are developed. These methods are presented in the Appendix.

In this chapter two problems of practical interest will be solved. First, the capacitances of circular metallic cylinders are tabulated. Secondly, the capacitances of metallic parallelepipeds of arbitrary dimensions are tabulated. In both cases there are no closed-form analytical solutions to the corresponding electrostatic problems, and the results are new. Special cases of these results, such as the capacitance of a cube, disk, or very long cylinder, will be compared with previously published results. The numerical results show that the formulas for calculating the capacitances, which have been derived in Section 3, are quite efficient.

4.2 Capacitance of a Circular Cylinder

Let $2L$ be the length and a be the radius of a metallic cylinder. Let $C_1 = C/(2L)$ and $\ell = La^{-1}$.

The capacitance per unit length C_1 is given in Figure 4.1 and Figure 4.2 as a function of ℓ, $0.1 \leq \ell \leq 10$. The capacitance C was calculated using formula (3.12) with $n = 0$ and $n = 1$. It turned out that for $\ell \geq 5$, $n = 0$ this formula gives a value which agrees within 1% with the capacitance of a hollow metallic tube with the same geometry. Numerical calculation of the capacitance of such a tube was given in [43]. For $1 \leq \ell \leq 5$, $n = 0$ the difference (i.e., the relative error) is at most 3%. For $\ell \geq 1$ and $n = 1$ the difference is at most 1%, while for $0.1 \leq \ell \leq 1$, $n = 1$ the difference is at most 3%. For $\ell \leq 0.1$ the asymptotic formula holds

$$C_1 = 4\varepsilon_e \ell^{-1} \tag{4.1}$$

with the relative error at most 3%. This formula follows from the known formula $C = 8a\varepsilon_e$ for the capacitance of the metallic disk of radius a and the definition $C_1 = C/(2L)$. As $\ell \to 0$ the accuracy of formula (4.1) increases.

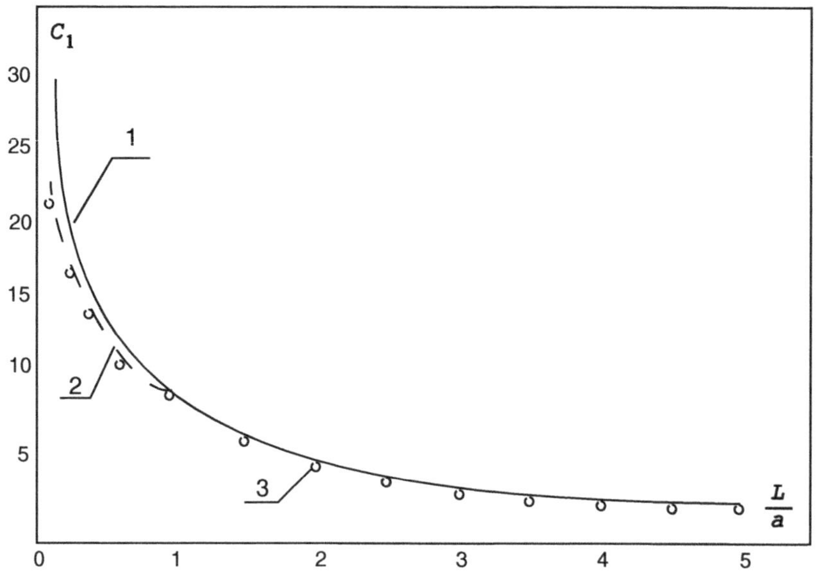

Fig. 4.1

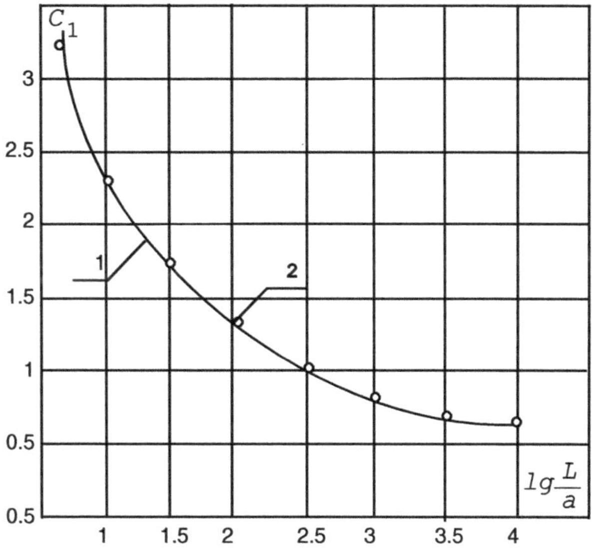

Fig. 4.2

For $\ell \geq 10$ the formula

$$C_1 = 4\pi\,\varepsilon_e\big(\Omega^{-1} + 0.71\Omega^{-3}\big), \quad \Omega \equiv 2\big[\ln(4\ell) - 1\big] \tag{4.2}$$

holds [43] with error at most 1%. For $\ell \geq 4$ formula (4.2) holds with the error at most 3.5%. For $0.1 \leq \ell \leq 4$ the formula

$$C_1 = \frac{2\pi^2\varepsilon_e}{\ln(16\ell^{-1})} \tag{4.3}$$

holds with error at most 3.5%. Thus formulas (4.1)–(4.3) give C_1 for any ℓ with the error at most 3.5%. An unexpected observation is that

$$\frac{C_{1\,\text{tube}}}{C_{1\text{cylinder}}} = \frac{\pi^2}{2\ln(16\ell^{-1})} = \frac{4.93}{\ln(16\ell^{-1})}, \quad \ell \ll 1. \tag{4.4}$$

This formula follows from (4.1) and (4.3). Formula (4.3) is the asymptotic formula for the capacitance of the tube for $\ell \ll 1$. For $\ell = 0.1$ the ratio (4.4) is equal to 0.98. This ratio is equal to 0.5 for $\ell^{-1} = 1250$. Thus *the capacitance per unit length of the metallic cylinder is nearly equal to that of the tube for* $\ell \geq 0.1$.

4.3 Capacitances of Parallelepipeds

Let a parallelepiped have edges

$$A_1 \le A_2 \le A_3, \tag{4.5}$$

let V denote its volume, set

$$\lambda = V^{1/3}\big(A_1 A_2 A_3\big)^{1/3}, \tag{4.6}$$

and let $C_\lambda = C(A_1 A_2 A_3)$ be its capacitance. Let

$$a_j = A_j \lambda^{-1}, \quad 1 \le j \le 3; \quad a_1 \le a_2 \le a_3, \quad a_1 a_2 a_3 = 1. \tag{4.7}$$

It is clear that

$$C_\lambda = \lambda \cdot C, \tag{4.8}$$

where C is the capacitance of the parallelepiped with sides a_1, a_2, a_3 and unit volume.

Therefore it is sufficient to tabulate $C(a_1, a_2, a_3)$, where a_j, $1 \le j \le 3$ satisfy (4.7).

Some long calculations (see [95]), which are based on formula (3.12) with $n = 0$, lead to the formula

$$\frac{C_\lambda}{4\pi\varepsilon_0} \approx \frac{S^2}{J}, \tag{4.9}$$

where

$$S = 2\big(A_1 A_2 + A_1 A_3 + A_2 A_3\big) \tag{4.10}$$

and

$$\begin{aligned} J = {} & \frac{4}{3}\sum_{i=l}^{3}\left[d\left(D^2 - \frac{S}{2} - \frac{3V}{A_i}\right) - A_i^3\right] \ln \frac{D - A_i}{D + A_i} \\ & + \frac{4}{3}\sum_{i=l}^{3}\sum_{j\ne i} \frac{V^2}{A_i^2 A_j}\left(3 + \frac{V}{A_i A_j^2}\right) \ln \frac{\overline{D^2 - A_i^2} + A_j}{\overline{D^2 - A_i^2} - A_j} \\ & - \frac{8}{3}\sum_{i=l}^{3}\left(D^2 - A_i^2 - \frac{2V}{A_i}\right)\overline{D^2 - A_i^2} \\ & - \frac{8}{3}SD + \frac{16}{3}\left[d\left(D^2 - \frac{S}{2}\right) + 3V\right] - \frac{8}{3}\sum_{i=l}^{3} A_i\big(A_i^2 + 3S\big) \operatorname{arctg} \frac{V}{A_i^2 D}, \end{aligned} \tag{4.11}$$

where

$$D = \left(\sum_{i=l}^{3} A_i^2 \right)^{1/2}; \quad d = \sum_{i=l}^{3} A_i; \quad S = 2V \sum_{i=l}^{3} \frac{1}{A_i}; \quad V = \prod_{i=l}^{3} A_i.$$

Let us describe a way to tabulate

$$\tilde{C} \equiv \frac{C}{4\pi\varepsilon_e}. \tag{4.12}$$

It follows from (4.7) that

$$0 \le a_1 \le 1. \tag{4.13}$$

Let

$$a_1 = kn^{-1}, \quad 1 \le k \le n \tag{4.14}$$

where n is an integer which defines the table. Let

$$a_2 = jn^{-1}, \quad j \ge k. \tag{4.15}$$

Then

$$a_3 = \frac{1}{a_1 a_2} = \frac{n^2}{kj}, \quad k \le j. \tag{4.16}$$

From (4.7) it follows that $jn^{-1} \le n^2(kj)^{-1}$. Thus

$$\frac{k}{n} \le \frac{j}{n} \le \sqrt{n/k}. \tag{4.17}$$

Therefore

$$a_1 \le a_2 \le \frac{1}{\sqrt{a_1}}. \tag{4.18}$$

For fixed a_1 and a_2, the parameter a_3 is uniquely determined by (4.16). This means that $\tilde{C}$ can be tabulated as a function of a_1 and a_2. In Table 4.1 the results are given for $n = 10$. In the horizontal line the values of a_1 are given. In the vertical line the values of a_2 are given. At the intersections the values of $\tilde{C}(a_1, a_2)$ are given. If zero stands at the intersection, this means that for the given a_1 the chosen a_2 is not allowed by (4.18).

Let us formulate an algorithm for calculating C_λ for an arbitrary parallelepiped.

Step 1. Order the sides of the parallelepiped as shown in (4.5) and calculate λ from (4.6) and a_1 and a_2 from (4.7).

Table 4.1 The capacitances $\tilde{C} = C/(4\pi\varepsilon_e)$ of the unit parallelepiped.

	a_1									
a_2	0.1	0.2	0.3	0.4	0.5	0.6	0.7	0.8	0.9	1.0
0.1000	7.00313									
0.2000	4.12588	2.47336								
0.3000	3.08985	1.88108	1.44955							
0.4000	2.54667	1.57289	1.22690	1.04998						
0.5000	2.21009	1.38371	1.09186	0.94404	0.85669					
0.6000	1.98066	1.25629	1.00224	0.87489	0.80064	0.75381				
0.7000	1.81434	1.16528	0.93938	0.82736	0.76294	0.72320	0.69733			
0.8000	1.68855	1.09767	0.89367	0.79366	0.73693	0.70237	0.68067	0.66708		
0.9000	1.59040	1.04604	0.85968	0.76936	0.71883	0.68859	0.67007	0.65894	0.65278	
1.0000	1.51203	1.00586	0.83405	0.75174	0.70633	0.67963	0.66373	0.65463	0.65011	0.6488
1.1000	1.44832	0.97417	0.81461	0.73906	0.69794	0.67461	0.66050	0.65312		
1.2000	1.39582	0.94897	0.79989	0.73010	0.69264	0.67145				
1.3000	1.35207	0.92885	0.78885	0.72404	0.68974					
1.4000	1.31531	0.91277	0.78074	0.72029	0.68872					
1.5000	1.28423	0.89998	0.77499	0.71837						
1.6000	1.25784	0.88990	0.77118							
1.7000	1.23534	0.88206	0.76896							
1.8000	1.21614	0.87611	0.76808							
1.9000	1.19975	0.87173								
2.0000	1.18577	0.86878								
2.1000	1.17387	0.86698								
2.2000	1.16380	0.96620								
2.3000	1.15532									
2.4000	1.14825									
2.5000	1.14243									
2.6000	1.13771									
2.70000	1.13399									
2.8000	1.13115									
2.9000	1.12911									
3.0000	1.12780									
3.1000	1.12714									

Step 2. Find the numbers closest to a_1 and a_2 in the horizontal and vertical line of Table 4.1 respectively. Find $C(a_1, a_2)$ in this table.

Step 3. Find C_λ from (4.8) and (4.12).

Example 4.1 Let $A_1 = 1$, $A_2 = 2$, $A_3 = 4$. Then $V = 8$, $\lambda = 2$, $a_1 = 0.5$, $a_2 = 1$, $\tilde{C} = 0.70633$. Thus $C_\lambda = 8\pi\varepsilon_e \cdot 0.70633 \simeq 17.7514\varepsilon_e$.

Example 4.2 Let $A_1 = A_2 = A_3 = 1$, i.e., we have a unit cube, $a_1 = a_2 = a_3 = 1$, $V = 1$, $\lambda = 1$. From Table 4.1 one find $C = 4\pi\varepsilon_e \cdot 0.649$.

References [43] and [79] mention about 17 papers dealing with the test problem of calculating the capacitance of a cube. The best results reported in [79] and obtained by means of some complicated calculations with harmonic polynomials with the symmetry group of a cube, state that the capacitance C of the unit cube satisfies the following estimates:

$$0.632 < \frac{C}{4\pi\varepsilon_e} < 0.710, \quad \frac{C}{4\pi\varepsilon_e} \approx 0.646. \tag{4.19}$$

From (3.12) and (3.15) it follows that the value $C/(4\pi\varepsilon_e) = 0.649$ is not only an approximation to $C/(4\pi\varepsilon_e)$ but also a lower bound. One can see that for a cube formula (3.12) gave a good result even for $n = 0$.

Example 4.3 Let $A_1 = 0$, $A_2 = 2$, $A_3 = 5$. This is the case of a thin rectangular metallic plate. Since the smallest $a_1 = 0.1$ in Table 4.1, we take $A_1 = 0.1$, $A_2 = 2$, $A_3 = 5$ and find $C = 4\pi\varepsilon_e \cdot 1.18577$. This agrees with the value given in [33].

Example 4.4 Consider the square thin plate: $A_1 = 0.1$, $A_2 = A_3 = 1$. Let $a_1 = 0.1$, $a_2 = a_3 = 3.16$. Then $a_1a_2a_3 = 1$ and from Table 4.1 one finds $C/(4\pi\varepsilon_e) = 1.12714$. For the capacitance of the thin plate with the unit side one finds $C^{(1)}/(4\pi\varepsilon_e) = 1.12714/3.16 = 0.3566$. This agrees with the value 0.360 given in [43].

Remark 4.1 *Table 4.1 shows that among all parallelepiped with the fixed volume the cube has the minimal capacitance. This can be proved, but the proof (see [68]) is not elementary. The error in the calculation of the capacitances in Table 4.1 is at most 2%.*

4.4 Interaction Between Conductors

Let two conducting balls of radius a be charged to potential V each. Then $Q = C_{11}V + C_{12}V$, $Q = C_{21}V + C_{22}V$ and by symmetry $C_{11} = C_{22}$, $C_{12} = C_{21}$. Let us join these balls. The electrostatic equilibrium will be preserved since the potentials of the balls are the same. Let $\tilde{C}$ denote the capacitance of the joined balls. Then $\tilde{C} = 2Q/V = 2(C_{11} + C_{12})$. Let C be the capacitance of a single ball. Then $\tilde{C}/(2C) = (C_{11} + C_{12})/C$. Let d be the distance between the centers of the balls. Then the numerical results [43] give $\tilde{C}/(2C) = 0.75$ if $2ad^{-1} = 0.5$; $\tilde{C}/(2C) = 0.91$ if $2ad^{-1} = 0.2$; $\tilde{C}/(2C) = 0.71$ if $2ad^{-1} = 0.9$. Therefore one makes an error of at most 25% if one neglects the interaction of the conductors if $a \leq 0.25d$ and one makes an error of at most 10% if $a \leq 0.1d$.

Chapter 5

Calculating Polarizability Tensors

5.1 Calculating the Polarizability Tensor of a Solid Body

1. If a solid conductor is placed in an exterior homogeneous electrostatic field E, then the induced charge distribution $\sigma(t)$ appears on its surface. Therefore the conductor acquires the dipole moment

$$P_i = \int_\Gamma t_i \sigma(t) dt, \tag{5.1}$$

where t_i is the ith coordinate of the radius vector t of the point t at the surface Γ of the conductor. Since the equations of electrostatics are linear, there is a linear relation between P and E:

$$P_i = \alpha_{ij} \varepsilon_e V E_j \tag{5.2}$$

(summation over the repeated indices is understood), where V is the volume of the conductor, ε_e is the dielectric permittivity of the exterior medium, the matrix α_{ij} is called the polarizability tensor. The dipole moment is interesting in many applications, especially in scattering theory (see Chapter 7).

A more general definition of the dipole moment is as follows. Let $\phi_0 = -(E, x)$ be the potential of the exterior homogeneous field, $\phi = \phi_0 + u$ be the potential of the total field. If the obstacle is finite, then

$$u \sim \frac{(P, x)}{4\pi\varepsilon_e |x|^3} \text{ as } |x| \longrightarrow \infty. \tag{5.3}$$

We assume here that the obstacle is electroneutral, that is, its total charge is zero. The vector P is called the dipole moment induced on the obstacle by the exterior field E.

2. Let the obstacle be a homogeneous body with dielectric constant ε. Put

$$\gamma = \frac{\varepsilon - \varepsilon_e}{\varepsilon + \varepsilon_e}. \tag{5.4}$$

The polarizability tensor is defined by the formula

$$P_i = \alpha_{ij}(\gamma)\varepsilon_e V E_j. \tag{5.5}$$

If $\varepsilon = \infty$ then $\gamma = 1, \alpha_{ij}(1) = \alpha_{ij}$ where α_{ij} is the polarizability tensor of the perfect conductor with the same shape. If $\varepsilon = 0$, then $\gamma = -1$, $\alpha_{ij}(-1) := \beta_{ij}$, where β_{ij} is the magnetic polarizability tensor (the polarizability tensor of the insulator). Our aim is to give approximate analytical formulas for calculating $\alpha_{ij}(\gamma)$.

Let us introduce some notations. Let

$$b_{ij}^{(0)} = V\delta_{ij}, \quad \delta_{ij} = \begin{cases} 1, & i = j, \\ 0, & i \neq j, \end{cases} \tag{5.6}$$

$$b_{ij}^{(1)} = \int_\Gamma \int_\Gamma \frac{N_i(t)N_j(s)}{r_{st}} ds\, dt, \tag{5.7}$$

where $N_i(t)$ is the ith component of the outer unit normal to Γ at the point t,

$$\begin{aligned} b_{ij}^{(m)} = {} & \int_\Gamma \int_\Gamma ds\, dt N_i(t)N_j(s) \underbrace{\int_\Gamma \cdots \int_\Gamma}_{m-1} \\ & \times \frac{1}{r_{st_{m-1}}} \psi(t_1, t)\psi(t_2, t_1) \cdots \psi(t_{m-1}, t_{m-2}) dt_1 \cdots dt_{m-1}, \end{aligned} \tag{5.8}$$

where

$$\psi(t, s) = \frac{\partial}{\partial N_t} \frac{1}{r_{st}}.$$

Define

$$\alpha_{ij}^{(n)}(\gamma) = \frac{2}{V} \sum_{m=0}^{n} \frac{(-1)^m}{(2\pi)^m} \frac{\gamma^{n+2} - \gamma^{m+1}}{\gamma - 1} b_{ij}^{(m)}, \quad n > 0. \tag{5.9}$$

In particular

$$\alpha_{ij}^{(1)}(\gamma) = 2(\gamma + \gamma^2)\delta_{ij} - \frac{\gamma^2}{\pi V} b_{ij}^{(1)}, \tag{5.10}$$

$$\alpha_{ij}^{(1)} = 4\delta_{ij} - \frac{1}{\pi V} b_{ij}^{(1)}, \tag{5.11}$$

$$\beta_{ij}^{(1)} = -\frac{1}{\pi V} b_{ij}^{(1)}. \tag{5.12}$$

Note that $b_{ij}^{(m)}$ depends only on the geometry of the body.

Theorem 5.1 *The following estimate holds*

$$\left|\alpha_{ij}(\gamma) - \alpha_{ij}^{(n)}(\gamma)\right| \le cq^n, \quad 0 < q < 1, \quad -1 \le \gamma \le 1, \tag{5.13}$$

where $c > 0$ and q are constants which depend only on the shape of Γ and on γ.

Remark 5.1 *From (5.9) for $\varepsilon = \infty$ (i.e., $\gamma = 1$) it follows that*

$$\alpha_{ij}^{(n)} = \frac{2}{V} \sum_{m=0}^{n} \frac{(-1)^m}{(2\pi)^m} (n+1-m) b_{ij}^{(m)}, \tag{5.14}$$

and for $\varepsilon = 0$ (i.e., $\gamma = -1$) it follows that

$$\beta_{ij}^{(n)} = \frac{1}{V} \sum_{m=0}^{n} \frac{(-1)^{n+m-1} - 1}{(2\pi)^m} b_{ij}^{(m)}. \tag{5.15}$$

Proof of Theorem 5.1. Let us define

$$P_i^{(n)} = \int_\Gamma t_i \sigma_n dt := \alpha_{ij}^{(n)} V \epsilon_e E_j, \tag{5.16}$$

where $\alpha_{ij}^{(n)}$ is calculated below in formula (5.28), σ_n is defined in (2.4) with $\sigma_0 = -2\gamma\varepsilon_e(\partial\phi_0/\partial N)$,

$$\left|\sigma_n - \sigma\right| \le cq^n, \quad 0 < q < 1 \tag{5.17}$$

where $c > 0$ and q depend on Γ and γ. From (2.4) it follows that

$$\sigma_n = \sum_{m=0}^{n} (-1)^m \gamma^m A^m (2\gamma(E, N)) \varepsilon_e. \tag{5.18}$$

From (5.16) and (5.18) one obtains

$$P_i^{(n)} = \frac{2}{V} \sum_{m=0}^{n} \frac{(-1)^m \gamma^{m+1}}{(2\pi)^m} \int_\Gamma t_j B^m(N_j) dt \, V \varepsilon_e E_j, \tag{5.19}$$

where

$$B \equiv 2\pi A. \tag{5.20}$$

Therefore

$$\alpha_{ij}^{(n)}(\gamma) = \frac{2}{V}\sum_{m=0}^{n}\frac{(-1)^m\gamma^{m+1}}{(2\pi)^m}J_{ij}^{(m)}, \tag{5.21}$$

where

$$J_{ij}^{(m)} = \int_\Gamma t_i B^m(N_j)dt. \tag{5.22}$$

Let us prove that

$$J_{ij}^{(m)} = b_{ij}^{(m)} - 2\pi J_{ij}^{(m-1)}, \tag{5.23}$$

where $b_{ij}^{(m)}$ is defined in (5.8). We have

$$J_{ij}^{(0)} = \int_\Gamma t_i N_j(t)dt = \int_D \frac{\partial x_i}{\partial x_j}dx = V\delta_{ij} = b_{ij}^{(0)}, \tag{5.24}$$

and

$$\begin{aligned} J_{ij}^{(1)} &= \int_\Gamma s_i B(N_j)ds = \int_\Gamma dt\, N_j(t)\int_\Gamma s_i \frac{\partial}{\partial N_s}\frac{1}{r_{st}}ds \\ &= \int_\Gamma dt\, N_j(t)\left(\int_\Gamma \frac{\partial s_i}{\partial N_s}\frac{ds}{r_{st}} - 2\pi t_i\right) = \int_\Gamma\int_\Gamma \frac{N_i(s)N_j(t)}{r_{st}}ds\,dt - 2\pi V\delta_{ij} \\ &= b_{ij}^{(1)} - 2\pi J_{ij}^{(0)}. \end{aligned} \tag{5.25}$$

In a similar manner, one obtains

$$\begin{aligned} J_{ij}^{(m)} &= \int_\Gamma ds\, s_i B^m(N_j) = \int_\Gamma dt\, N_j(t)\int_\Gamma dt_1\psi(t_1,t) \\ &\quad \cdots\int_\Gamma dt_{m-1}\psi(t_{m-1},t_{m-2})\left[\int_\Gamma \frac{N_i(s)ds}{r_{st_{m-1}}} - 2\pi\left(t_{m-1}\right)_i\right] \\ &= b_{ij}^{(m)} - 2\pi J_{ij}^{(m-1)}. \end{aligned} \tag{5.26}$$

From (5.26) it follows that

$$J_{ij}^{(m)} = \sum_{k=0}^{m} b_{ij}^{(k)}(2\pi)^{m-k}(-1)^{m-k}. \tag{5.27}$$

Using (5.27) and (5.21) one finds that

$$\begin{aligned}\alpha_{ij}^{(n)}(\gamma) &= \frac{2}{V}\sum_{m=0}^{n}\frac{(-1)^m\gamma^{m+1}}{(2\pi)^m}\sum_{k=0}^{m} b_{ij}^{(k)}(2\pi)^{m-k}(-1)^{m-k} \\ &= \frac{2}{V}\sum_{k=0}^{n} b_{ij}^{(k)}\frac{(-1)^k}{(2\pi)^k}\,\frac{\gamma^{n+2}-\gamma^{k+1}}{\gamma-1}.\end{aligned} \tag{5.28}$$

Estimate (5.13) follows from (5.17). Theorem 5.1 is proved. □

5.2 Polarizability Tensors of Thin Metallic Screens

Let F be a thin metallic screen. Its polarizability tensor is defined as

$$P_i = \alpha_{ij}E_j\varepsilon_e, \quad P_i = \int_F t_i\sigma(t)dt, \tag{5.29}$$

where $\sigma(t)$ is the distribution of the charge induced by the exterior homogeneous electrostatic field E. Let e_i, $1 \le i \le 3$, be the orthonormal unit vectors of the coordinate system, let $E = e_j$, and let $\phi_0 = -x_j$ be the potential corresponding to E. Then

$$P_i = \alpha_{ij}\varepsilon_e. \tag{5.30}$$

Let $\sigma_n(t)$ be the approximate charge distribution constructed in (2.36). Then

$$P_i^{(n)} = \int_\Gamma t_i\sigma_n(t)dt \equiv \alpha_{ij}^{(n)}\varepsilon_e. \tag{5.31}$$

Thus

$$\alpha_{ij}^{(n)} = \varepsilon_e^{-1}\int_\Gamma t_i\sigma_n(t)dt. \tag{5.32}$$

Note that the index j is implicitly present in the right-hand side of (5.32) because $\sigma_n(t)$ is constructed for the initial field $E = e_j$, or for the initial potential $\phi_0 = -x_j$. Thus, calculating the polarizability tensor is reduced to finding σ_n according to Theorem 2.8 and to the calculation of the six integrals in (5.32), $1 \le i \le j \le 3$. The number of the integrals is six (and not nine) because $\alpha_{ij}^{(n)} = \alpha_{ji}^{(n)}$.

Let F be a plane plate. Let e_3 be orthogonal to F. Then $\alpha_{i3} = \alpha_{3i} = 0$ and the polarizability tensor is defined by the three numbers α_{11}, α_{22} and $\alpha_{12} = \alpha_{21}$.

5.3 Polarizability Tensors of Flaky-Homogeneous Bodies or a System of Bodies

1. The integral equation for the surface charge densities, induced by the initial field, is given in Theorem 2.3. The nth approximation for the polarizability tensor of the flaky-homogeneous body is rather cumbersome. Therefore only the first approximation will be considered. Let A_{ij} be the polarizability tensor

$$P_i = A_{ij} E_j \varepsilon_e. \tag{5.33}$$

There is no factor V in this definition of A_{ij} because if the body is nonhomogeneous the matrix $\alpha_{ij} = A_{ij} V^{-1}$ does not depend solely on the geometry of the body. For the dipole moment of the flaky-homogeneous body one has the formula

$$P_i = \sum_{j=1}^{p} \int_{\Gamma_j} t_i \sigma_j(t) dt. \tag{5.34}$$

Substituting $\sigma_j^{(n)}$ from Theorem 2.3 in (5.34) in place of σ_j yields the n–th approximation to P_i,

$$P_i^{(n)} = \sum_{j=1}^{p} \int_{\Gamma_j} t_i \sigma_j^{(n)}(t) dt \equiv A_{ij}^{(n)} E_j \varepsilon_e. \tag{5.35}$$

Let us take $n = 1$. From (5.35) and Theorem 2.3, it follows that ($E = -\nabla \phi_0$)

$$\begin{aligned} P_i^{(1)} = \sum_{j=1}^{p} \varepsilon_e \int_{\Gamma_j} t_i \Big\{ 2\gamma_j N_q(t) E_q - 2\gamma_j^2 A_j \big(E_q N_q(t)\big) \\ - 2\gamma_j \gamma_m \sum_{m \neq j, m=1}^{p} T_{jm}\big(E_q N_q(t)\big) \Big\} dt \quad (5.36) \\ = \Big\{ \sum_{j=1}^{p} \alpha_{iq}^{(1)}(\gamma_j) V_j + \sum_{j=1}^{p} \sum_{m \neq j, m=1}^{p} \alpha_{iq}^{(j,m)} \Big\} E_q \varepsilon_e, \end{aligned}$$

where V_j is the volume of the body inside Γ_j,

$$\alpha_{iq}^{(1)}(\gamma_j) = 2\delta_{iq}\big(\gamma_j + \gamma_j^2\big) - \frac{\gamma_j^2}{\pi V_j} b_{iq}^{(1)}, \tag{5.37}$$

γ_j is given in (1.59) (compare (5.37)) and (5.10)), and

$$\alpha_{iq}^{(j,m)} = \begin{cases} -\frac{\gamma_j\gamma_m}{\pi} b_{iq}^{(j,m)}, & j > m \\ -\frac{\gamma_j\gamma_m}{\pi} b_{iq}^{(j,m)} + 4\gamma_j\gamma_m V_m \delta_{iq}, & j < m, \end{cases} \tag{5.38}$$

where

$$b_{iq}^{(j,m)} = \int_{\Gamma_j} \int_{\Gamma_m} \frac{N_i(t)N_q(s)}{r_{st}} ds\, dt. \tag{5.39}$$

These formulas and their proof are quite similar to formulas (5.8)–(5.12).

From (5.35)–(5.39) one finds

$$A_{iq}^{(1)} = \sum_{j=1}^{p} \alpha_{iq}^{(1)}(\gamma_j)V_j + \sum_{j=1}^{p} \sum_{m\neq j, m=1}^{p} \alpha_{iq}^{(j,m)}, \tag{5.40}$$

where $\alpha_{iq}^{(1)}(\gamma_j)$ and $\alpha_{iq}^{(j,m)}$ are defined in (5.37) and (5.38) respectively.

2. Let us derive an approximate formula for the polarizability tensor of a system of bodies. We use Theorem 2.4 in the same manner as Theorem 2.3 was used. Let us define the polarizability tensor of a system of bodies by

$$P_i = B_{ij} E_j \varepsilon_e. \tag{5.41}$$

Then, using the argument given in Section 5.1, one finds

$$B_{iq}^{(1)} = \sum_{j=1}^{n} \alpha_{iq}^{(1)}(k_j)V_j + \sum_{j=1}^{p} \sum_{m\neq j, m=1}^{p} \tilde{\alpha}_{iq}^{(j,m)}, \tag{5.42}$$

where k_j is defined in (1.56), $\alpha_{iq}^{(1)}(k_j)$ is defined in (5.37) with k_j in place of γ_j,

$$\tilde{\alpha}_{iq}^{(j,m)} = -\frac{k_j k_m}{\pi} b_{iq}^{(j,m)} \tag{5.43}$$

and $b_{iq}^{(j,m)}$ is defined in (5.39).

If the jth body is a perfect conductor then $k_j = 1$.

5.4 Variational Principles for Polarizability Tensors

1. The purpose of this section is to give variational principles for polarizability tensors and to show how some two-sided estimates for polarizability tensors can be obtained from these principles.

Let $E = e_j$, where e_j is the coordinate unit vector, $\phi_0 = -x_j$, $E = -\nabla\phi_0$. Suppose that the body is a perfect conductor. Then the induced surface charge distribution $\sigma_j(t)$ satisfies the equation

$$\int_\Gamma \frac{\sigma_j(t)dt}{4\pi\varepsilon_e r_{st}} = U_j + s_j, \quad U_j = \text{const}, \tag{5.44}$$

and the electroneutrality condition

$$\int_\Gamma \sigma_j dt = 0. \tag{5.45}$$

The quantity U_j is the potential of the conductor. The induced dipole moment of the conductor is

$$P_i = \alpha_{iq}\varepsilon_e V E_q = \alpha_{ij}\varepsilon_e V = \int_\Gamma t_i\sigma_j(t)dt, \tag{5.46}$$

because $E_q = \delta_{jq}$. Therefore

$$V\alpha_{ij} = \varepsilon_e^{-1}\int_\Gamma t_i\sigma_j(t)dt, \quad \alpha_{ij} = \alpha_{ji}. \tag{5.47}$$

Note that (5.45) and (5.44) imply

$$\int_\Gamma U_j\sigma_j dt = 0. \tag{5.48}$$

From (5.44), (5.48), and (3.26) it follows that

$$V\alpha_{ij} = 4\pi \ st \frac{\int_\Gamma t_i\phi_j dt \int_\Gamma t_j\phi_i dt}{\int_\Gamma\int_\Gamma \frac{\phi_i(t)\phi_j(s)ds\,dt}{r_{st}}}, \tag{5.49}$$

where the admissible functions satisfy (5.45). For $i = j$ the st in (5.49) can be replaced by max.

$$V\alpha_{jj} = \max 4\pi\left(\int_\Gamma t_j\phi_j dt\right)^2\left(\int_\Gamma\int_\Gamma \frac{\phi_j(t)\phi_j(s)ds\,dt}{r_{st}}\right)^{-1}, \tag{5.50}$$

where again ϕ_j satisfies (5.45). Principle (5.50) allows one to find lower bounds for the diagonal elements of the polarizability tensor.

2. In order to find upper bounds for these elements we need another variational principle. The energy U of the electrostatic field of the conductor is

$$U = \frac{\varepsilon_e}{2}\int_{D_e} |\nabla\phi_j|^2 dx, \tag{5.51}$$

where ϕ_j is the secondary potential corresponding to the initial field $E = e_j$, D_e is the exterior domain with the boundary Γ, and $D = D_i$ is the conductor with the boundary Γ.

On the other hand the same energy is equal to

$$U = \frac{\varepsilon_e V}{2} \alpha_{jj}. \tag{5.52}$$

Indeed, if P is the dipole moment, then $U = \frac{1}{2}(P, E) = \frac{\varepsilon_e V}{2} \alpha_{im} E_m E_i$ and since $E_m = \delta_{jm}$ one obtains (5.52). Thus

$$V \alpha_{jj} = \min \int_{D_e} |\nabla u|^2 dx, \tag{5.53}$$

where the admissible functions $u \in C^1(D_e)$ satisfy the condition

$$u|_\Gamma = U_j + s_j, \quad U_j = \text{const}. \tag{5.54}$$

The minimum in (5.53) is attained at the solution of the problem

$$\triangle\, \phi = 0 \text{ in } D_e, \quad u|_\Gamma = U_j + s_j, \quad \int_\Gamma \frac{\partial \phi}{\partial N} dt = 0, \quad \phi(\infty) = 0. \tag{5.55}$$

The variational principle (5.53)-(5.54) allows one to obtain upper bounds for α_{jj}.

Example 5.1 Let Γ be a sphere with radius a. By symmetry one concludes that $\alpha_{ij} = \alpha \delta_{ij}$, where $\alpha > 0$ is a scalar. Let $\phi_j(t) = Y_{ij}(t)$, where Y_{ij} are the spherical harmonics, $Y_{11} = \cos\theta, Y_{12} = \sin\theta \cos\phi$, $Y_{13} = \sin\theta \sin\phi$, and $t = (1, \theta, \phi)$. From (5.49) one finds that $\alpha_{ij} = 0$ for $i \neq j$. For $i = j$ it follows from (5.50) that $\alpha_{jj} = \alpha = 3$. In this example we obtained the exact value of α because of the symmetry.

3. Suppose that $V \to 0$ and the body tends to a thin screen F with the edge L. Then the variational principles (5.49), (5.50), and (5.53)-(5.54) remain valid but the admissible functions should satisfy the edge condition. The tensor

$$\lim_{V \to 0} \alpha_{ij} V = \tilde{\alpha}_{ij} \tag{5.56}$$

is the polarizability tensor of the screen F. Therefore the derivation of the variational principles for the electric polarizability tensor of the metallic screen has no new points.

4. Let us derive some variational principles for the magnetic polarizability tensor β_{ij}. This tensor is defined as follows.

Consider the boundary value problem

$$\triangle \phi = 0 \text{ in } D_e, \quad -\frac{\partial \phi}{\partial N_e} = -\frac{\partial t_j}{\partial N} = -N_j(t), \quad \phi(\infty) = 0. \tag{5.57}$$

This problem is a mathematical formulation of the physical problem of finding the magnetic field around a superconductor (i.e., a body D inside which the magnetic induction $B = 0$). On the surface Γ of this body $B_N|_\Gamma = 0$. Outside the body $\operatorname{div} B = 0$, $\operatorname{curl} H = 0$, $B = \mu_0 H$ in D_e where μ_0 is the magnetic permittivity of the exterior medium. If $H = e_j - \nabla\phi = \nabla(x_j - \phi)$ then the condition $B_N|_\Gamma = 0$ can be written as

$$\left.\frac{\partial(x_j - \phi)}{\partial N_e}\right|_\Gamma = 0, \quad \text{or} \quad -\frac{\partial \phi}{\partial N_e} = -N_j \text{ on } \Gamma, \tag{5.58}$$

which is the same condition as in (5.57). Let

$$\phi = \phi_j = \int_\Gamma \frac{\sigma_j(t)dt}{4\pi\mu_0 r_{xt}}. \tag{5.59}$$

Then from (5.58) it follows that

$$\sigma_j = A\sigma_j - 2\mu_0 N_j(t), \quad A\sigma = \int_\Gamma \frac{\partial}{\partial N_s}\frac{1}{2\pi r_{st}}\sigma(t)dt \tag{5.60}$$

and

$$\sigma_j = \left(\frac{\partial \phi_j}{\partial N_i} - \frac{\partial \phi_j}{\partial N_e}\right)\mu_0. \tag{5.61}$$

The magnetic polarizability tensor is defined by the equation

$$V\beta_{pj} = \mu_0^{-1}\int_\Gamma t_p \sigma_j(t)dt, \tag{5.62}$$

where V is the volume of the body D.

If we substitute (5.61) into (5.62), we obtain

$$\begin{aligned} V\beta_{pj} &= \int_\Gamma t_p\left(\frac{\partial \phi_j}{\partial N_i} - \frac{\partial \phi_j}{\partial N_e}\right)dt = \int_\Gamma \frac{\partial t_p}{\partial N}\phi_j dt - \int_\Gamma t_p \frac{\partial t_j}{\partial N}dt \\ &= \int_\Gamma \frac{\partial \phi_p}{\partial N_e}\phi_j dt - \delta_{pj}V = -\int_{D_e} \nabla\phi_p \nabla\phi_j dx - \delta_{pj}V. \end{aligned} \tag{5.63}$$

In particular

$$V\beta_{jj} + V = -\int_{D_e} |\nabla\phi_j|^2 dx, \tag{5.64}$$

$$V\beta_{pj} = \int_\Gamma \frac{\partial t_p}{\partial N}\phi_j dt - V\delta_{pj}. \tag{5.65}$$

The operator $-\partial/\partial N_e$ is nonnegative-definite on the set of functions on Γ which are restrictions on Γ of harmonic functions defined in D_e and vanishing at infinity. This follows from the Green formula

$$-\int_\Gamma V\frac{\partial u}{\partial N_e}dt = \int_{D_e} \nabla u\,\nabla v\,dx = -\int_\Gamma u\frac{\partial v}{\partial N_e}dt. \tag{5.66}$$

Therefore formulas (5.58) and (3.26) yield

$$-V\beta_{pj} = st\frac{\int_\Gamma N_p(t)u_j(t)dt\int_\Gamma N_j(t)u_p(t)dt}{-\int_\Gamma \frac{\partial u_p}{\partial N_e}u_j dt} + V\delta_{pj}, \tag{5.67}$$

where the admissible functions $u_j(t)$ are harmonic in D_e and $u_j(\infty) = 0$. If $p = j$ then st in (5.67) can be replaced by max, obtaining

$$-(V + V\beta_{jj}) = \max\left\{\left(\int_\Gamma N_j(t)u_j(t)dt\right)^2\left(-\int_\Gamma \frac{\partial u_j}{\partial N_e}u_j dt\right)^{-1}\right\}, \tag{5.68}$$

or

$$-(V + V\beta_{jj}) = \max\left\{\left(\int_\Gamma N_j u_j dt\right)^2\left(\int_{D_e}|\nabla u_j|^2 dx\right)^{-1}\right\}. \tag{5.69}$$

The maximum in (5.68), (5.69) is attained at the solution to (5.57).

Remark 5.2 *Formulas* (5.68), (5.69) *remain valid if the admissible functions u are not necessarily harmonic in D_e but are arbitrary functions $u \in C^1(D_e)$, $u(\infty) = 0$.*

Proof. From (5.64) and (5.69) it follows that (5.69) can be written as

$$\int_{D_e}|\nabla u|^2 dx\int_{D_e}|\nabla\phi_j|^2 dx \geq \left(\int_{D_e}\nabla\phi_j\nabla u_j dx\right)^2 = \left(\int_\Gamma N_j(t)u_j dt\right)^2. \tag{5.70}$$

The equality in (5.70) follows from Green's formula.

Inequality (5.70) is just the Cauchy inequality and is valid for any u, ϕ_j such that $\nabla u \in L^2(D_e)$, $\nabla\phi_j \in L^2(D_e)$. □

Exercise 5.1 Prove that

$$-2\pi V\beta_{pj} = \delta_{pj} + st\frac{\int_\Gamma\int_\Gamma N_p(t)\sigma_j(s)\frac{ds\,dt}{r_{st}}\int_\Gamma\int_\Gamma N_j(t)\sigma_p(s)\frac{ds\,dt}{r_{st}}}{\int_\Gamma\int_\Gamma\{\sigma_p(t) - A\sigma_p(t)\}\frac{\sigma_j(s)ds\,dt}{r_{st}}},$$

where A is defined in (5.60) and the admissible functions $\sigma_j(t) \in C(\Gamma)$.

Remark 5.3 *Principle* (5.69) *allows one to obtain lower bounds for* β_{jj}. *In order to obtain some upper bounds for* β_{jj} *the variational principle*

$$-V - V\beta_{jj} = \min \int_{D_e} \left|q_j\right|^2 dx, \tag{5.71}$$

where q_j *are arbitrary vector fields such that the integral* (5.71) *converges and*

$$\operatorname{div} q_j = 0 \text{ in } D_e, \quad \big(q_j, N\big) = N_j(t) \text{ on } \Gamma. \tag{5.72}$$

Proof. In order to prove principle (5.71)-(5.72), note that if q satisfies (5.72), then

$$\int_{D_e} \left|q - \nabla\phi_j\right|^2 dx = \int_{D_e} |q|^2 dx + \int_{D_e} \left|\nabla\phi_j\right|^2 dx - 2\int_{D_e} q\nabla\,\phi_j dx \tag{5.73}$$

and

$$\begin{aligned}\int_{D_e} q\nabla\phi_j dx &= \int_{D_e} \operatorname{div}\big(q\phi_j\big) dx - \int_{D_e} \phi_j \operatorname{div} q\, dx = -\int_\Gamma (q, N)\phi_j dt \\ &= -\int_\Gamma N_j\phi_j dt = -\int_\Gamma \frac{\partial\phi_j}{\partial N_e} u_j dt = \int_{D_e} \left|\nabla\phi_j\right|^2 dx.\end{aligned} \tag{5.74}$$

From (5.73), (5.74) and (5.64) it follows that

$$\begin{aligned}\int_{D_e} \left|q - \nabla\phi_j\right|^2 dx &= \int_{D_e} |q|^2 dx - \int_{D_e} \left|\nabla\phi_j\right|^2 dx \\ &= \int_{D_e} |q|^2 dx + V + V\,\beta_{jj}\end{aligned} \tag{5.75}$$

provided that q satisfies (5.72). Principle (5.71) follows from (5.75). The minimum in (5.71) is attained at $q_j = \nabla\phi_j$, where ϕ_j is the solution to (5.57). □

5. Magnetic polarizability of screens. In connection with magnetic polarizability, the screen is a model of a thin superconductor or a perfect magnetic film. The latter case is of interest because thin magnetic films are parts of the memory elements of computers.

Let us denote the magnetic polarizability tensor of the screen by

$$\tilde{\beta}_{ij} = \lim_{V\to 0} V\beta_{ij}. \tag{5.76}$$

This definition is similar to (5.56). The new point, in comparison with Section 5.3, is: if Γ is an unclosed surface one cannot look for a solution to problem (5.57) of the form (5.59). Indeed the normal derivative of the potential of a single layer (5.59) has a jump when x crosses Γ, while the boundary condition in (5.57) shows that the normal derivative is continuous when x crosses Γ. Therefore in the case when the body is an unclosed thin surface F let us look for the solution of (5.57) of the form

$$\psi_j = \int_F \eta_j(t) \frac{\partial}{\partial N_t} \frac{1}{4\pi\mu_0 r_{xt}} dt. \tag{5.77}$$

It is known [38] that $\partial\psi_j/\partial N$ is continuous when x crosses F provided that the surface F is smooth. We have

$$\psi_j \sim \frac{(M_j, x)}{4\pi\mu_0 |x|^3}, \quad |x| \longrightarrow \infty, \tag{5.78}$$

where

$$M_j \equiv \int_F \eta_j(t) N(t) dt. \tag{5.79}$$

The vector M is the induced magnetic moment. In particular,

$$M_{jj} = \int_F \eta_j(t) N_j(t) dt. \tag{5.80}$$

Since the initial field H corresponds to the potential $\phi_0 = -x_j$, we have

$$M_{jj} = \mu_0 \tilde{\beta}_{jj} H_j = \mu_0 \tilde{\beta}_{jj}. \tag{5.81}$$

Thus

$$\tilde{\beta}_{jj} = \mu_0^{-1} \int_F \eta_j(t) N_j(t) dt, \tag{5.82}$$

and

$$\tilde{\beta}_{pj} = \tilde{\beta}_{jp} = \mu_0^{-1} \int_F \eta_j(t) N_p(t) dt. \tag{5.83}$$

Let us consider the boundary condition

$$-\frac{\partial \psi_j}{\partial N} = -N_j \text{ on } \Gamma \tag{5.84}$$

as an equation for η_j. The function ψ_j must satisfy the edge condition which can be formulated for this problem as

$$\lim_{\rho \to 0} \int_{S_\rho} \psi_j \frac{\partial \psi_j}{\partial N} ds = 0, \tag{5.85}$$

where S_ρ is the surface of the torus generated by a disk of radius ρ whose center moves along the edge L of F so that the disk is perpendicular to L. Condition (5.85) allows one to integrate over F as if F were a closed surface. Namely $\int_F = \int_{F_+} + \int_{F_-}$, where $F_+(F_-)$ is the upper (lower) side of F. It does not matter which of the two sides is chosen as the upper one. As $V \to 0$, it follows from (5.69) that

$$-\tilde{\beta}_{jj} = \max \left\{ \left(\int_F N_j(t) u_j(t) dt \right)^2 \left(\int_{D_e} |\nabla u_j|^2 dx \right)^{-1} \right\}, \tag{5.86}$$

where the maximum is taken over the set of harmonic functions satisfying the edge condition (5.85). For example, one can take the admissible functions of the form (5.77). The surface F is the surface of discontinuity for the admissible functions.

Passing to the limit $V \to 0$ in (5.71) yields

$$-\tilde{\beta}_{jj} = \min \int_{D_e} |q_j|^2 dx, \tag{5.87}$$

where q_j satisfies (5.72) and the edge condition (1.18).

Principles (5.86) *and* (5.87) *allow one to obtain lower and upper bounds for* $\tilde{\beta}_{jj}$ *respectively.*

From (5.53), (5.84), and (3.26) it follows that

$$-\beta_{pj} = st \frac{\int_F N_p(t)\eta_j(t)dt \int_F N_j(t)\eta_p(t)dt}{-\int_F \frac{\partial}{\partial N_t}\{\int_F \eta_p(s) \frac{\partial}{\partial N_s} \frac{1}{4\pi r_{st}} ds\}\eta_j(t)dt}, \tag{5.88}$$

where the admissible functions $\eta_j(t)$ should satisfy the edge condition (1.17). Principle (5.88) holds also for closed surfaces, in which case $\tilde{\beta}_{pj} = V\beta_{pj}$.

The integral in the denominator of (5.88) can be transformed by means of the identity [39]:

$$\frac{\partial}{\partial N_t} \int_F \eta(s) \frac{\partial}{\partial N_s} \frac{1}{r_{st}} ds = \int_F \left([N_s, \hat{\nabla}_s \eta], [N_t, \hat{\nabla}_s r_{st}^{-1}] \right) ds, \tag{5.89}$$

where $\hat{\nabla}$ is the surface gradient and $[a, b]$ is the vector product.

6. Polarizability tensors for plane screens. Let the x_3-axis be perpendicular to the screen. If $\Gamma = F$ in (5.49), (5.50) and F is a plane domain on the $(x_1, x_2)-$plane, then $\tilde{\alpha}_{i3} = \tilde{\alpha}_{3i} = 0$, for $1 \leq i \leq 3$.

Similarly from (5.83) and (5.86) it follows that only $\tilde{\beta} \equiv \tilde{\beta}_{33} \neq 0$ if $\Gamma = F$ is a plane screen. From (5.86) it follows that

$$-\tilde{\beta} = \max\left\{\left(\int_F u\,dt\right)^2\left(\int_{D_e}|\nabla u|^2 dx\right)^{-1}\right\}, \tag{5.90}$$

where the admissible functions u satisfy the edge condition, vanish at infinity, and are harmonic.

From (5.87) it follows that

$$-\tilde{\beta} = \min\int_{D_e}|q|^2 dx, \tag{5.91}$$

where the admissible vectors satisfy the conditions

$$\operatorname{div} q = 0 \text{ in } D_e, \quad q_3|_F = 1. \tag{5.92}$$

Exercise 5.2 Derive from (5.67) that

$$-\tilde{\beta} = \max\left\{\left(\int_F u(t)dt\right)^2\left(\int_F\int_F \frac{\hat{\nabla}_t u(t)\hat{\nabla}_s u(s)ds\,dt}{4\pi\, r_{st}}\right)^{-1}\right\} \tag{5.93}$$

7. The variational principles, i.e., principles involving a maximum or minimum, were derived only for the diagonal elements of the polarizability tensors. Nevertheless they allow one to obtain two-sided estimates for any elements of the tensors.

To do so one can use the transformation properties of tensors and take into account that any element of a selfadjoint matrix is a linear combination of its diagonal elements in the coordinates in which the matrix is diagonal.

Chapter 6

Iterative Methods: Mathematical Results

6.1 Iterative Methods of Solving the Fredholm Equations of the Second Kind at a Characteristic Value

The aim of this chapter is to provide in abstract setting some results which justify the iterative processes given in Chapter 2.

1. Let A be a linear compact operator on a Hilbert space H, λ_n, ϕ_n its characteristic values and eigenelements, $\phi_n = \lambda_n A\phi_n$, $|\lambda_1| < |\lambda_2| \le |\lambda_3| \le \cdots$. Let $G_1 \equiv \{\psi : (I - \bar{\lambda}_1 A^*)\psi = 0\}$ and $G_1^{\perp}$ be its orthogonal complement in H. The equation

$$g - \lambda_1 Ag = f \tag{6.1}$$

is solvable if and only if $f \in G_1^{\perp}$.

$$\text{Main assumption: } \lambda_1 \text{ is semisimple.} \tag{6.2}$$

This means that the pole $\lambda = \lambda_1$ of the resolvent $(I - \lambda A)^{-1}$ is simple. This also means that the root subspace of A corresponding to λ_1 coincides with the eigensubspace of A corresponding to λ_1. The root subspace is defined as follows. Let $\phi = \lambda_1 A\phi$. Consider the equations

$$\phi^{(j+1)} - \lambda_1 A\phi^{(j+1)} = \phi^{(j)}, \quad j \ge 0, \quad \phi^{(0)} = \phi. \tag{6.3}$$

Only a finite number r of these equations are solvable ([44]). If (6.3) has no solution for $j = 0$ then λ_1 is semisimple. If (6.3) is solvable for $0 \le j \le r$ and is not solvable for $j = r + 1$ then the set $\{\phi, \phi^{(1)}, \dots, \phi^{(r)}\}$ is called the Jordan chain of length $r + 1$ associated with the pair (λ_1, ϕ). The elements $\phi^{(1)}, \dots, \phi^{(r)}$ are called root vectors of A corresponding to λ_1. The linear span of all eigenvectors and root vectors corresponding to λ_1 is called the root space corresponding to λ_1. The linear span of eigenvectors corresponding to λ_1 is called the eigenspace corresponding to λ_1.

If the root space is one-dimensional then λ_1 is called simple. If the root space coincides with the eigenspace but has dimension greater than one then λ_1 is called semisimple. It can be proved that λ_1 is semisimple if and only if λ_1 is a simple pole of $(I-\lambda A)^{-1}$ ([44]). It can also be proved that λ_1 is semisimple iff

$$\left(I-\lambda_1 A\right)^2\phi = 0 \Longrightarrow \left(I-\lambda_1 A\right)\phi = 0. \tag{6.4}$$

Lemma 6.1 *If λ_1 is semisimple then equation* (6.1) *has at most one solution in $G_1^{\perp}$.*

Proof. It is sufficient to prove that the homogeneous equation (6.1) has only trivial solutions in $G_1^{\perp}$. Suppose $\phi = \lambda_1 A\phi$, $\phi \in G_1^{\perp}$, $\phi \neq 0$. Since $G_1^{\perp} = R(I-\lambda_1 A)$, where $R(A)$ denotes the range of A, and since $G_1^{\perp}$ is closed, because A is compact, the condition $\phi \in G_1^{\perp}$ implies that there exists an f such that $\phi = (I-\lambda_1 A)f$. Therefore $(I-\lambda_1 A)^2 f = 0$, and from (6.4) it follows that $(I-\lambda_1 A)f = 0$, i.e., $\phi = 0$. □

Remark 6.1 *Equation* (6.1) *with semisimple λ_1 is important because most of the basic equations of electrostatic, magnetostatics, elastostatics, and hydrodynamics of the ideal incompressible fluids are of this type. In practice f in* (6.1) *belongs to $G_1^{\perp}$ so that* (6.1) *is solvable. On the other hand, λ_1 is a characteristic value so that the resolvent $(I-\lambda A)^{-1}$ does not exist at $\lambda = \lambda_1$. Therefore solving equation* (6.1) *is an ill-posed problem: small perturbations of f can produce large perturbations in the solution or make equation* (6.1) *unsolvable. The theorems below show how to handle this difficulty and how to construct a stable approximation to the solution of* (6.1).

Let $\{\phi_j\}$ be an orthonormal basis of $N(I-\lambda_1 A) \equiv \{\phi : (I-\lambda_1 A)\phi = 0\}$ and let $\{\psi_j\}$ be an orthonormal basis of $G_1 = N(I-\bar{\lambda}_1 A^*)$, $1 \le j \le m$. Let P be the orthogonal projection of H onto G. Define

$$B_\gamma g := Ag + \gamma \sum_{j=1}^{m} (g, \psi_j)\psi_j \tag{6.5}$$

and

$$r_\gamma := \min\left(|\lambda_2|, \left|\lambda_1 \left(1+\gamma\lambda_1\right)^{-1}\right|\right), \tag{6.6}$$

where γ is an arbitrary number which will be so chosen that $r_\gamma = |\lambda_2|$ e.g., $\gamma = -\lambda_1^{-1}$ and $(\cdot,\cdot)$ denotes the inner product in H.

Consider the equation

$$g = \lambda_1 B_\gamma g + f, \quad f \in G_1^\perp. \tag{6.7}$$

It is clear that equations (6.7) and (6.1) are equivalent on $G_1^\perp$ because the sum in (6.5) vanishes if $g \in G_1^\perp$. Therefore every solution $g \in G_1^\perp$ of (6.7) is a solution of (6.1) and vice versa.

Theorem 6.1 *The operator B_γ defined in (6.5) has no characteristic values in the disk $|\lambda| < r_\gamma$. If $|\lambda_1(1+\gamma\lambda_1)^{-1}| > |\lambda_2|$, then the iterative process*

$$g_{n+1} = \lambda_1 B_\gamma g_n + F, \quad g_0 = F \equiv \lambda_1 Af - f, \quad F \in G_1^\perp \tag{6.8}$$

converges as a geometric series with ratio q, $0 < q < |\lambda_1\lambda_2^{-1}|$, to an element $g = \phi - f$, where $\phi \in N(I - \lambda_1 A)$ and $P\phi = Pf$. If $\dim G_1 = 1, \phi \in N(I-\lambda_1 A)$, $\psi \in G_1$ and $\|\psi\| = \|\phi\| = 1$, then $\phi = \phi(f,\psi)/(\phi,\psi)$. Process (6.8) is stable: the sequence

$$h_{n+1} = \lambda_1 B_\gamma h_n + F + \varepsilon_n, \quad h_0 = F, \quad \|\varepsilon_n\| < \varepsilon \tag{6.9}$$

satisfies the estimate

$$\limsup_{n\to\infty} \|g - h_n\| = O(\varepsilon), \tag{6.10}$$

where

$$g = \lim_{n\to\infty} g_n. \tag{6.11}$$

Theorem 6.2 *If $\dim G_1 = 1$, then the iterative process*

$$f_{n+1} = \lambda_1 A f_n, \quad f_0 = f \tag{6.12}$$

converges as a geometrical series with ratio $q = |\lambda_1\lambda_2^{-1}|$ to the element $a\phi, \phi \in N(I-\lambda_1 A), a = (f,\psi)/(\phi,\psi)$. Here $f \in H$ is arbitrary.

Proof of Theorem 6.1. If $g = \lambda B_\gamma g$, then $(g,\psi_j) = \lambda\lambda_1^{-1}(g,\psi_j) + \lambda\gamma(g,\psi_j)$, or $(g,\psi_j)(1 - \lambda\lambda_1^{-1} - \lambda\gamma) = 0$. If for some $j, 1 \le j \le m, (g,\psi_j) \ne 0$, then $\lambda = \lambda_1(1+\lambda_1\gamma)^{-1}$. If $(g,\psi_j) = 0$ for all $1 \le j \le m$, then $B_\gamma g = Ag, g = \lambda Ag$, i.e., $\lambda \in \sigma_1(A)$, where $\sigma_1(A)$ is the set of all characteristic values of A except λ_1 . The value λ_1 is excluded because if $(g,\psi_j) = 0, 1 \le j \le m$, then $g = 0$, since λ_1 is semisimple. Therefore the disk $|\lambda| < r_\gamma$ does not contain any characteristic values of B_γ. Our argument shows that

$\sigma(B_\gamma) \subset \{\sigma(A)\} \cup \{\lambda_1(1+\lambda_1\gamma)^{-1}\}$. If $g = \lambda B_\gamma g, g \in G_1^\perp$ then $g = \lambda A g$. Let us show that every $\lambda \in \sigma_1(A)$ belongs to $\sigma(B_\gamma)$. It is sufficient to prove that if $g = \lambda_n A g, n > 1$ then $g \in G_1^\perp$. In order to prove this, we start with the identity $(g, \psi_j) = \lambda_n (Ag, \psi_j) = \lambda_n (g, A^* \psi_j) = \lambda_n \lambda_1^{-1}(g, \psi_j)$. Thus $(g, \psi_j)(1 - \lambda_n \lambda_1^{-1}) = 0, 1 \leq j \leq m$. Since $\lambda_n \lambda_1^{-1} \neq 1$ it follows that $g \in G_1^\perp$. We have proved that every $\lambda \in \sigma_1(A)$ belongs to $\sigma(B_\gamma)$ and moreover the eigenvectors of A corresponding to $\lambda_n, n > 1$, are the eigenvectors of B corresponding to λ_n.

Let us prove that process (6.6) converges. If γ is chosen so that $|\lambda_1(1+\gamma\lambda_1)^{-1}| > |\lambda_2|$ then there are no characteristic values of B_γ in the disk $|\lambda| < |\lambda_2|$. Therefore process (6.8) converges as the geometric series with ratio $0 < q < |\lambda_1 \lambda_2^{-1}|$. Since $F \in G_1^\perp$ implies that $AF = B_\gamma F$, one can see that $g := \sum_{j=0}^\infty \lambda_1^j B_\gamma^j F = F + \lambda_1 B_\gamma g$ and $B_\gamma g = Ag$. Therefore $g + f = \lambda_1 A f + \lambda_1 B_\gamma g = \lambda_1 A f + \lambda_1 A g = \lambda_1 A(g+f)$. This means that $h := g + f \in N(I - \lambda_1 A)$. Since $Pg = 0$ we have $Ph = Pf$. If $\dim G_1 = 1$ then $\dim N(I - \lambda_1 A) = 1$. Let $\phi \in N(I - \lambda_1 A), \psi \in G_1, \|\phi\| = \|\psi\| = 1$. Then $h = c\phi, (h, \psi) = c(\phi, \psi)$, i.e., $c = (h,\psi)/(\phi,\psi) = (f,\psi)/(\phi,\psi)$. Note that $(\phi, \psi) \neq 0$ because λ_1 is semisimple, and $(g, \psi) = 0$ because $g \in G_1^\perp$. Let us prove (6.10). We have

$$h_n = \sum_{j=0}^{n} (\lambda_1 B_\gamma)^j F + \sum_{j=0}^{n-1} (\lambda_j B_\gamma)^j \varepsilon_{n-1-j}, \quad \|\lambda_1 B_\gamma\| \leq q < 1,$$

$$\|g - h_n\| \leq \varepsilon \sum_{j=0}^{n-1} q^j + \sum_{j=n+1}^{\infty} q^j \|F\| \leq \frac{\varepsilon + \|F\| q^{n+1}}{1-q}.$$

This implies (6.10). □

Proof of Theorem 6.2. First let us formulate and prove a lemma.

Lemma 6.2 *Let $f(\lambda)$ be a function of the complex variable λ with values in the set of linear bounded operators on a Banach space. Let $f(\lambda)$ be analytic in the disk $|\lambda| < r$ and meromorphic in the disk $|\lambda| < r + \varepsilon, \varepsilon > 0$. Suppose that λ_1 is a simple pole of $f(\lambda)$, $Res_{\lambda=\lambda_1} f(\lambda) = c$ and $f(\lambda) = \sum_{n=0}^\infty a_n \lambda^n$ for $|\lambda| < r$. If there are no other poles in the disk $|\lambda| < r + \varepsilon$, then*

$$\lim_{n\to\infty} \lambda_1^{n+1} a_n = -c. \tag{6.13}$$

Proof of Lemma 6.2. The function $f(\lambda) - c(\lambda - \lambda_1)^{-1}$ is analytic in the disk $|\lambda| < r + \varepsilon$. Therefore $f(\lambda) - c(\lambda - \lambda_1)^{-1} = \sum_{n=0}^\infty b_n \lambda^n, |\lambda| < r + \varepsilon$.

For $|\lambda| < r$ the identity $\sum_{n=0}^{\infty} b_n \lambda^n = \sum_{n=0}^{\infty}(a_n + c\lambda_1^{-n-1})\lambda^n$ holds. This identity can be analytically continued into the disk $|\lambda| < r + \varepsilon$. Thus $a_n + c\lambda_1^{-(n+1)} \to 0$ as $n \to \infty$. This implies (6.13). □

Let us prove Theorem 6.2: The function $(I - \lambda A)^{-1} f = \Sigma_{j=0}^{\infty} \lambda^j A^j f$ is analytic in the disk $|\lambda| < |\lambda_1|$, has a simple pole at $\lambda = \lambda_1$, and has no other poles in the disk $|\lambda| < |\lambda_2|$. Lemma 6.2 says that $\lim_{n\to\infty} \lambda_1^{n+1} A^n f = -c$ with the rate of convergence $O(|\lambda_1 \lambda_2^{-1}|^n)$. Since $f_n = \lambda_1^n A^n f$ we conclude that $\lim_{n\to\infty} f_n = h$ exists and $h = \lambda_1 A h$. If $\dim N(I - \lambda_1 A) = 1$, then $h = a\phi, \phi \in N(I - \lambda_1 A)$. Note that

$$(f_{n+1}, \psi) = \lambda_1 (A f_n, \psi) = (f_n, \psi) = \cdots = (f, \psi).$$

Therefore $a(\phi, \psi) = (f, \psi), a = (f, \psi)/(\phi, \psi)$. Theorem 6.2 is proved. □

Remark 6.2 *Process* (6.12) *is unstable in the sense that the process*

$$h_{n+1} = \lambda_1 A h_n + \varepsilon_n, \quad \|\varepsilon_n\| < \varepsilon, \quad h_0 = f \tag{6.14}$$

can diverge because $\lambda_1 \in \sigma(A)$, *where* $\sigma(A)$ *is the set of characteristic values of* A.

Let $\phi \in N(I - \lambda_1 A), \phi = \psi + h$, where $\psi \in G_1, h \in G_1^{\perp}$. From $\phi = \lambda_1 A \phi$ it follows that

$$h = \lambda_1 A h + F, \quad F \equiv \lambda_1 A \psi - \psi, \quad F \in G_1^{\perp}. \tag{6.15}$$

A stable iterative process for solution of (6.15) is given in Theorem 6.1, namely the process (6.8). In order to use it one must know a basis of G_1. In the case of electrostatics this basis is known explicitly (e.g., $\psi = 1$ in the case of a single conductor). In the general case one can find numerically an approximation to a basis of G_1. If $\{\psi_j\}$, $1 \le j \le m$, is an orthonormal basis of G_1 and $\|\psi_{j\varepsilon} - \psi_j\| < \varepsilon$, then the operator $B_{\gamma,\varepsilon} = \lambda_1 A + \gamma \sum_{j=1}^{m} (\cdot, \psi_{j\varepsilon}) \psi_{j\varepsilon}$ has no characteristic values in the disk $|\lambda| < |\lambda_1| + \delta$ where $\delta = \delta(\varepsilon) > 0$ and $\delta(\varepsilon) \to (|\lambda_2| - |\lambda_1|)$ as $\varepsilon \to 0$ provided that γ is chosen so that $|\lambda_1(1 + \gamma\lambda_1)^{-1}| > |\lambda_2|$. This follows from the uniform convergence $\|B_{\gamma,\varepsilon} - B_\gamma\| \to 0$ as $\varepsilon \to 0$.

Remark 6.3 *One can use the following general principle in order to construct a stable iterative process which converges to* $\phi \in N(I - \lambda_1 A)$:

Suppose that a convergent iterative process for solutions of the equation $Bg = f$ *is known for the exact data* f. *Then it is possible to construct a stable iterative process for solving this equation with perturbed (noisy) data* f_δ, $\|f_\delta - f\| < \delta$.

Indeed, let $S_n f$ be the nth approximation of the iterative process. We assume that each S_n is a continuous operator. We have

$$\|S_n f_\delta - g\| \leq \|S_n f_\delta - S_n f\| + \|S_n f - g\|. \tag{6.16}$$

Here g solves the equation $Bg = f$. By our assumption

$$\|S_n f - g\| \equiv a(n) \longrightarrow 0 \text{ as } n \longrightarrow \infty \tag{6.17}$$

and

$$\|S_n f_\delta - S_n f\| \equiv b(\delta, n), \quad b(\delta, n) \longrightarrow 0 \text{ as } \delta \longrightarrow 0. \tag{6.18}$$

The last limit is not uniform in n. Let us find for any given $\delta > 0$ such $n(\delta)$ that

$$b(\delta, n) + a(n) = \min := \alpha(\delta) \tag{6.19}$$

It follows from (6.17), (6.18) that

$$n(\delta) \longrightarrow \infty \text{ as } \delta \longrightarrow 0, \quad \alpha(\delta) \longrightarrow 0 \text{ as } \delta \longrightarrow 0. \tag{6.20}$$

Therefore

$$\|S_{n(\delta)} f_\delta - g\| \longrightarrow 0 \text{ as } \delta \longrightarrow 0. \tag{6.21}$$

Let us summarize this observation.

Proposition 6.1 *If a convergent iterative process for solution of the equation $Bg = f$ is known, $g_n = S_n f$ is the nth approximation of this process and each operator S_n is continuous, then $\|S_{n(\delta)} f_\delta - g\| \to 0$ as $\delta \to 0$ provided that $n(\delta)$ is chosen from* (6.19) *and $\|f_\delta - f\| \leq \delta$.*

In practice, if B is the linear operator $I - A$, then

$$S_n = \sum_{j=0}^{n} A^j, \quad \|S_n\| \leq \frac{\|A\|^{n+1} - 1}{\|A\| - 1}$$

if $\|A\| > 1$, and $\|S_n\| \leq n + 1$ if $\|A\| \leq 1$. This gives an explicit estimate for $b(\delta, n)$ (e.g., if $\|A\| \leq 1$ then $b(\delta, n) \leq \delta(n + 1)$). To estimate $a(n)$ in (6.19) one must use specific information about A. For example, under the assumptions of Theorem 6.2 one has $a(n) \leq c|\lambda_1 \lambda_2^{-1}|^n$.

2. The spectral radius of a linear bounded operator A on a Banach space is defined as $r(A) = \lim_{n\to\infty} \|A^n\|^{1/n}$. This limit always exists ([44]). If $|\lambda| > r(A)$, then $(A - \lambda I)^{-1}$ exists and is bounded. Let us assume that

$$r(A) = 1, \quad 1 \notin \sigma(A). \tag{6.22}$$

It is clear that the equation

$$g = Ag + f \tag{6.23}$$

is equivalent to the equation

$$g = Bg + f(1+t)^{-1}, \quad t \neq -1, \quad B \equiv (A + tI)(1+t)^{-1}. \tag{6.24}$$

Consider the iterative process

$$g_{n+1} = Bg_n + f(1+t)^{-1}, \quad g_0 = f(1+t)^{-1}, \quad t > 0. \tag{6.25}$$

Theorem 6.3 *If* (6.22) *holds, then the solution* g *of equation* (6.23) *can be obtained by the iterative process* (6.25)*:* $g = \lim_{n\to\infty} g_n$*. The process converges as a geometric series.*

Proof of Theorem 6.3. The equation

$$g = \lambda Bg + g_0 \tag{6.26}$$

coincides with (6.24) if $\lambda = 1$ and can be solved by iterations for sufficiently small $|\lambda|, |\lambda| < \delta$. Its solution

$$g(\lambda) = \sum_{n=0}^{\infty} \lambda^n B^n g_0 \tag{6.27}$$

is analytic in the disk $|\lambda| < \delta$. If $g(\lambda)$ has no singular points in the disk $|\lambda| \leq R$, then the series (6.27) converges in this disk. If $R > 1$ then the series converges for $\lambda = 1$ at the rate of the geometric series with ratio R^{-1}.

Let us prove that for some $R > 1$ the function $g(\lambda)$ is analytic in the disk $|\lambda| \leq R$. Let us rewrite (6.26)

$$g = zAg + bf, \quad z = \frac{\lambda}{1+t-\lambda t}, \quad b = \frac{1}{1+t-\lambda t}. \tag{6.28}$$

The solution of (6.28) is analytic in a domain Δ of the complex plane z. This domain includes the disk $|z| < 1$ and a neighborhood of the point $z = 1$. For any $t > 0$ one can find $R > 1$ such that the disk $|\lambda| \leq R$ is mapped by the function $z = \lambda(1+t-\lambda t)^{-1}$ onto a disk $K_r \subset \Delta$. This implies the conclusion of Theorem 6.3. Indeed, the function $z = \lambda(1+t-\lambda t)^{-1}$ is analytic in the disk $|\lambda| \leq R$, $1 < R < 1 + t^{-1}$, and maps this disk onto $K_r \subset \Delta$. The solution $g(z(\lambda))$ to (6.26) is analytic in the disk $|\lambda| \leq R$. Let us to show that for some $1 < R < 1 + t^{-1}$ the function $z = \lambda(1+t-\lambda t)^{-1}$ maps the disk $|\lambda| \leq R$ into Δ. Since $z(\lambda)$ is linear fractional it maps disks onto disks. Note that $z(\bar{\lambda}) = \overline{z(\lambda)}$ where the bars denote complex conjugation. Therefore the circle $|\lambda| = R$ is mapped onto the circle K_r with the diameter $[z(-R), z(R)]$, $r = [z(R) - z(-R)]/2$, and the center lies on the real axis at the point $[z(R) + z(-R)]/2$. Hence $K_r \subset \Delta$ provided that $z(-R) > -1$, $|z(R) - 1| < \alpha$ where $\alpha > 0$ is sufficiently small. We have $z(-R) = -R(1 + t + Rt)^{-1} > -1$ if $t > (R-1)/(R+1)$. If $t > 0$ is fixed, then $z(-R) > -1$ when $R < (1+t)/(1-t)$ for $t < 1, z(-R) > -1$ for any $R > 0$ when $t \geq 1$. On the other hand, $z(R) = R(1+t-tR)^{-1} < 1+\alpha$ if $R < (1+\alpha)(1+t)/(1+t(1+\alpha)) = 1 + \alpha[1 + t(1+\alpha)]^{-1}$. Therefore there exists $R > 1$ which satisfies the last inequality. We have proved that for some $R \in (1, 1+t^{-1})$ the function $z(\lambda)$ maps the disk $|\lambda| \leq R$ onto the disk $K_r \subset \Delta$. This completes the proof of Theorem 6.3. One can choose $t > 0$ so that R will be maximal and the rate of convergence of the process (6.25) will be fastest in this case. □

3. Let us formulate a well known theorem whose proof is left to the reader. Let A be a linear bounded operator on a Banach space X and $\sigma(A)$ be its characteristic set (i.e., the image of the spectrum of A under the mapping $z \to z^{-1}$).

Theorem 6.4 *If $\sigma(A) \subset \{\lambda : |\lambda| > 1\}$, then for every $f \in X$ the equation*

$$g = Ag + f \tag{6.29}$$

has a unique solution g, given by the iterative process

$$g_{n+1} = Ag_n + f, \quad g = \lim_{n\to\infty} g_n \tag{6.30}$$

for any initial approximation g_0. If there are points of $\sigma(A)$ in the disk $|\lambda| < 1$ then there exists a set $E \subset X$ such that E is of the second category and the process (6.30) *diverges if $f \in E$ and $g_0 = 0$.*

The set E is said to be of the second category if it is not a countable union of nowhere dense sets.

6.2 Iterative Processes for Solving Some Operator Equations

Let A be a selfadjoint linear operator on a Hilbert space $H, \|A\| = 1$. Consider the equation

$$g = Ag + f. \tag{6.31}$$

The following theorem is proved in [53].

Theorem 6.5 *Suppose that* -1 *is not a characteristic value of* A *and* (6.31) *is solvable. Then the iterative process*

$$g_{n+1} = Ag_n + f, \tag{6.32}$$

converges to a solution of (6.31) *for any* $g_0 \in H$.

Proof of Theorem 6.5. Let H_1 be the eigenspace of A corresponding to $\lambda = 1$ and let P_1 be the projection on H_1. If g is a solution to (6.31) then $g' = g - P_1 g$ is also a solution to (6.31) and $g' \perp H_1$. Let us prove that $g_n \to g' + P_1 g_0$ as $n \to \infty$. Let $0 < \delta < 1$ and $P_2 := \int_{-1+\delta}^{1-\delta} dE_\lambda$, $P_3 = I - P_1 - P_2$, where $A = \int_{-1}^{1} \lambda dE_\lambda$ is the spectral representation of A. The operator P_3 is an orthoprojection and since -1 is not a characteristic value of A, one has

$$\left\| P_3 f \right\| \longrightarrow 0 \text{ as } \delta \longrightarrow 0 \text{ for any fixed } f \in H. \tag{6.33}$$

Since $AP_j = P_j A$ and $P_i P_j = 0$ for $i \neq j$ one can rewrite (6.32) as

$$P_1 g_{n+1} = AP_1 g_n + P_1 f, \tag{6.34}$$

$$P_2 g_{n+1} = AP_2 g_n + P_2 f, \tag{6.35}$$

$$P_3 g_{n+1} = AP_3 g_n + P_3 f. \tag{6.36}$$

Since (6.31) is solvable, $P_1 f = 0$ and (6.34) shows that $P_1 g_n = P_1 g_0$. Let $H_2 = P_2 H$. The process (6.35) can be considered as an iterative process for the restriction A_2 of A to H_2. Since $\|A_\varepsilon\| \leq 1 - \delta$ the process (6.35) converges to $h := P_2 g'$ which is the solution to the equation $h = Ah + P_2 f$. Thus $\|P_2 g_n - P_2 g'\| < \varepsilon$ for $n > n(\varepsilon)$. Furthermore,

$$\begin{aligned} \left\| P_3 (g_n - g') \right\| &= \left\| A(P_3 g_{n-1} - P_3 g') \right\| \leq \left\| P_3 (g_{n-1} - g') \right\| \\ &\leq \cdots \leq \left\| P_3 (g_0 - g') \right\| < \varepsilon \end{aligned}$$

provided that δ is sufficiently small (see (6.33)).

Now one has

$$\begin{aligned} \left\| g_n - (g' + P_1 g_0) \right\| &\leq \left\| P_1 (g_n - g' - g_0) \right\| + \left\| P_2 (g_n - g') \right\| \\ &\quad + \left\| P_3 (g_n - g') \right\| < 2\varepsilon \end{aligned}$$

provided that $n > n(\varepsilon)$ and δ is sufficiently small. This completes the proof. $\square$

The following result is discussed in [53], [113] and in [107].

Theorem 6.6 *Every solvable linear equation with a bounded operator in a Hilbert space H can be solved by an iterative process.*

Proof of Theorem 6.6. Let the equation

$$Bg = f \tag{6.37}$$

be solvable in H and B be a linear bounded operator. The equation

$$Ag \equiv B^*Bg = B^*f \tag{6.38}$$

is equivalent to (6.37). Indeed, (6.37) implies (6.38). On the other hand, since (6.37) is solvable $f = Bh$ and (6.38) can be written as $B^*B(g-h) = 0$. Multiplying this by $g - h$ yields $B(g-h) = 0$, i.e., $Bg = Bh = f$. That is, (6.38) implies (6.37).

Equation (6.38) can be written as

$$g = (I - kA)g + F, \quad F = kB^*f, \tag{6.39}$$

where $k > 0$ is a constant. Suppose that

$$0 < k < 2\|A\|^{-1}. \tag{6.40}$$

Then the operator $I - kA$ is selfadjoint, -1 is not an eigenvalue of it and $\|I - kA\| \leq 1$. By Theorem 6.5, equation (6.39) is solvable by the iterative process

$$g_{n+1} = (I - kA)g_n + F, \tag{6.41}$$

with an arbitrary initial element $g_0 \in H$. $\square$

Remark 6.4 *Assume $0 < m \leq A \leq M$. Then*

$$\|I - kA\| = \frac{M - m}{M + m} \quad \textit{if} \quad k = \frac{2}{m + M}.$$

The following observation is useful ([143]).

Remark 6.5 *Let $B \geq 0$ be a linear operator on a Hilbert space H such that equation (6.37) is solvable. Then the iterative process*

$$g_{n+1} + Bg_{n+1} = g_n + f, \tag{6.42}$$

converges to a solution of (6.37) for any initial element $g_0 \in H$.

Proof of Remark 6.5. We have $g_{n+1} = (I+B)^{-1}g_n + (I+B)^{-1}f$. For the operator $A = (I+B)^{-1}$ the assumptions of Theorem 6.5 hold and Remark 6.5 follows from this theorem. □

6.3 Iterative Processes for Solving the Exterior and Interior Boundary Value Problems

1. Let D be a bounded domain with a smooth boundary Γ, and D_e be the exterior domain. Consider the problems

$$\triangle\, u = 0 \quad \text{in } D, \quad u|_\Gamma = f, \tag{6.43}$$

$$\triangle\, u = 0 \quad \text{in } D_e, \quad \left.\frac{\partial u}{\partial N_e}\right|_\Gamma = f, \quad u(\infty) = 0, \tag{6.44}$$

$$\triangle\, u = 0 \quad \text{in } D_e, \quad u|_\Gamma = f, \quad u(\infty) = 0, \tag{6.45}$$

$$\triangle\, u = 0 \quad \text{in } D, \quad \frac{\partial u}{\partial N_i} = f, \quad \int_\Gamma f\, dt = 0. \tag{6.46}$$

Define

$$v = \int_\Gamma \frac{\sigma(t)dt}{4\pi r_{xt}}, \quad w = \int_\Gamma \mu(t) \frac{\partial}{\partial N_t} \frac{1}{4\pi r_{xt}} dt. \tag{6.47}$$

One has:

$$w_{ie} = \frac{A^*\mu \mp \mu}{2}, \quad A^*\mu = \int_\Gamma \mu(t) \frac{\partial}{\partial N_t} \frac{1}{4\pi r_{xt}} dt. \tag{6.48}$$

$$\frac{\partial v}{\partial N_{ie}} = \frac{A\sigma \pm \sigma}{2}, \quad A\sigma = \int_\Gamma \sigma(t) \frac{\partial}{\partial N_s} \frac{1}{2\pi r_{st}} dt, \tag{6.49}$$

and $\partial w/\partial N_i = \partial w/\partial N_e$, provided that Γ is smooth. In (6.48) and (6.49) the upper (lower) signs correspond to the upper (lower) subscript $i(e)$.

Let $u = w(\mu)$ in (6.43) and $u = v(\sigma)$ in (6.44). Then, using (6.48)–(6.49), one gets:

$$\mu = A^*\mu - 2f, \tag{6.50}$$

$$\sigma = A\sigma - 2f. \tag{6.51}$$

It is known [38] that A and A^* have no characteristic values in the disk $|\lambda| < 1$ and only one characteristic value $\lambda = -1$ on the circle $\lambda = 1$. The operators A and A^* are compact in $C(\Gamma)$ and $H = L^2(\Gamma)$ if Γ is smooth. We have

Proposition 6.2 *Theorem 6.3 is applicable to equations* (6.50) *and* (6.51).

Remark 6.6 *Setting* $t = 1, B = A^*$ *in* (6.25) *yields the classical Neumann process for solving the interior Dirichlet problem which reduces to equation* (6.51).

From (6.46) and (6.49) it follows that problem (6.46) can be reduced to the integral equation

$$\sigma = -A\sigma + 2f, \qquad \int_\Gamma f\,dt = 0. \tag{6.52}$$

Equation (6.52) was discussed in detail in Chapter 2. Theorem 6.1 was basic in this discussion and the crucial assumption (6.2) is fulfilled for the operator A defined in (6.49). For this operator $\lambda_1 = -1$ and this λ_1 is *simple*, i.e., $\dim N(I + A) = 1$ and the function $\psi = 1$ belongs to $N(I + A^*) = G_1$. Condition (6.52) means that $f \in G_1^\perp$. Therefore equation (6.52) can be solved by the iterative process:

$$\sigma_{n+1} = -A\sigma_n + 2f, \quad \sigma_0 = 2f \tag{6.53}$$

or its modification (6.8) which guarantees the stability of the calculations with respect to small errors.

2. Let us discuss problem (6.45). If $u = w(\mu)$, then $\mu = -A^*\mu + 2f$. This equation may have no solutions and is not equivalent to problem (6.45) because the solution to (6.45) is not necessarily representable by a double-layer potential w. Therefore let us look for a solution to (6.45) of the form

$$u = \frac{a}{|x|} + \int_\Gamma \mu(t) \frac{\partial}{\partial N_t} \frac{1}{4\pi r_{xt}} dt, \quad a = \text{const} \tag{6.54}$$

From (6.54) and (6.48) it follows that

$$\mu = -A^*\mu + 2\left(f - \frac{a}{|s|}\right), \tag{6.55}$$

where s is a point on the boundary Γ.

Because the equation $\mu = -A^*\mu$ has a non-trivial solution, let us consider the equation

$$\nu = M\nu + 2\left(f - \frac{a}{|s|}\right), \quad M\nu \equiv -A^*\nu + \int_\Gamma \nu\, dt. \tag{6.56}$$

Proposition 6.3 *The operator M has no characteristic values in the disk $|\lambda| \le 1$, so that the iterative process*

$$\nu_{n+1} = M\nu_n + 2\left(f - \frac{a}{|s|}\right), \tag{6.57}$$

converges (in $C(\Gamma)$) to the solution of equation (6.56) for an arbitrary initial approximation $\nu_0 \in C(\lambda)$. Moreover, one can choose a so that equation (6.56) and (6.55) are equivalent, i.e., so that

$$\int_\Gamma \nu\, dt = 0. \tag{6.58}$$

This will be true if

$$a = \int_\Gamma Qf\, dt \left(\int_\Gamma Q\left(\frac{1}{|s|}\right) ds\right)^{-1}, \quad Q := (I - M)^{-1}. \tag{6.59}$$

Proof of Proposition 6.3. First let us prove that the disk $|\lambda| \le 1$ contains no characteristic values of M. Let

$$\nu = \lambda M\nu = -\lambda A^*\nu + \lambda \int_\Gamma \nu\, dt, \tag{6.60}$$

and

$$u(x) = \int_\Gamma \nu(t) \frac{\partial}{\partial N_t} \frac{1}{4\pi r_{xt}} dt. \tag{6.61}$$

Then from (6.60), (6.61), and (6.48) it follows that

$$(1 + \lambda)u_e = (1 - \lambda)u_i + \lambda \int_\Gamma (u_e - u_i) dt. \tag{6.62}$$

Multiplying (6.62) by $\overline{\frac{\partial u}{\partial N}} = \overline{\frac{\partial u}{\partial N_e}} = \overline{\frac{\partial u}{\partial N_i}}$ one obtains

$$\frac{1+\lambda}{1-\lambda} \int_\Gamma u_e \overline{\frac{\partial u}{\partial N_e}} dt = \int_\Gamma u_i \overline{\frac{\partial u}{\partial N_i}} dt + \frac{\lambda}{1-\lambda} \int_\Gamma (u_e - u_i) dt \int_\Gamma \overline{\frac{\partial u}{\partial N}} dt. \tag{6.63}$$

By Green's formula

$$\begin{aligned} &\int_\Gamma u_e \overline{\frac{\partial u}{\partial N_e}} dt = -\int_{D_e} |\nabla u|^2 dx \le 0, \\ &\int_\Gamma u_i \overline{\frac{\partial u}{\partial N_i}} dt = \int_D |\nabla u|^2 dx \ge 0, \quad \int_\Gamma \frac{\partial u}{\partial N_i} dt = 0. \end{aligned} \tag{6.64}$$

From (6.64) and (6.63) it follows that $(1+\lambda)(1-\lambda)^{-1} \le 0$. Hence λ is real and $|\lambda| \ge 1$. It remains to be proved that $\lambda = \pm 1$ is not a characteristic value of M. If $\lambda = -1$ then (6.63) shows that

$$\int_\Gamma u_i \overline{\frac{\partial u}{\partial N_i}} dt = \int_D |\nabla u|^2 dx = 0. \tag{6.65}$$

Therefore u is constant in D, $\partial u/\partial N_i = 0 = \partial u/\partial N_e$. Hence $u = 0$ in D_e and $\nu = u_e - u_i =$ const. Without loss of generality, suppose $\nu = 1$ is a solution to (6.60):

$$1 = A^*1 - S, \quad S := \operatorname{meas}\Gamma. \tag{6.66}$$

Let ν_0 be the electrostatic density, i.e.,

$$\nu_0 = -A\nu_0, \quad \int_\Gamma \nu_0 dt > 0. \tag{6.67}$$

Multiplying (6.66) by ν_0 and integrating over Γ one obtains

$$\int_\Gamma \nu_0 dt = (\nu_0, A^*1) - S \int_\Gamma \nu_0 dt,$$

or

$$(1+S) \int_\Gamma \nu_0 dt = -\int_\Gamma \nu_0 dt. \tag{6.68}$$

This is a contradiction. Therefore $\nu = 1$ is not a solution to (6.60).

If $\lambda = 1$ then $\nu = -A^*\nu + \int_\Gamma v\, dt$. The solvability condition is

$$\int_\Gamma \nu\, dt \int_\Gamma \nu_0 dt = 0. \tag{6.69}$$

Thus

$$\int_\Gamma \nu\, dt = 0 \tag{6.70}$$

and

$$\nu = -A^*\nu, \tag{6.71}$$

so that $\nu = \text{const} \neq 0$. This contradicts (6.70). Therefore $\lambda = 1$ is not a characteristic value of M. The other statements of Proposition 6.3 are obvious. □

Remark 6.7 *In practice, in order to find a from formula* (6.59) *one can use the processes*

$$h_{n+1} = Mh_n + \frac{1}{|s|}, \quad \lim_{n\to\infty} h_n = Q\left(\frac{1}{|s|}\right) \tag{6.72}$$

and

$$v_{n+1} = Mv_n + f, \quad \lim_{n\to\infty} v_n = Q(f), \tag{6.73}$$

and then find a from (6.59).

3. Consider the third boundary value problem

$$\Delta\, u = 0 \quad \text{in } D_e, \quad -\frac{\partial u}{\partial N_e} + hu|_\Gamma = f, \quad u(\infty) = 0, \tag{6.74}$$

$$\Delta\, u = 0 \quad \text{in } D, \quad \frac{\partial u}{\partial N_i} + hu|_\Gamma = f, \tag{6.75}$$

$$h = h_1 + ih_2, \quad h_1 \geq 0, \quad h_2 \leq 0, \quad |h_1| + |h_2| > 0. \tag{6.76}$$

It is easy to prove that under the assumption (6.76) problems (6.74) and (6.75) have at most one solution.

Let us look for the solution of (6.74) and (6.75) of the form

$$v = \int_\Gamma \frac{g(t)dt}{4\pi r_{xt}}. \tag{6.77}$$

Then problems (6.74) is reduced to the equation

$$g = Ag - Tg + 2f, \tag{6.78}$$

where A is defined in (6.49) and

$$Tg = h \int_\Gamma \frac{g(t)dt}{2\pi r_{st}} := hT_1 g. \tag{6.79}$$

The problem (6.75) is reduced to the equation

$$g = -Ag - Tg + 2f. \tag{6.80}$$

Consider the problem

$$g + Tg = \lambda A g. \tag{6.81}$$

Theorem 6.7 *If (6.76) holds then all the eigenvalues of (6.81) satisfy the inequality $|\lambda| > 1$ and they are real if $h > 0$. Moreover the equation*

$$g + Tg = \lambda A g + F, \quad \lambda = \pm 1 \tag{6.82}$$

can be solved by the iterative process

$$g_{n+1} + Tg_{n+1} = \lambda A g_n + F, \tag{6.83}$$

where $g_0 \in H = L^2(\Gamma)$ is arbitrary. This method converges as a geometric series.

Remark 6.8 *The iterative process*

$$g_{n+1} + Tg_n = Ag_n + F, \tag{6.84}$$

with an arbitrary $g_0 \in H$ converges if $0 < h < k$, where

$$k := \min \left\{ \int_D |\nabla u|^2 dx \left(\int_\Gamma |u|^2 dt \right)^{-1} \right\}. \tag{6.85}$$

Proof of Theorem 6.7. Let us rewrite (6.81) as

$$(1-\lambda)\frac{\partial v}{\partial N_i} + 2hv = (1+\lambda)\frac{\partial v}{\partial N_e}, \tag{6.86}$$

where v is defined in (6.77). Multiplying (6.86) by $\bar{v}$ and integrating over Γ yields

$$\frac{1-\lambda}{1+\lambda}\mathcal{A} + h\frac{\mathcal{B}}{1+\lambda} = \mathcal{C}, \tag{6.87}$$

where

$$\mathcal{A} = \int_\Gamma \frac{\partial v}{\partial N_i}\bar{v}\, dt > 0, \quad \mathcal{B} = 2\int_\Gamma |v|^2 dt > 0, \tag{6.88}$$

$$\mathcal{C} = \int_\Gamma \frac{\partial v}{\partial N_e}\bar{v}\, dt < 0. \tag{6.89}$$

If $\mathcal{A}$, $\mathcal{B}$, or $\mathcal{C}$ is zero then $v = 0$. Let $\lambda = a + ib$. Taking real and imaginary parts of (6.87) yields

$$\frac{(1 - a^2 - b^2)\mathcal{A} + [h_1(1+a) + h_2 b]\mathcal{B}}{(1+a)^2 + b^2} = \mathcal{C} < 0, \tag{6.90}$$

and

$$\frac{-2b\mathcal{A} + [h_2(1+a) - h_1 b]\mathcal{B}}{(1+a)^2 + b^2} = 0, \tag{6.91}$$

Hence

$$(1 - |\lambda|^2)\mathcal{A} + \big[h_1(1+a) + h_2 b\big]\mathcal{B} < 0, \tag{6.92}$$

$$h_2 = \frac{h_1\mathcal{B} + 2\mathcal{A}}{1+a} b. \tag{6.93}$$

Suppose that $|\lambda| \le 1$. Since $h_2 \le 0, h_1 \le 0$, and $|a| \le |\lambda| \le 1$, it follows from (6.92) that $b \le 0$. Thus $h_2 b > 0$. Therefore (6.92) cannot be valid. This contradiction proves that $|\lambda| > 1$. If $h = h_1 > 0, h_2 = 0$, then $b = 0$, i.e., all the eigenvalues are real-valued. In order to prove that the process (6.83) converges, let us consider the equation

$$g = \lambda G g + Q, \tag{$*$}$$

where $G := (I+T)^{-1}A, Q := (I+T)^{-1}F$. The operator $(I+T)^{-1}$ exists and is bounded because T is compact and $(I+T)f = 0$ implies that $f = 0$. The latter conclusion follows immediately from the positive definiteness of the operator $Re(I+T) = I + h_1 T_1, T_1 > 0, h_1 \ge 0$. The operator G has no characteristic values in the disk $|\lambda| \le 1$ (as was proved above). Therefore the iterative process

$$g_{n+1} = G g_n + Q, \tag{6.94}$$

with an arbitrary $g_0 \in H$, converges at the rate of a geometric series to the solution of the equation (*). The process (6.94) is equivalent to (6.83) and equation (6.82) is equivalent to (*). Theorem 6.7 is proved. □

Proof of Remark 6.8. Consider the equation

$$g = \lambda(-T + A)g. \tag{6.95}$$

If the characteristic values $|\lambda_j| > 1$ then process (6.84) converges. Let us find when $|\lambda_j| > 1$. Let us rewrite (6.95) as

$$(1-\lambda)\frac{\partial v}{\partial N_i} + 2\lambda h v = (1+\lambda)\frac{\partial v}{\partial N_e}. \tag{6.96}$$

From (6.96) it follows that

$$(1-\mu)\mathcal{A} + \mu\, h\mathcal{B} = (1+\mu)\mathcal{C}, \tag{6.97}$$

where $\mathcal{A}$, $\mathcal{B}$, $\mathcal{C}$ are defined in (6.88) and (6.89).

If $h > 0$ then as in the proof of Theorem 6.7 one can show that if $\lambda = a + ib$ then

$$(1-a)\mathcal{A} + ah\mathcal{B} = (1+a)\mathcal{C}, \tag{6.98}$$

$$-b\mathcal{A} + bh\mathcal{B} = b\mathcal{C}. \tag{6.99}$$

If $b \neq 0$ then from (6.99) and (6.98) it follows that $\mathcal{A} = \mathcal{C}$. This is a contradiction because of (6.88), (6.89). Thus $b = 0, \lambda = a$, and

$$\frac{1-a}{1+a}\mathcal{A} + \frac{a}{1+a}h\mathcal{B} < 0. \tag{6.100}$$

Suppose that $|a| < 1$. Then (6.100) cannot hold for $0 \leq a \leq 1$. If $-1 < a < 0$ then (6.100) can be written as

$$\frac{\int_D |\nabla u|^2 dx}{\int_\Gamma |u|^2 dt} < \frac{2|a|h}{1-|a|}\,\frac{1-|a|}{1+|a|} = \frac{2|a|h}{1+|a|} < h. \tag{6.101}$$

Since $|a| < 1$, one has $2|a|h/(1+|a|) < h$. Therefore (6.101) cannot hold if $k > h$, where k is defined in (6.85).

If $a = -1$ then (6.98) shows that

$$2\mathcal{A} = h\mathcal{B}, \quad \int_D |\nabla u|^2 dx \left(\int_\Gamma |u|^2 dt \right)^{-1} = h. \tag{6.102}$$

If $k > h$ the equality (6.102) cannot hold. This argument proves that if $k > h > 0$ then the process (6.84) converges.

If $h > k$ then equation (6.98) does not lead to a contradiction even if $|a| < 1$. In this case it is not known if the process (6.84) diverges for some F. □

4. Consider the problem

$$\Delta\, u = 0 \text{ in } D, \quad u|_\Gamma = f_1, \quad \left.\frac{\partial u}{\partial N_i}\right|_{\Gamma_2} = f_2, \quad \Gamma_1 \cup \Gamma_2 = \Gamma, \tag{6.103}$$

$\Gamma_1 \cap \Gamma_2 = \emptyset, \Gamma_1 \neq \emptyset$, where $\emptyset$ denotes the empty set. This problem was studied probably for the first time by Zaremba (1910). It has at most one solution. Numerical approaches to this problem have been studied recently by many authors and by means of various techniques (see [161] and the bibliography in this paper). In this section a simple approach taken from [91] is discussed. Consider the problem

$$\triangle\, v_h = 0 \quad \text{in } D, \quad \frac{\partial v_h}{\partial N_i} + h(s) v_h|_\Gamma = F, \tag{6.104}$$

$$F = \begin{cases} hf_1 & \text{on } \Gamma_1 \\ f_2 & \text{on } \Gamma_2 \end{cases}, \quad h(s) = \begin{cases} h & \text{on } \Gamma_1 \\ 0 & \text{on } \Gamma_2 \end{cases}, \quad h = \text{const} > 0. \tag{6.105}$$

The idea is to first solve (6.104) by an iterative process and then to show that $v_h \to u$ as $h \to +\infty$ and to establish the estimates

$$\|u - v_h\|_{H_1} \le ch^{-1}, \quad \|u - v_h\|_{\tilde{H}_2} \le ch^{-1}. \tag{6.106}$$

Here and below $c > 0$ denotes various constants, $H_1 = W_2^1$ is the Sobolev space [44], and $\tilde{H}_2 = W_2^2(\tilde{D})$, where $\tilde{D} \subset D$ is any fixed strictly inner subdomain of D, i.e., $\text{dist}(\bar{\tilde{D}}, \partial D) > 0$ where ∂D is the boundary of D and $\bar{\tilde{D}}$ is the closure of $\tilde{D}$.

Theorem 6.8 *The solution of* (6.104) *exists, is unique, and satisfies* (6.106) *where u is the solution of* (6.103). *Furthermore the solution of* (6.104) *can be calculated by means of the iterative process described in Theorem 6.7.*

Proof of Theorem 6.8. Let $w_h = v_h - u$. Then

$$\triangle\, w_h = 0 \quad \text{in } D, \quad \left.\frac{\partial w_h}{\partial N}\right|_{\Gamma_2} = 0, \quad \left.\frac{\partial w_h}{\partial N} + h w_h\right|_{\Gamma_1} = -\left.\frac{\partial u}{\partial N}\right|_{\Gamma_1}.$$

From this it follows that

$$\int_\Gamma w_h \frac{\partial w_h}{\partial N} dt + h \int_{\Gamma_1} |w_h|^2 dt = -\int_{\Gamma_1} w_h \frac{\partial u}{\partial N} dt.$$

Therefore

$$\int_D |\nabla w_h|^2 dx + h \int_{\Gamma_1} |w_h|^2 dt \le c\|w_h\|_{L^2(\Gamma_1)}, \quad c = \left\|\frac{\partial u}{\partial N}\right\|_{L^2(\Gamma_1)}.$$

Thus

$$\|w_h\|_{L^2(\Gamma_1)} \le ch^{-1}, \quad \int_D |\nabla w_h|^2 dx \le c^2 h^{-1}. \tag{6.107}$$

From (6.107) and the inequality

$$\|w_h\|_{L^2(D)} \le C_1\Big(\|\nabla w_h\|_{L^2(D)} + \|w_h\|_{L^2(\Gamma_1)}\Big) \qquad (6.108)$$

where $C_1 = C_1(D, \Gamma_1)$, the first estimate (6.106) follows. The second estimate (6.106) follows from the inequality

$$\|w_h\|_{\tilde{H}_2} \le C_2\Big(\| \Delta w\|_{L^2(D)} + \|w\|_{L^2(D)}\Big)$$

which is valid for any function $w \in W_2^2(D)$ and any $\tilde{D} \subset D$ which is a strictly inner subdomain of D [44].

It remains to be proved that problem (6.104) can be solved by an iterative process.

To this end one can use a generalization of Theorem 6.7. Define T as in (6.70) with $h = h(s)$, where $h(s)$ is defined in (6.105). Alternatively, one may assume that $0 < m \le h(s) \le M$ and that h is a piecewise-continuous function. The conclusion and the proof of Theorem 6.7 remain valid. The only new point in the proof is the invertibility of the operator $I + T$. This new point is discussed in the following lemma. □

Lemma 6.3 *Under the above assumptions on $h(s)$, the operator $(I+T)^{-1}$ is bounded and defined on all of $H = L^2(\Gamma)$.*

Proof of Lemma 6.3. Since T is compact it is sufficient to prove that $(^*) f + Tf = 0$ implies $f = 0$. If $0 < m \le h(s) \le M$ and $h^{-1/2}f = g$ then $g + Sg = 0$, where $S = h^{1/2}T_1h^{1/2}$ and T_1 is defined in (6.79). Therefore $S \ge 0$ and $I + S \ge I$. Thus $g = 0$ and $f = 0$. If $h(s)$ is defined in (6.105) then $(^*)$ shows that $f = 0$ on Γ_2 and

$$f(s) + h\int_{\Gamma_1} \frac{f(t)dt}{2\pi r_{st}} = 0, \quad s \in \Gamma_1, \quad h > 0. \qquad (6.109)$$

Since the kernel r_{st}^{-1} is positive semidefinite, (6.109) implies that $f = 0$ on Γ_1. This completes the proof. □

6.4 An Iterative Process for Solving the Fredholm Integral Equations of the First Kind with Pointwise Positive Kernel

In Section 2.4 a problem of practical interest was discussed, reduced to equation (2.35), and solved by means of the iterative process (2.36). Here we give a theoretical justification of this process in a general setting.

Consider the equation

$$Kf = \int_D K(x,y)f(y)dy = g(x), \quad x \in D \subset R^r, \tag{6.110}$$

where D is a bounded domain, the operator $K : L^2(D) \to L^2(D)$ is compact and

$$K(x,y) > 0 \tag{6.111}$$

almost everywhere. Suppose there exists a function $h(x) > 0$ such that $Kh \leq c$ and $\int_D a(x)dx < \infty$, where $a(x) := h(x)/(Kh(x))$. Let $\phi = fa^{-1}(x)$ and $H_\pm = L^2(D, a^{\pm 1}(x))$, $\|f\|_\pm^2 = \int_D |f|^2 a^{\pm 1}(x)dx$. Let us rewrite (6.110) as

$$K_1\phi = g, \quad K_1\phi \equiv \int_D K(x,y)a(y)\phi(y)dy = Ka\,\phi. \tag{6.112}$$

Let

$$Q = I - K_1, \quad K_1 f_j = \lambda_j f_j, \quad \lambda_1 > |\lambda_2| \geq |\lambda_3| \geq \cdots \tag{6.113}$$

The first eigenvalue of the integral operators with pointwise positive kernels is positive and simple, i.e., the corresponding eigenspace is one dimensional (Perron-Frobenius theorem for matrices, Jentsch theorem for integral operators, Krein-Rutman theorem for abstract operators [164]). Let us assume that

$$g(x) \in H_+, \tag{6.114}$$

$$0 < c_1(\Delta) \leq \int_\Delta K(x,y)a(y)dy \leq c_2(\Delta), \quad x \in D, \tag{6.115}$$

where $\Delta \subset D$, meas $\Delta > 0$,

$$\text{equation (6.112) is solvable in } H_+, \tag{6.116}$$

$$\text{the eigenfunctions } \{f_j\} \text{ form a Riesz basis of } H_+, \tag{6.117}$$

$$|\arg \lambda_j| \leq \frac{\pi}{3}, \quad \lambda_j \neq 0. \tag{6.118}$$

Theorem 6.9 *If the above assumptions (6.111)–(6.118) hold, then the iterative process*

$$\phi_{n+1} = Q\phi_n + g, \quad \phi_0 = g \tag{6.119}$$

converges in H_+ *to a solution* ϕ *of (6.112), the function* $f = a\phi$ *solves (6.110), and* $f \in H_-$.

Remark 6.9 *A complete minimal system* $\{f_j\} \subset H$ *forms a Riesz basis of the Hilbert space* H *if for any numbers* $c_1, \ldots, c_n$ *and any* n *the inequality*

$$a\sum_{j=1}^{n} |c_j|^2 \le \left\| \sum_{j=1}^{n} c_j f_j \right\|^2 \le b\sum_{j=1}^{n} |c_j|^2, \quad a > 0 \tag{6.120}$$

holds, where a *and* b *do not depend on* n.

Proof of Theorem 6.9. Let ϕ be a solution to (6.112), $g_n = \phi - \phi_n$. Then $g_n = Q^n g$. Let $g \sum_{j=1}^{\infty} c_j f_j$. Then

$$g_n = \sum_{j=1}^{\infty} (1 - \lambda_j)^n c_j f_j, \text{ and } |\lambda_j| < 1 \text{ if } j \ge 2.$$

From (6.118) it follows that $|1 - \lambda_j| < 1$. Indeed, if $\lambda = r\exp(i\psi), r < 1, |\psi| \le \pi/3$, then $|1 - \lambda|^2 = 1 + r^2 - 2r\cos\psi \le 1 + r^2 - r < 1$. Hence $|1 - \lambda_j|^n \to 0$ as $n \to \infty$. Therefore $\|g_n\|^2 \le b\sum_{j=1}^{\infty} |1 - \lambda_j|^{2n} |c_j|^2 \to 0$ as $n \to \infty$. This means that $\|\phi_n - \phi\|_{H_+} \to 0$ as $n \to \infty$. The rest is obvious. □

Example 6.1 Let $\Gamma = \{x : |x| = 1\}$, $m = 2$. Equation (6.110) is of the form

$$Af = \int_{-\pi}^{\pi} \ln \left| \frac{1}{2\sin\frac{\phi - \phi'}{2}} \right| f(\phi')d\phi' = g(\phi), \quad -\pi \le \phi \le \pi. \tag{6.121}$$

Since $\int_0^{\pi} \ln \sin x \, dx = -\pi \ln 2$ one has

$$\int_{\pi}^{\pi} \ln \left| \frac{1}{2\sin\frac{\phi - \phi'}{2}} \right| d\phi' = 0.$$

Therefore $f_0 = (2\pi)^{-1}$ is the solution of the homogeneous equation (6.121).

In this example equation (6.121), if solvable, is equivalent to the equation

$$Bf = -\int_{-\pi}^{\pi} \ln \left| \sin\{(\phi - \phi')/2\} \right| f(\phi')d\phi' = g(\phi), \quad -\pi \le \phi \le \pi \tag{6.122}$$

with the pointwise positive and selfadjoint kernel, provided that one looks for a solution of (6.121) which satisfies the condition $\int_0^{2\pi} f\,du = 0$. In this example $a(x) = (2\pi \ln 2)^{-1} := a$, $B_1 = aB$, $f = a\psi$ and (6.119) takes the form

$$\begin{aligned}\psi_{n+1}(\phi) = \psi_n(\phi) + (2\pi \ln 2)^{-1} \int_{-\pi}^{\pi} \ln \left| \sin\{(\phi - \phi')/2\} \right| \\ \times \psi_n(\phi')d\phi' + g(\phi)\psi_0 = g(\phi).\end{aligned} \tag{6.123}$$

Let $g(\phi) = \cos\phi$. Since

$$-\ln \left| \sin \frac{\phi' - \phi}{2} \right| = \ln 2 + \sum_{m=1}^{\infty} \frac{\cos\{m(\phi' - \phi)\}}{m} \tag{6.124}$$

one has

$$B\cos\phi = \pi\cos\phi; \quad -B_1\cos\phi = -(2\ln 2)^{-1}\cos\phi. \tag{6.125}$$

With this in mind, one concludes from (6.123) that

$$\begin{aligned}\psi_1 &= \cos\phi\big(2 - (\ln 4)^{-1}\big) = c_1\cos\phi, \\ \psi_2 &= \big(1 + c_1\big)\cos\phi - c_1(\ln 4)^{-1}\cos\phi \equiv c_2\cos\phi,\end{aligned}$$

and

$$\psi_{n+1} = c_{n+1}\cos\phi, \quad c_{n+1} = \big(1 + c_n\big) - c_n(\ln 4)^{-1}. \tag{6.126}$$

Thus

$$c_{n+1} = qc_n + 1, \quad c_0 = 1, \quad q = 1 - (\ln 4)^{-1} = 0.28. \tag{6.127}$$

Therefore

$$\begin{aligned}\lim c_n = c = \ln 4 &= 2\ln 2, \\ \psi = \lim \psi_n &= 2\ln 2\cos\phi, \\ \big|\psi - \psi_n\big| &\le (1-q)^{-1}q^{n+1}, \\ f &= \pi^{-1}\cos\phi.\end{aligned}$$

Chapter 7

Wave Scattering by Small Bodies

7.1 Introduction

Wave scattering by small bodies is of great interest in theory and applications. An incomplete list of problems for which wave scattering by small bodies is of prime importance includes: radio wave scattering by rain and hail, light scattering by cosmic dust, light scattering in colloidal solutions, light propagation in muddy water, wave scattering in a medium consisting of many small particles, ultrasound mammography, finding small cracks and holes in metals and other materials, detecting mines and other subsurface inhomogeneities from the scattered field, measured on the surface, etc. We will show that the skin effect for thin wires and radiation from small holes are a particular examples of the the theory of wave scattering by small bodies. The number of examples is practically unlimited. The theory was originated by Rayleigh (1871) who contributed to this field until his death (1919). Rayleigh understood that the main term in the scattering amplitude in the problem of wave scattering by a small body with diameter much less than the wavelength of the incident field is the dipole radiation. J. J. Thomson (1893) realized that for a small perfect conductor the magnetic dipole radiation is of the same order as the electric one. Some efforts were made in order to develop an algorithm for finding the expansion of the scattered field in powers of ka, where k is the wave number and a is the characteristic dimension of the scatterer, $ka \ll 1$ ([151], [50]). Since in many cases the first term of this expansion already provides a good approximation we will only discuss this first approximation. The general idea of our presentation is very simple. First, it will be shown that a low-frequency approximation to the scattering matrix can be calculated if the electric and magnetic polarizability tensors for the scatterer are known. In this chapter an explicit formula for the scattering matrix, S-matrix is derived. The entries of this matrix are expressed in terms of the polarizability tensors, for which approximate analytical formulas are derived in Chapter 5. These formulas allow one to calculate the polarizability tensors with any desired accuracy. Therefore we have derived explicit approximate analytical formulas for the S-matrix, which allow one to compute this matrix with any desired accuracy. Using these formulas one can write computer codes for calculating the scattering matrix for small bodies of arbitrary shapes. Exact solutions in closed form for the exterior problems of potential theory for bounded bodies in the three dimensional space are not known, except for ellipsoids.

The other important point which should be emphasized is that we study dependence of the scattering matrix on the boundary condition.

We study wave scattering by many small bodies. Two cases are considered: first, when the number r of these bodies is of order 10, not very large, and second, when this number is very large, say, of order 10^23, so that one has a medium consisting of many small bodies. In the first case, the smallness of the bodies allows one to reduce the problem to a linear algebraic system (see, e.g., equation (7.71) below), rather than to a system of integral equations, as in the case of wave scattering by many bodies which are not small. The scattering amplitude in the case of small bodies of arbitrary shapes is determined by finitely many numbers, which have physical meaning. In the second case, one derives an integro-differential equation (see equation (7.81)), or an integral equation in the simplest case, (see equation (7.62)), for the self-consistent field in the medium consisting of many small particles (see [146], [113]).

7.2 Scalar Wave Scattering: The Single-Body Problem

1. Consider the problem

$$(\nabla^2 + k^2)v = 0 \text{ in } D' := D_e, \tag{7.1}$$

$$\frac{\partial v}{\partial N} - hv|_\Gamma = \left(- \frac{\partial u_0}{\partial N} + hu_0 \right)\bigg|_\Gamma, \tag{7.2}$$

$$|x|\left(\frac{\partial v}{\partial |x|} - ikv \right) \longrightarrow 0 \quad \text{as } |x| \longrightarrow \infty, \tag{7.3}$$

where u_0 is the incident field, D' is the exterior domain with smooth boundary Γ, $h = \text{const}, h = h_1 + ih_2$, $h_2 \leq 0$, $h_1 \geq 0$, $k > 0$, $D = R^3 \setminus D'$ is the interior bounded domain. Let us look for a solution of (7.1)–(7.3) of the form

$$v(x) = \int_\Gamma g(x, s, k)\sigma(s)ds, \quad g = \frac{\exp(ik|x-s|)}{4\pi|x-s|}. \tag{7.4}$$

The scattering amplitude $f(n, k)$ is defined by the formula

$$v \sim \frac{\exp(ik|x|)}{|x|} f(n, k), \quad |x| \longrightarrow \infty, \quad n = x|x|^{-1}. \tag{7.5}$$

From (7.4) and (7.5) it follows that

$$f(n, k) = (4\pi)^{-1} \int_\Gamma \exp\{-ik(n, s)\}\sigma(s)ds. \tag{7.6}$$

Substituting (7.4) into (7.2) yields

$$\sigma = A(k)\sigma - hT(k)\sigma - 2hu_0 + 2\frac{\partial u_0}{\partial N}, \tag{7.7}$$

where

$$A(k)\sigma = 2\int_\Gamma \frac{\partial}{\partial N_s} g(s, t, k)\sigma(t)dt, \tag{7.8}$$

$$T(k)\sigma = 2\int_\Gamma g(s, t, k)\sigma(t)dt. \tag{7.9}$$

Let us expand σ, $A(k)$, $T(k)$, and u_0 in powers of k.

$$\sigma = \sigma_0 + ik\sigma_1 + \frac{(ik)^2}{2}\sigma_2 + \cdots, \tag{7.10}$$

$$A(k) = A + ikA_1 + \frac{(ik)^2}{2}A_2 + \cdots, \tag{7.11}$$

$$T(k) = T + ikT_1 + \frac{(ik)^2}{2}T_2 + \cdots, \tag{7.12}$$

$$u_0 = u_{00} + iku_{01} + \frac{(ik)^2}{2}u_{02} + \cdots, \tag{7.13}$$

From (7.10)–(7.13) and (7.7) it follows that

$$\sigma_0 = A\sigma_0 - hT\sigma_0 - 2hu_{00} + 2\frac{\partial u_{00}}{\partial N}, \tag{7.14}$$

$$\sigma_1 = A\sigma_1 - hT\sigma_1 + A_1\sigma_0 - hT_1\sigma_0 - 2hu_{01} + 2\frac{\partial u_{01}}{\partial N}, \tag{7.15}$$

$$\begin{aligned} \sigma_2 = A\sigma_2 - hT\sigma_2 + A_2\sigma_0 + 2A_1\sigma_1 - hT_2\sigma_0 \\ - 2hT_1\sigma_1 - 2hu_{02} + 2\frac{\partial u_{02}}{\partial N}. \end{aligned} \tag{7.16}$$

From (7.15) and (7.6) it follows that

$$\begin{aligned} 4\pi f &= \int_\Gamma \left[1 - ik(n,s) + \frac{(ik)^2}{2}(n,s)^2 + \cdots\right]\left[\sigma_0 + ik\sigma_1 + \frac{(ik)^2}{2}\sigma_2 + \cdots\right] ds \\ &= \int_\Gamma \sigma_0 ds + ik\left[\int_\Gamma \sigma_1 ds - \int_\Gamma (n,s)\sigma_0 ds\right] \\ &\quad + \frac{(ik)^2}{2}\left[\int_\Gamma \sigma_2 ds - 2\int_\Gamma (n,s)\sigma_1 ds + \int_\Gamma \sigma_0 (n,s)^2 ds\right] + \cdots \end{aligned} \tag{7.17}$$

Let us assume that

$$u_0 = \exp\{ik(\nu, x)\}. \tag{7.18}$$

Then

$$\begin{aligned} u_{00} = 1, \quad u_{01} = (\nu, s), \quad u_{02} = (\nu, s)^2 \\ \frac{\partial u_{00}}{\partial N} = 0, \quad \frac{\partial u_{01}}{\partial N} = (\nu, N), \quad \frac{\partial u_{02}}{\partial N} = 2(\nu, s)(\nu, N). \end{aligned} \tag{7.19}$$

We note that the following formulas hold

$$A\sigma = \int_\Gamma \frac{\partial}{\partial N_s}\frac{1}{2\pi r_{st}}\sigma(t)dt, \quad A_1\sigma = 0, \tag{7.20}$$

$$-\int_\Gamma A\sigma dt = \int_\Gamma \sigma\, dt, \tag{7.21}$$

$$T\sigma = \int_\Gamma \frac{\sigma\, dt}{2\pi r_{st}}, \tag{7.22}$$

$$T_1\sigma = (2\pi)^{-1}\int_\Gamma \sigma(t)dt. \tag{7.23}$$

Let us integrate (7.14) over Γ and take into account (7.19)–(7.21). This yields

$$2\int_\Gamma \sigma_0 dt = -2hS - h\int_\Gamma\int_\Gamma \frac{\sigma_0(t)dt\,ds}{2\pi r_{st}}, \quad S = \operatorname{meas}\Gamma,$$

or

$$\int_\Gamma \sigma_0 dt = -hS - \frac{h}{4\pi}\int_\Gamma\int_\Gamma \frac{\sigma_0(t)dt\,ds}{r_{st}}. \tag{7.24}$$

The exact value of σ_0 should be found from the integral equation (7.14). An approximate value of $\int_\Gamma \sigma_0 dt$ can be found from (7.24) if one uses the approximation

$$\int_\Gamma \frac{ds}{r_{st}} \approx \frac{1}{S}\int_\Gamma dt \int_\Gamma \frac{ds}{r_{st}} = JS^{-1}, \quad J \equiv \int_\Gamma\int_\Gamma r_{st}^{-1} ds\,dt. \tag{7.25}$$

From (7.25) and (7.24) it follows that

$$\int_\Gamma \sigma_0 dt \approx -\frac{hS}{1 + hJ(4\pi S)^{-1}}. \tag{7.26}$$

In Chapter 3 the approximate formula

$$C \approx C^{(0)} = 4\pi S^2 J^{-1}, \quad \varepsilon_0 = 1 \tag{7.27}$$

was given. Combining (7.26) and (7.27) yields

$$\int_\Gamma \sigma_0 dt = -hS\left(1 + hSC^{-1}\right)^{-1}. \tag{7.28}$$

Therefore

$$f(n,k) \approx -hS\left(1 + hSC^{-1}\right)^{-1}\frac{u_{00}}{4\pi}. \tag{7.29}$$

If $h = \infty$, i.e., the scatterer is a perfect conductor, then

$$f = -\frac{C}{4\pi}u_{00}. \tag{7.30}$$

From (7.29) and (7.30) it follows that *the scattering from a small body of arbitrary shape under the Dirichlet boundary condition (i.e., acoustically soft body, $h = \infty$) or under the impedance boundary condition ($h < \infty$) is isotropic and the scattering amplitude is of order a, where a is the characteristic dimension of the scatterer.*

Note that if the scatterer is not too prolate then $C \sim a$. We also assumed above that h is not too small, e.g., $hS > C^{-1}$.

The scattering amplitude $f = -\frac{Cu_{00}}{4\pi}(1 + ikaf_1 + O((ka)^2))$, where $ka << 1$ and f_1 is a real number, because of the property $f(-k) = \overline{f(k)}$, where the overbar stands for complex conjugate. Therefore, the differential cross-section is $|f|^2 = \frac{C^2|u_{00}|^2}{16\pi^2}(1 + O((ka)^2))$. If $ka < 0.1$, then the first term (7.30) is practically the dominant term since the next term is of order of 10^{-2} of the main term in the formula for the differential cross-section, which is measured in experiments. A similar remark holds in relation to (7.39) below.

3. Consider now the case when $h = 0$, i.e., the case of the acoustically rigid body. We shall see that *in this case the scattering is anisotropic, is defined by the magnetic polarizability tensor and the scattering amplitude is of order* k^2a^3. If $h = 0$ then (7.14) takes the form $\sigma_0 = A\sigma_0$ and therefore $\sigma_0 = 0$ since 1 is not an eigenvalue of A. Equation (7.15) takes the form

$$\sigma_1 = A\sigma_1 + 2\frac{\partial u_{01}}{\partial N}. \tag{7.31}$$

Integrating (7.31) over Γ and using (7.21) yields

$$\int_\Gamma \sigma_1 dt = \int_\Gamma \frac{\partial u_{01}}{\partial N} dt = \int_D \Delta u_{01} dx = 0, \tag{7.32}$$

since $\Delta u_{00} = 0$ and $\Delta u_{01} = 0$. The latter equations follow from the equation

$$(\Delta + k^2)u_0 = 0$$

and the asymptotic expansion (7.13).

Thus, in the case $h = 0$ formula (7.17) takes the form

$$4\pi f(n,k) = -\frac{k^2}{2}\int_\Gamma \sigma_2 ds + k^2 \int_\Gamma (n,s)\sigma_1 ds. \tag{7.33}$$

For the initial field (7.18) it follows from (7.31) that

$$\left(n, \int_\Gamma s\sigma_1 ds\right) = -V\beta_{pq}\nu_q n_p. \tag{7.34}$$

Here and below one should sum over the repeating indices, V denotes the volume of the scatterer and β_{pq} is the magnetic polarizability tensor defined in Chapter 5 as $V\beta_{pq} = \mu_0^{-1}\int_\Gamma s_p\sigma_q(s)ds$, where σ_q is the solution of the equation $\sigma_q = A\sigma_q - 2N_q$ and N is the unit outer normal to Γ. In order to calculate the term $\int_\Gamma \sigma_2 ds$, let us rewrite (7.16) for $h = 0$ as

$$\sigma_2 = A\sigma_2 + 2\frac{\partial u_{02}}{\partial N}. \tag{7.35}$$

Here we have used (7.20) and took into account that $\sigma_0 = 0$. From (7.19) and (7.35) it follows that

$$\sigma_2 = A\sigma_2 + 4(\nu, s)(\nu, N). \tag{7.36}$$

Integrating (7.36) over Γ and taking into account (7.21) yields

$$\begin{aligned}\int_\Gamma \sigma_2 dt &= 2\int_\Gamma (\nu, s)(\nu, N)ds = 2\left(\nu, \int_\Gamma N(\nu, s)ds\right)\\ &= 2\left(\nu, \int_D \nabla(\nu, x)dx\right) = 2(\nu, \nu)V = 2V.\end{aligned} \tag{7.37}$$

From (7.33), (7.34), and (7.37) it follows that if $h = 0$ and the initial field is given by (7.18) then the scattering amplitude is:

$$f(n, \nu, k) = -\frac{k^2 V}{4\pi} - \frac{k^2 V}{4\pi}\beta_{pq}\nu_q n_p, \quad f \sim k^2 a^3. \tag{7.38}$$

The scattering is *anisotropic* in this case (i.e., in the case $h = 0$, i.e., in the case of acoustically hard obstacle).

4. Let us derive the following formula for the scattering amplitude in the case $h = 0$ for an arbitrary initial field u_0:

$$f(n, k) = \frac{ikV}{4\pi}\beta_{pq}\frac{\partial u_0}{\partial x_q}n_p + \frac{V\Delta u_0}{4\pi}. \tag{7.39}$$

The initial field satisfies the equation $(\Delta + k^2)u_0 = 0$, so $\Delta u_0 = -k^2 u_0$ and one has $\operatorname{grad} u_0 = O(k)$ for $k \to 0$. *The main assumption is the smallness of the scatterer.* We want to derive formula (7.39) for two reasons. First, the initial field u_0 is not assumed to be a plane wave. Second, we want to isolate the dependence of the scattering amplitude on the size of the body from its dependence on the wave number k. When the initial field is $u_0 = \exp\{ik(\nu, x)\}$ and the scatterer is placed at the origin, then the small parameter is ka, so that $k \to 0$ is equivalent to $x \to 0$. But if we consider the many-body problem then the phase difference should be taken into account. For example, if k is small but x is large then $ik(\nu, x)$ is not necessarily small, and such a situation occurs in the many-body problem if the distance between some of the bodies is larger than the wavelength.

As in Section 7.3 we consider the problem (7.1)–(7.3) with $h = 0$ and look for the solution of the form (7.4). The integral equation for σ takes the form (7.7) with $h = 0$. If a is very small we can rewrite this equation as

$$\sigma = A\sigma + 2\frac{\partial u_0}{\partial N}, \tag{7.40}$$

where A is defined in (7.20) and the error is $O(a)$.

Let us rewrite formula (7.6) as

$$f(n, k) = (4\pi)^{-1} \int_\Gamma \sigma(s) ds - ik(4\pi)^{-1} \int_\Gamma (n, s)\sigma(s) ds, \tag{7.41}$$

where the terms of the order $O(k^2a^2)$ are omitted because $ka \ll 1$. We expand the initial field u_0 in the Taylor series with respect to x, assuming that the origin is placed inside the scatterer. This yields

$$u_0(x, k) = u_{00} + (\nu, x) + \frac{1}{2}(Bx, x) + O(a^3), \tag{7.42}$$

where $u_{00} = u_0(0, k)$,

$$\nu = \nabla u_0(x, k)|_{x=0}, \quad (B)_{mj} = b_{mj} = \left.\frac{\partial^2 u_0(x, k)}{\partial x_m \partial x_j}\right|_{x=0} \tag{7.43}$$

Therefore

$$\frac{\partial u_0}{\partial N} = (\nu, N) + (Bs, N), \tag{7.44}$$

where $s \in \Gamma$. Integrating (7.40) over Γ and taking into account (7.21), one obtains

$$\int_\Gamma \sigma\, ds = \int_\Gamma (Bs, N) ds = V trB = V\Delta u_0|_{x=0}, \tag{7.45}$$

where tr is the trace and the formula $\int_\Gamma (\nu, N)ds = 0$ was used. Furthermore, one obtains

$$-ik(4\pi)^{-1} n_q \int_\Gamma s_q \sigma(s) ds = -ik(4\pi)^{-1} n_p \nu_q \beta_{pq} V, \tag{7.46}$$

where $\sigma = A\sigma + 2\nu_p N_p$, one sums up over the repeated indices, $\int_\Gamma s_q\sigma(s)ds = -V\nu_p\beta_{pq}$, $\beta_{pq} = \beta_{qp}$ is the magnetic polarizability tensor defined in Chapter 5 as

$$V\beta_{pq} = \int_\Gamma s_q \sigma_p(s) ds, \tag{7.47}$$

where σ_p is the solution of the equation

$$\sigma_p = A\sigma_p - 2N_p. \tag{7.48}$$

Formula (7.39) follows from (7.45), (7.46), and (7.43). In calculating the integral in (7.46) one can neglect the term (Bs, N) in the right-hand side of (7.44) because this term is of order $O(a)$, while $(\nu, N) = O(1)$.

7.3 Scalar Wave Scattering: The Many-Body Problem

1. Consider scattering by r bodies. Let

$$D = \bigsqcup_{j=1}^{r} D_j, \quad \Gamma = \bigsqcup_{j=1}^{r} \Gamma_j, \quad D_j \cap D_i = \emptyset, \quad i \neq j, \quad \Omega = R^3 \setminus D, \tag{7.49}$$

where $\emptyset$ denotes the empty set, $R^3 \setminus D$ denotes the complement of D in R^3, and Γ_j is the boundary of D_j. Let

$$h|_{\Gamma_j} = h_j = h_{1j} + ih_{2j}, \quad h_{1j} \geq 0, \quad h_{2j} \leq 0, \quad |h_j| > 0, \tag{7.50}$$

$$a = \max_{1 \leq j \leq r} a_j, \tag{7.51}$$

$$d = \min_{i,j} d_{ij}, \quad i \neq j \tag{7.52}$$

$$\ell = \max_{i,j} d_{ij}, \quad \int_\Gamma \equiv \sum_{j=1}^{r} \int_{\Gamma_j}, \tag{7.53}$$

where d_{ij} is the distance between D_i and D_j

Consider the problem (7.1)–(7.3), which looks formally identical in the cases $r = 1$ and $r > 1$. As in Section 7.2 we look for a solution of the form (7.4) and define the scattering amplitude by formula (7.5). The scattering amplitude can be written as in (7.6), $\sigma = (\sigma_1, \dots, \sigma_r)$ but an important difference between cases $r = 1$ and $r > 1$ is that if $r = 1$ then $|s| \sim a$ in (7.6), while if $r > 1$ the magnitude $|s|$ can be large.

Let us denote by s_j some point inside D_j and rewrite (7.6) for the case $r > 1$ as

$$f(n,k) = (4\pi)^{-1} \sum_{j=i}^{r} \int_{\Gamma_j} \exp\Big\{ -ik\big(n, s - s_j\big)\Big\} \sigma_j(s)ds \exp\Big\{ -ik\big(n, s_j\big)\Big\}. \tag{7.54}$$

In (7.54) the magnitudes $|s - s_j| \sim a$ if $s \in \Gamma_j$ and $|s - s_j| \sim d_{ij}$ if $s \in \Gamma_i$, $i \neq j$. The integral equations for σ_j, $1 \leq j \leq r$, can be obtained by substituting (7.4) into the boundary condition (7.2). This yields

$$\begin{aligned} \sigma_j &= A_j(k)\sigma_j - h_j T_j(k)\sigma_j + \sum{}' A_{jp}(k)\sigma_p - h_j \sum{}' T_{jp}(k)\sigma_p \\ &\quad + 2\frac{\partial u_0}{\partial N} - 2h_j u_0, \quad 1 \leq j \leq r, \quad \sum{}' := \sum_{p=1, p\neq j}^{r}, \end{aligned} \tag{7.55}$$

where

$$A_{jp}\sigma_p = \int_{\Gamma_p} \frac{\partial}{\partial N_{s_j}} \frac{\exp(ikr_{s_j t_p})}{2\pi r_{s_j t_p}} \sigma_p\big(t_p\big)dt_p, \tag{7.56}$$

$$T_{jp}\sigma_p = \int_{\Gamma_p} \frac{\exp(ikr_{s_j t_p})}{2\pi r_{s_j t_p}} \sigma_p\big(t_p\big)dt_p. \tag{7.57}$$

Suppose that

$$d \gg a. \tag{7.58}$$

If one neglects the terms A_{jp} and T_{jp} for $j \neq p$ in (7.55) then for σ_j one obtains the same equations as for a single body in Section 7.2. Therefore for $h_j \neq 0$ the scattering amplitude can be calculated from the formula

$$f(n,k) = -\frac{1}{4\pi} \sum_{j=1}^{r} \exp\Big\{ -ik\big(n, s_j\big)\Big\} \frac{h_j S_j}{1 + h_j S_j C_j^{-1}} u_{0j}, \tag{7.59}$$

where C_j is the capacitance of the jth body, S_j is the area of its surface, $u_{0j} = u_0(s_j, k)$ (see formula (7.29)). If we assume that every small body is affected by the self-consistent field u in the medium consisting of many small bodies, then (7.59) takes the form

$$f(n,k) = -\frac{1}{4\pi} \int \exp\big\{ -ik(n,y)\big\} q(y)u(y,k)dy, \tag{7.60}$$

where $q(y)$ is the "effective potential" which is defined as

$$q(y) = N(y)\frac{hS}{1 + hSC^{-1}}. \tag{7.61}$$

Here $N(y)$ is the number of the small bodies (particles) per unit volume and $hS(1 + hSC^{-1})$ is the average value of $h_jS_j(1 + h_jS_jC_j^{-1})^{-1}$ in a neighborhood of the point y. The integral in (7.60) is taken over the domain where $N(y) \neq 0$. The self-consistent field u satisfies the equation

$$u(x, k) = u_0(x, k) - \int \frac{\exp(ik|x - s|)}{4\pi|x - s|} q(y)u(y, k)dy, \tag{7.62}$$

which is obtained by taking the limit $r \to \infty$ in the formula

$$u(x, k) = u_0(x, k) - \sum_{j=1}^{r} \frac{\exp(ik|x - s_j|)}{4\pi|x - s_j|} \frac{h_jS_j}{1 + hSC^{-1}} u(s_j, k). \tag{7.63}$$

Equation (7.62) can be written as the Schrödinger equation

$$\left[\nabla^2 + k^2 - q(x)\right]u(x, k) = 0, \tag{7.64}$$

$$u - u_0 \sim \frac{\exp(ik|x|)}{4\pi|x|} f(n, k) \text{ as } |x| \longrightarrow \infty. \tag{7.65}$$

2. If the number r of the small scatterers is not very large ($r \sim 10$) then the scattering amplitude and the scattered field can be found from a linear system of algebraic equations. The matrix of the system has dominant main diagonal so that the system is easily solvable by iterations. In order to prove this statement let us look for the solution of the problem (7.1)–(7.3) (with $\Gamma = \bigsqcup_{j=1}^{r} \Gamma_j$) of the form

$$v = \sum_{j=1}^{r} \int_{\Gamma_j} \frac{\exp(ik|x - s|)}{4\pi|x - s|} \sigma_j(s)ds. \tag{7.66}$$

In general, in order to find v one derives a system of integral equations for finding σ_j, $1 \leq j \leq r$. In our case, when $ka \ll 1$, the scattering amplitude depends just on finitely many numbers Q_j:

$$\begin{aligned} f(n, k) &= \frac{1}{4\pi} \sum_{j=1}^{r} \int_{\Gamma_j} \exp\{-ik(n, s)\}\sigma_j(s)ds \\ &= \frac{1}{4\pi} \sum_{j=1}^{r} \exp\left\{-ik(n, s_j)\right\}Q_j + O(ka), \end{aligned} \tag{7.67}$$

where

$$Q_j = \int_{\Gamma_j} \sigma_j(t)dt, \tag{7.68}$$

and we assume that $Q_j \neq 0$, $1 \leq j \leq r$ This is the case when $h_j \neq 0$. Consider, for example, the Dirichlet boundary condition ($h_j = \infty$) (acoustically soft particles). Then from (7.66) and (7.2) it follows that

$$\begin{aligned} &\int_{\Gamma_m} \frac{\exp(ik|x_m - s|)}{4\pi|x_m - s|}\sigma_m(s)ds \\ &+ \sum_{m\neq j, j=1}^{r} \int_{\Gamma_j} \frac{\exp(ik|x_m - s|)}{4\pi|x_m - s|}\sigma_j ds = -u_0(x_m, k), \quad 1 \leq m \leq r. \end{aligned} \tag{7.69}$$

If $ka \ll 1$, then this system can be written with the accuracy $O(ka)$ as

$$\int_{\Gamma_m} \frac{\sigma_m ds}{4\pi|x_m - s|} = -u_0(x_m, k) - \sum_{m\neq j, j=1}^{r} \frac{\exp(ik|x_m - s_j|)}{4\pi|x_m - s_j|}Q_j, \quad 1 \leq m \leq r. \tag{7.70}$$

Equation (7.70) can be considered as an equation for the electrostatic charge distribution σ_m on the surface Γ_m of the perfect conductor charged to the potential given by the right-hand side of (7.70). Therefore the total charge on Γ_m is

$$Q_m = \int_{\Gamma_m} \sigma_m ds = C_m \left\{ -u_0(x_m, k) - \sum_{m\neq j, j=1}^{r} \frac{\exp(ik|x_m - s_j|)}{4\pi|x_m - s_j|}Q_j \right\},$$

where C_m is the electrical capacitance of the perfect conductor with boundary Γ_m. The above system of equations for $Q = (Q_1, \ldots, Q_r)$ can be written as

$$AQ = b, \tag{7.71}$$

where

$$\begin{aligned} &A = (a_{mj}), \quad a_{mj} := \delta_{mj} + C_m \frac{\exp(ik|x_m - s_j|)}{4\pi|x_m - s_j|}, \\ &b_m = -C_m u_0(x_m, k), \quad \delta_{mj} = \begin{cases} 1, & m = j, \\ 0, & m \neq j. \end{cases} \end{aligned} \tag{7.72}$$

If the particles are not too prolate then $C_m \sim a$. The matrix A will have dominant main diagonal if

$$(4\pi)^{-1} r a\, d^{-1} < 1, \tag{7.73}$$

where d is defined in (7.52). If condition (7.73) holds then the system (7.71) can be solved by iterations and the scattering amplitude can be found from formula (7.67). The scattered field v can be found from the formula

$$v = \sum_{j=1}^{r} \frac{\exp(ik|x - s_j|)}{4\pi|x - s_j|} Q_j \tag{7.74}$$

with the accuracy $O(ka)$. If $h \neq 0$ then the scattering amplitude can be calculated from (7.67) and (7.68), and the linear algebraic system for Q_m can be obtained from (7.55). To this end let us integrate (7.55) over Γ_j, yielding

$$Q_j = -Q_j - \frac{h_j J_j}{2\pi S_j} Q_j + \sum_{p \neq j} d_{jp} Q_p - 2h_j S_j u_0(s_j). \tag{7.75}$$

Here we have used arguments similar to those given in Section 7.2, subsection 2, and the following notations:

$$J_j = \int_{\Gamma_j} \int_{\Gamma_j} \frac{ds\, dt}{r_{st}}, \quad d_{jp} = \int_{\Gamma_j} ds \left\{ \frac{\partial}{\partial N_s} \frac{\exp(ikr_{st_p})}{2\pi r_{st_p}} - h_j \frac{\exp(ikr_{st_p})}{2\pi r_{st_p}} \right\}.$$

Equation (7.75) can be written as

$$\tilde{A} Q = \tilde{b}, \tag{7.76}$$

$$(\tilde{A}_{jp}) = \tilde{a}_{jp} = \delta_{jp}\left(1 + \frac{h_j J_j}{4\pi S_j}\right) - \tilde{d}_{jp}, \quad \tilde{d}_{jp} = \frac{d_{jp}}{2}, \quad \tilde{b}_j = -h_j S_j u_0(s_j). \tag{7.77}$$

The linear system (7.76) can be solved by iterations if

$$1 + \frac{h_j J_j}{4\pi S_j} > \sum_{p \neq j, p=1}^{r} |\tilde{d}_{jp}|, \quad 1 \leq j \leq r. \tag{7.78}$$

If $h_j = 0$, $1 \leq j \leq r$, then $Q_j = 0$ and formula (7.67) for the scattering amplitude becomes more complicated. This was shown in Section 7.2. If we consider each of the small bodies as being affected by the self-consistent field u, then from (7.67) and (7.39) it follows that

$$f(n,k) = (4\pi)^{-1}\sum_{j=1}^{r} \exp\Big\{-ik(n,s_j)\Big\}\Big\{ikV_j\beta_{pq}^{(j)}\frac{\partial u}{\partial x_q}n_p + V \Delta u\Big\}, \quad (7.79)$$

where V_j is the volume of the jth body and $\beta_{pq}^{(j)}$ is its magnetic polarizability tensor. The same argument leads to the following formula for the self-consistent field in the medium:

$$u = u_0 + \sum_{j=1}^{r}\frac{\exp(ik|x-s_j|)}{4\pi|x-s_j|}\Big\{ikV_j\beta_{pq}^{(j)}\frac{\partial u(s_j,k)}{\partial y_q}n_p + V_j\Delta u(s_j,k)\Big\}, \quad (7.80)$$

where s_j is the radius vector of the jth body.

If one passes to the limit as $r \to \infty$ in (7.71) then the integro-differential equation for the field u takes the form

$$u(x,k) = u_0(x,k) + \int \frac{\exp(ik|x-y|)}{4\pi|x-y|}\Big(ikB_{pq}(y)\frac{\partial u}{\partial y_q}\frac{x_p-y_p}{|x-y|} + b(y)\Delta u\Big)dy, \quad (7.81)$$

where one must sum over the repeating indices, the integral is taken over the domain where $b(y) \neq 0$, $b(y)$ is the average volume of the bodies near the point y, and $B_{pq}(y)$ is the average magnetic polarizability tensor. That is, if $K_h(y)$ is the ball of radius h centered at y, then

$$b(y) = \lim_{h\to 0}\frac{\sum V_j}{|K_h(y)|}, \quad B_{pq}(y) = \lim_{h\to 0}\frac{\sum V_j\beta_{pq}^{(j)}}{|K_h(y)|}, \quad (7.82)$$

where $|K_h(y)|$ is the volume of $K_h(y)$ and $\sum$ denotes the sum over the bodies which are located in the ball $K_h(y)$. The vector $(x_p - y_p)/|x-y|$ in formula (7.81) replaces n_p in formula (7.80).

7.4 Electromagnetic Wave Scattering

1. Let us consider the scattering by a single homogeneous body D with characteristic dimension a. Let ε, μ, σ be its dielectric permeability, magnetic permeability, and conductivity, ε_0, μ_0, $\sigma_0 = 0$ be the corresponding parameters of the exterior medium, $\varepsilon' = \varepsilon + i\sigma\omega^{-1}$, ω be the frequency of the initial field, λ_0 be its wavelength and $k_0 = 2\pi\lambda_0^{-1}$. Let $\lambda = \lambda_0(|\varepsilon'\mu|)^{-1/2}$ be the wavelength in the body, and $\delta = (\frac{2}{\omega\gamma\delta})^{1/2}$ be the depth of the skin layer, where we assume that $\varepsilon << \sigma$. We consider wave scattering under

the following assumptions, which will be discussed separately:

$$|\varepsilon'| \gg 1, \quad \delta \gg a, \quad k_0 a \ll 1, \tag{7.83}$$

$$|\varepsilon'| \gg 1, \quad \delta \ll a, \quad k_0 a \ll 1, \tag{7.84}$$

$$\left|(\varepsilon' - \varepsilon_0)\varepsilon_0^{-1}\right| + \left|(\mu - \mu_0)\mu_0^{-1}\right| \ll 1. \tag{7.85}$$

Assumption (7.83) corresponds to a small dielectric body. Assumption (7.84) corresponds to a small well-conducting body. Assumption (7.85) corresponds to the case when the body does not differ much from the exterior medium. This assumption does not require the body to be small. Our aim is to derive explicit analytical approximate formulas for the scattering amplitude and for the scattering matrix.

2. The basic equations are

$$\operatorname{curl} E = i\omega\mu H, \quad \operatorname{curl} H = -i\omega\varepsilon' E \text{ in } D, \tag{7.86}$$

$$\operatorname{curl} E = i\omega\mu_0 H, \quad \operatorname{curl} H = -i\omega\varepsilon_0 E + j_0 \text{ in } D', \tag{7.87}$$

where D' is the exterior domain with respect to D. The boundary conditions are

$$N \times E \text{ and } \mu H \cdot N \text{ are continuous when crossing } \Gamma, \tag{7.88}$$

where Γ is the boundary of D and N is the outward pointing unit normal at the boundary.

If $\sigma = \infty$ then

$$N \times E = 0 \text{ on } \Gamma, \tag{7.88'}$$

This case can occur only under assumption (7.84). In (7.87) j_0 is the initial current source. Let

$$A_0 = \int G(x,y) j_0 dy, \quad G = \frac{\exp(ik_0|x-y|)}{4\pi|x-y|}, \quad k_0^2 = \omega^2\varepsilon_0\mu_0, \quad \int = \int_{R^3} \tag{7.89}$$

and

$$E_0 = \frac{1}{-i\omega\varepsilon_0}\left(\operatorname{curl}\operatorname{curl} A_0 - j_0\right), \quad H_0 = \operatorname{curl} A_0, \tag{7.90}$$

The total field can be found from the formulas

$$E = E_0 + E_1, \quad H = H_0 + H_1 \tag{7.91}$$

where

$$E_1 = \frac{1}{-i\omega\varepsilon_0}\operatorname{curl}\operatorname{curl} A - \operatorname{curl} F,$$
$$H_1 = \frac{1}{-i\omega\mu_0}\operatorname{curl}\operatorname{curl} F + \operatorname{curl} A, \tag{7.92}$$

and

$$A = \int_\Gamma G(x,s) N \times H_1 ds, \quad F = -\int_\Gamma G(x,s) N \times E_1 ds. \tag{7.93}$$

Remark 7.1 *Let us assume* (7.84). *If one tries to calculate the scattering using the approximations* $N \times E_1 = -N \times E_0$ *on* Γ *and* $N \times H_1 = 0$ *on* Γ, *this leads to wrong results (for example one can take the spherical scatterer* D *and use the known explicit solution to check the above statement). Therefore the above approximations, which are used in geometrical optics, are not valid for our low-frequency problem.*

3. Let $n = x|x|^{-1}$, $|x| = r$ and

$$f = f_E(n,k) = \lim_{|x|\to\infty, x|x|^{-1}=n} |x| \exp\big(-ik|x|\big) E_1. \tag{7.94}$$

Let us prove, assuming (7.84), that

$$f = \frac{k_0^2}{4\pi\varepsilon_0}\big[n,[P,n]\big] + \frac{k_0^2}{4\pi}\left(\frac{\mu_0}{\varepsilon_0}\right)^{1/2}[M,n], \tag{7.95}$$

where P and M are the electric and magnetic dipole moments induced on the body by the initial field and $[A,B] = A \times B$ is the vector product. Let us consider the vector potential in the far-field region and keep the first two terms of its expansion in powers of ka:

$$\begin{aligned} A &= \int_D j(y) G(x,y) dy \\ &= \frac{\exp(ik_0|x|)}{4\pi|x|}\int_D dy j(y) \exp\{-ik_0(n,y)\} \\ &= \frac{\exp(ik_0|x|)}{4\pi|x|}\left\{\int_D j(y)dy - ik_0\int_D (n,y) j(y) dy + \cdots\right\} \\ &= \frac{\exp(ik_0|x|)}{4\pi|x|}\{-i\omega P - ik_0 M \times n\}, \end{aligned} \tag{7.96}$$

where

$$P = \int_D y\rho(y)dy, \quad M = \frac{1}{2}\int_D [y, j]dy, \tag{7.97}$$

and $\rho = (i\omega)^{-1}\operatorname{div} j$. Indeed, using the condition $(j, N) = 0$ on Γ, one gets:

$$\begin{aligned} -i\omega P &= -i\omega \int_D y\rho(y)dy = -\int_D y \operatorname{div} j\, dy \\ &= -\int_\Gamma (j, N) y\, ds + \int_D j\, dy = \int_D j\, dy. \end{aligned} \tag{7.98}$$

Furthermore,

$$\int_D (n, y) j\, dy = \frac{1}{2}\int_D \big([y, j] \times n + j(n, y) + y(n, j)\big)dy = M \times n, \tag{7.99}$$

where we have used the relations:

$$\int_D \big(j(n, y) + y(n, j)\big)dy + \int_D y(n, y) \operatorname{div} j\, dy = \int_\Gamma y(j, N)(y, n)ds = 0, \tag{7.100}$$

and

$$\int_D y(n, y) \operatorname{div} j\, dy = i\omega \int_D y(n, y)\rho\, dy \approx 0.$$

In the far-field region, $j = 0$ and $E_1 = (-i\omega\varepsilon_0)^{-1} \operatorname{curl}\operatorname{curl} A$. Therefore from (7.96) and (7.94) it follows that

$$\begin{aligned} f &= -\big(4\pi i\omega\varepsilon_0\big)^{-1} ik_0 n \times \Big[ik_0 n \times \big\{-i\omega P - ik_0 M \times n\big\}\Big] \\ &= \frac{k_0^2}{4\pi\varepsilon_0} n \times [P, n] + \frac{k_0^2}{4\pi}\left(\frac{\mu_0}{\varepsilon_0}\right)^{1/2} M \times n. \end{aligned} \tag{7.101}$$

If the domain D shrinks to a surface S, then (7.101) still holds with

$$P = \int_S s\sigma(s)ds, \quad M = \frac{1}{2}\int_S s \times j\, ds. \tag{7.102}$$

Algorithms and formulas for calculating P and M are given in Chapter 5.

Under the assumption (7.83) the magnetic dipole radiation can be neglected if $\mu = \mu_0$ because the eddy currents are negligible if $\delta \gg a$. Under the assumption (7.84) the magnetic dipole radiation is of the order of the electric dipole radiation even in the case $\mu = \mu_0$.

In general, the magnetic polarizability vector can be calculated from the formula

$$M_i = \tilde{\beta}_{ij} V \mu_0 H_j, \tag{7.103}$$

where

$$\tilde{\beta}_{ij} = \alpha_{ij}(-1) + \alpha_{ij}(\gamma_\mu), \quad \gamma_\mu = \frac{\mu - \mu_0}{\mu + \mu_0} \tag{7.104}$$

and $\alpha_{ij}(\gamma)$ is defined in Section 5.1. We denote

$$\alpha_{ij}(-1) := \beta_{ij}. \tag{7.105}$$

If $\mu = \mu_0$ then $\alpha_{ij}(\gamma_\mu) = 0$.

Remark 7.2 *Suppose that D is a metallic body. In this case the current can be calculated by the formula: $j = N \times H$, where H is the magnetic field on the surface of the body. Let H^1 denote the magnetic field on the surface Γ of the ideal magnetic insulator D, i.e., a body with $\mu = 0$. This field is the value on Γ of the solution of the problem*

$$\operatorname{curl} H = 0, \quad \operatorname{div} H = 0 \text{ in } D', \quad N \cdot H = 0 \text{ on } \Gamma, \quad H(\infty) = H^0, \tag{7.106}$$

where H^0 is a given constant field. In the quasistatic problem H^0 is the initial field at the point where the small body is placed. If $\delta \ll a$ then neither magnetic nor electric field can penetrate into the body and therefore the body behaves like a perfect magnetic insulator in the initial homogeneous magnetic field H^0. Under the assumption (7.83) a good approximation for $N \times H$ is $N \times H^1$. This approximation leads to the correct value of M. On the other hand, this approximation leads to a *wrong* value of P. Let us show this in the case when D is a ball of radius a. The magnetic field H^1 in this case is known explicitly:

$$H^1 = H^0 - \frac{a^3}{2|x|^3}\left\{3\frac{x(x, H^0)}{|x|^2} - H^0\right\}. \tag{7.107}$$

If Γ is a sphere of radius a and s is a point on Γ, then $N \times s = 0$. Therefore

$$N \times H^1 = \frac{3}{2}[N, H^0], \tag{7.108}$$

and

$$-i\omega P = \int_\Gamma j\,dy = \int_\Gamma N \times H^1 dy = \frac{3}{2}\int_\Gamma [N, H^0] ds = 0,$$

which is wrong. Thus one can calculate M using the approximation

$$j = N \times H^1 \tag{7.109}$$

if the body is metallic, but this approximation cannot be used for, calculating P.

4. Let us calculate f under the assumption (7.85). Equations (7.86) and (7.87) can be written as

$$\operatorname{curl} E = i\omega\mu_0 H + i\omega(\mu - \mu_0)\eta H, \tag{7.110}$$

$$\operatorname{curl} H = -i\omega\varepsilon_0 E + j_0 - i\omega(\varepsilon' - \varepsilon_0)\eta E \tag{7.111}$$

where

$$\eta = \begin{cases} 1, & x \in D, \\ 0, & x \notin D. \end{cases} \tag{7.112}$$

Let us set

$$j_e = -i\omega(\varepsilon' - \varepsilon_0)\eta E, \quad j_m = -i\omega(\mu - \mu_0)\eta H, \tag{7.113}$$

$$A = \int G(x,y) j_e dy, \quad F = \int G(x,y) j_m dy, \quad \int := \int_D . \tag{7.114}$$

Then the vectors E_1, H_1 defined in formula (7.91) can be found from the formulas

$$E_1 = -(i\omega\varepsilon_0)^{-1}(\operatorname{curl}\operatorname{curl} A - j_e) - \operatorname{curl} F, \tag{7.115}$$

$$H_1 = -(i\omega\mu_0)^{-1}(\operatorname{curl}\operatorname{curl} F - j_m) + \operatorname{curl} A. \tag{7.116}$$

From (7.113)–(7.116) and (7.91) one gets

$$\begin{aligned} E(x) = E_0(x) &+ \frac{\varepsilon' - \varepsilon_0}{\varepsilon_0} \operatorname{curl}\operatorname{curl} \int G(x,y) E\, dy \\ &- \frac{\varepsilon' - \varepsilon_0}{\varepsilon_0} \eta E + i\omega(\mu - \mu_0) \operatorname{curl} \int G(x,y) H\, dy, \end{aligned} \tag{7.117}$$

$$\begin{aligned} H = H_0(x) &+ \frac{\mu - \mu_0}{\mu_0} \operatorname{curl}\operatorname{curl} \int_D G(x,y) H\, dy \\ &- \frac{\mu - \mu_0}{\mu_0} \eta H - i\omega(\varepsilon' - \varepsilon_0) \operatorname{curl} \int G(x,y) E\, dy. \end{aligned} \tag{7.118}$$

The system (7.117)-(7.118) can be solved by iterations if

$$\left(\left|\frac{\mu-\mu_0}{\mu_0}\right|+\left|\frac{\varepsilon'-\varepsilon_0}{\varepsilon_0}\right|\right)(1+k_0^2a^2)\ll 1, \tag{7.119}$$

where a is the characteristic dimension of the domain D, say, half of its diameter. Indeed, under the assumption (7.119) the norm in $L^2(D)$ of the operator of system (7.117)-(7.118) is less than 1. Let us verify this statement. We have

$$\left\|\frac{\varepsilon'-\varepsilon_0}{\varepsilon_0}\eta E\right\|\le\left|\frac{\varepsilon'-\varepsilon_0}{\varepsilon_0}\right|\|E\|. \tag{7.120}$$

Here and below $\|\cdot\|$ denotes the $L^2(D)$ norm and c denotes various constants. Furthermore,

$$\left\|\operatorname{curl}\int G(x,y)E\,dy\right\|\le c\big(1+k_0a\big)\|E\|, \tag{7.121}$$

$$\left\|\operatorname{curl}\operatorname{curl}\int G(x,y)E\,dy\right\|\le c\big(1+k_0^2a^2\big)\|E\|. \tag{7.122}$$

Inequalities (7.121) and (7.122) can be proved as follows. Note that

$$4\pi G(x,y)=\frac{1}{|x-y|}+ik_0-\frac{k_0^2|x-y|}{2}+O\big(k_0^3|x-y|^2\big), \tag{7.123}$$

if $k_0|x-y|\ll 1$. Hence

$$4\pi D^2G=D^2\frac{1}{|x-y|}+O\left(\frac{k_0^2}{|x-y|}\right). \tag{7.124}$$

We have

$$\left\|\int\frac{E\,dy}{x-y}\right\|\le\|E\|\left(\int\int\frac{dx\,dy}{|x-y|^2}\right)^{1/2}=\|E\|O(a^2), \tag{7.125}$$

$$\left\|D_x\int\frac{E\,dy}{x-y}\right\|\le\|E\|O(a),$$

and

$$\left\|\int\frac{E\,dy}{|x-y|}\right\|_{W_2^2(D)}\le c\|E\|, \tag{7.126}$$

where $W^{\ell,p}(D) = W_p^{\ell}(D)$ are the Sobolev spaces ([29]). Inequality (7.126) is known in the theory of elliptic boundary-value problems (see, e.g., [44]). From the above estimates it follows that the norm of the oprator in (7.117)-(7.118) is less than one.

Let $Af := \int_D f(y)|x-y|^{-a}dy,\ D \in \mathbb{R}^n$. The operator $A : L^p(D) \to L^q(D)$ is bounded provided that $a = n(1-m), 0 < m \leq 1, 0 \leq \delta \leq p^{-1} - q^{-1} < m$. Moreover, $||A||_{L^p(D)\to L^q(D)} \leq c|D|^{m-\delta}$, where $|D|$ is the measure (volume) of D, $c = c(m,\delta,n) = const > 0$. (see, e.g.,[29]).

Inequality (7.119) holds even if the body is large ($k_0 a \gg 1$) provided that the quantity

$$\left|\frac{\varepsilon' - \varepsilon_0}{\varepsilon_0}\right| + \left|\frac{\mu - \mu_0}{\mu_0}\right|$$

is sufficiently small.

Let us set

$$g(n) = \int \exp\{-ik_0(n,y)\}dy, \tag{7.127}$$

iterate once system (7.117)-(7.118) and calculate the scattering amplitude. This yields

$$\begin{aligned} f =& -\frac{\varepsilon' - \varepsilon}{4\pi\varepsilon}k_0^2 n \times \left[n \times \int \exp\left(-ik_0 n \cdot y\right)E_0(y)dy\right] \\ & -\frac{k_0\omega(\mu-\mu_0)}{4\pi} n \times \int \exp\left(-ik_0 n \cdot y\right)H_0 dy. \end{aligned} \tag{7.128}$$

If E_0 and H_0 are the values of the electromagnetic field at the point where the small body D is situated, then an approximate formula for f can be written as

$$f = \left(-\frac{\varepsilon' - \varepsilon}{4\pi\varepsilon}k_0^2\, n \times \left[n \times E_0\right] - \frac{k_0\omega(\mu-\mu_0)}{4\pi}\, n \times H_0\right)g(n).$$

If D is a ball of radius a, then

$$g(n) = 4\pi a^3 \frac{\sin(k_0 a) - k_0 a \cos k_0 a}{(k_0 a)^3}. \tag{7.129}$$

If D is a cylinder with radius a and length $2L$, then

$$g(n) = 2L\frac{\sin(k_0 L\cos\theta)}{k_0 L\cos\theta}\,\frac{J_1(k_0 a\sin\theta)}{k_0 a\sin\theta}, \tag{7.130}$$

where θ is the angle between the axis of the cylinder and the unit vector n, and $J_1(x)$ is the Bessel function.

5. Many-body electromagnetic wave scattering can be developed along the lines of Section 7.3.

6. Let us derive the following formula for the scattering matrix for the electromagnetic wave scattering by a single body under the assumption (7.84):

$$S = \frac{k^2 V}{4\pi} \begin{pmatrix} \mu_0 \beta_{11} + \alpha_{22} \cos\theta - \alpha_{32} \sin\theta & \alpha_{21} \cos\theta - \alpha_{31} \sin\theta - \mu_0 \beta_{12} \\ \alpha_{12} - \mu_0 \beta_{21} \cos\theta + \mu_0 \beta_{31} \sin\theta & \alpha_{11} + \mu_0 \beta_{22} \cos\theta - \mu_0 \beta_{32} \sin\theta \end{pmatrix}, \tag{7.131}$$

where θ is the angle of scattering and β_{ij} and α_{ij} are the polarizability tensors defined in Chapter 5. In Chapter 5 approximate analytical formulas for calculating these tensors are given.

If assumption (7.83) holds then one can neglect terms involving β_{ij} in (7.131). Let us prove (7.131). Let the origin be inside D, the initial field be a plane wave propagating in the positive direction e_3 of the z-axis, n be a unit vector, and θ be the angle between e_3 and n (the angle of scattering). Let E_1, E_2 be the projections of the initial electric field onto the axes OX and OY, and f_1, f_2 be the projections of the scattered electric field onto the axes OX^1, OY^1. The axis OZ^1 is assumed to be in the direction of n. The plane (OZ, OY) coincides with the plane (OZ^1, OY^1) and is called the plane of scattering.

The scattering matrix is defined by the formula $f_E = SE$:

$$\begin{pmatrix} f_2 \\ f_1 \end{pmatrix} = \begin{pmatrix} S_2 & S_3 \\ S_4 & S_1 \end{pmatrix} \begin{pmatrix} E_2 \\ E_1 \end{pmatrix} \tag{7.132}$$

Formula(7.131) gives this matrix explicitly. All the elements of the smatrix are calculated by the same method. Let us derive in detail the formula for S_2. Let $e_j(e'_j)$ be the unit vectors of the above coordinate systems. Then $(e'_2, e_1) = 0$, $(e'_2, e_2) = \cos\theta$, $(e'_2, n) = -\sin\theta$, $f_2 = S_2 E_2 + S_3 E_1$. On the other hand,

$$f_2 = (f, e'_2) = \frac{k^2}{4\pi\varepsilon_0}\Big([n, [P, n]], e'_2\Big) + \frac{k^2}{4\pi}\left(\frac{\mu_0}{\varepsilon_0}\right)^{1/2} ([M, n], e'_2).$$

We have

$$\Big([n[P,n]],e_2'\Big) = (P - n(P,n), e_2') = (P, e_2') = \varepsilon_0 V \alpha_{ij} E_j (e_i, e_2')$$
$$= \varepsilon_0 V \Big\{ \big(\alpha_{21} E_1 + \alpha_{22} E_2\big) \cos\theta - \big(\alpha_{31} E_1 + \alpha_{32} E_2\big) \sin\theta \Big\}, \tag{7.133}$$

$$\big([M,n],e_2'\big) = \big([n,e_2'],M\big) = -\big(e_1,M\big) = -\mu_0 V\big(\beta H, e_1\big)$$
$$= -\mu_0 V\big(\beta_{11} H_1 + \beta_{12} H_2\big) = -\mu_0 V \left(\frac{\varepsilon_0}{\mu_0}\right)^{1/2} \big(-\beta_{11} E_2 + \beta_{12} E_1\big), \tag{7.134}$$

where the formulas

$$H_1 = -\left(\frac{\varepsilon_0}{\mu_0}\right)^{1/2} E_2, \quad H_2 = \left(\frac{\varepsilon_0}{\mu_0}\right)^{1/2} E_1 \tag{7.135}$$

were used. From (7.133) and (7.134) we find

$$S_2 = \frac{k^2 V}{4\pi}\big(\alpha_{22}\cos\theta - \alpha_{32}\sin\theta + \mu_0 \beta_{11}\big) \tag{7.136}$$

as the coefficient of E_2. Formulas for the other elements of the S-matrix can be obtained similarly.

Knowing the S-matrix for a single small body, one can find the refraction index tensor $n_{ij} = \delta_{ij} + 2\pi N k^{-2} S_{ij}(0)$ of the rarefied medium consisting of many small particles, the coefficient of absorption $\kappa = N\sigma = 4\pi N k^{-1} \mathrm{Im}\, S(0)$ and the crosssection $\sigma = 2\pi k^{-1} tr \mathrm{Im}\, S(0)$ for the anisotropic scattering. Here N is the number of the particles per unit volume, tr denotes the trace of a matrix, and Im denotes the imaginary part of a complex number.

7.5 Radiation from Small Apertures and the Skin Effect for Thin Wires

1. Let F be an aperture in an infinite conducting plane, α_0 be its coefficient of electrical polarizability, β_{ij}^0, $1 \le i,j \le r$, be its tensor of magnetic polarizability, the x_3-axis be perpendicular to the plane and $e_j, 1 \le j \le 3$, be the coordinate unit vectors. We assume that the electric field in the halfspace $x_3 < 0$ is $E_0' e_3$ and in the halfspace $x_3 > 0$ the electrostatic

potential $\phi \sim (P, x)/(4\pi\varepsilon_0|x|^3)$, $E = -\nabla\phi$. The electric dipole moment P can be calculated from the formula.

$$P = \alpha_0 \varepsilon_0 E_0' e_3. \tag{7.137}$$

The magnetic field in the half-space $x_3 < 0$ is $H_0' = H_{01}' e_1 + H_{02}' e_2$ and its asymptotic behavior in the half-space $x_3 > 0$ is given by $\psi \sim (M, x)/(4\pi\mu_0|x|^3)$, where ψ is the magnetostatic potential, M *is* is the magnetic dipole moment, $H = -\nabla\psi$ for $x_3 > 0$, and

$$M_i = \beta_{ij}^0 \mu_0 H_{0j}'. \tag{7.138}$$

Let $\tilde{\beta}$ and $\tilde{\alpha}_{ij}$ denote the magnetic polarizability coefficient and the electric polarizability tensor of the thin magnetic film and the thin metallic screen with the shape of F. The following theorem is a *duality principle in electrostatics.*

Theorem 7.1 *The following formulas hold*

$$\alpha_0 = -\tilde{\beta}/2, \quad \beta_{ij}^0 = -\tilde{\alpha}_{ij}/2. \tag{7.139}$$

Remark 7.3 *Formulas for calculating the values of* $\tilde{\beta}$ *and* $\tilde{\alpha}_{ij}$ *are given in Chapter 5. If one knows these values, one can find* α_0 *and* β_{ij}^0 *from* (7.139) *and* P *and* M *from* (7.137) *and* (7.138). *Knowing* P *and* M, *one can calculate the radiation from the aperture* F *from* (4.13).

Proof. [Proof of Theorem 7.1] Let us formulate *two principles*:

(A) Let there be an initial electrostatic field $\tilde{E}_0^{(2)} = E_0 e_3$ in the half-space $x_3 < 0$ bounded by the conducting plane $x_3 = 0$. If we cut an aperture F in the plane $x_3 = 0$ then the field $E(2)$ in the half-space $x_3 > 0$ can be calculated from the formula $E^{(2)} = H^{(1)} - H_0^{(1)}$, where $H^{(1)}$ is the magnetic field which is present when a magnetic plate F with $\mu = 0$ is placed in the initial field $H_0^{(1)} = -\frac{1}{2}\tilde{E}_0^{(2)} = -\frac{1}{2}E_0 e_3$.

(B) Let there be a magnetostatic field $H_0^{(2)}$ parallel to the plane $x_3 = 0$ in the half-space $x_3 < 0$ bounded by the plane $x_3 = 0$ with $\mu = 0$. If we cut an aperture F in the plane then the field $H^{(2)}$ in the half-space $x_3 > 0$ can be calculated from the formula $H^{(2)} = -(E^{(1)} - E_0^{(1)})$, where $E^{(1)}$ is the electric field which is present when the metallic plate F is placed in the initial field $E_0^{(1)} = \frac{1}{2}H_0^{(2)}$.

Formula (7.139) follows immediately from these principles and from the definition of α_0, $\tilde{\beta}$, $\tilde{\beta}^0_{ij}$, $\tilde{\alpha}_{ij}$. Both principles can be proved similarly. We give the proof of *principle* (A).

Let $S = R^2 \setminus F$. We have $E^{(2)} = -\nabla u$, where

$$u = \begin{cases} \phi, & x_3 > 0, \\ -E_0 x_3 + \phi, & x_3 < 0, \end{cases}$$

$\triangle\, \phi = 0$ outside S, $\phi|_S = 0$, $\phi(\infty) = 0$, and u, $\partial u/\partial x_3$ are continuous when crossing F, i.e., $(\partial\phi/\partial x_3)_+ = -E_0 + (\partial\phi/\partial x_3)_-$. By symmetry we have $\phi(\hat{x}, x_3) = \phi(\hat{x}, -x_3)$, $\hat{x} := (x_1, x_2)$. Hence $(\partial\phi/\partial x_3)_- = -(\partial\phi/\partial x_3)_+$, $(\partial\phi/\partial x_3)_+ = -\frac{1}{2}E_0$. Here $(\partial\phi/\partial x_3)$ are the limiting values of $\partial\phi/\partial x_3$ on F for $x_3 \to \pm 0$. So $\Delta\phi = 0$ for $x_3 > 0$, $\phi|_S = 0$, $\phi(\infty) = 0$, $(\partial\phi/\partial x_3)_+ = -\frac{1}{2}E_0$, and $E^{(2)} = -\nabla\phi$ for $x_3 > 0$. The field $H^{(1)} - H_0^{(1)} = -\nabla\psi$ for $x_3 > 0$, where $\triangle\, \psi = 0$, $\psi(\infty) = 0$, and by symmetry $\psi(\hat{x}, -x_3) = -\psi(\hat{x}, x_3)$. The magnetostatic potential $v = \frac{1}{2}E_0 x_0 + \psi$ satisfies the condition $(\partial v/\partial x_3)|_F = 0$, where N is the outward pointing normal to F. Hence $(\partial\psi/\partial x_3)_+ = -\frac{1}{2}E_0$. As ψ is odd with respect to x_3, we conclude that $\psi|_{x_3=0} = 0$, $\psi|_S = 0$. Hence ϕ, ψ are the solutions of the same boundary value problem in the half-space $x_3 > 0$. The solution of this problem is unique. Hence $\phi \equiv \psi$ for $x_3 > 0$. This means that $E^{(2)} = H^{(1)} - H_0^{(1)}$ for $x_3 > 0$. Principle (A) is proved. □

Example 7.1 For disk with radius a we have $\tilde{\beta} = -(8/3)a^3$, $\tilde{\alpha} = (16/3)a^3\delta_{ij}$, $1 \le i, j \le 2$, $\alpha_0 = (4/3)a^3$, $\beta^0_{ij} = -(8/3)a^3\delta_{ij}$, $1 \le i, j \le 2$, in *SI* units.

2. In Chapter 5 some two-sided variational estimates of $\tilde{\beta}$ and $\tilde{\alpha}_{ij}$ were given. In the special case in which F is a plane aperture one can give another variational estimate of $\tilde{\beta}$. Actually we will derive the estimate for $\alpha_0 = -\tilde{\beta}/2$.

Let $S = R^2 \setminus F$, be the complement of F in the plane, and let

$$g(s) = \int_F r_{st}^{-1}\, dt, \quad \alpha = (2\pi)^{-1} \int_F \int_F r_{st}^{-1}\, ds\, dt. \tag{7.140}$$

Then the following variational principle holds:

$$\alpha - \alpha_0 = \max \frac{(\int_S g(t) u(t) dt)^2}{2\pi \int_S \int_S \frac{u(s)u(t)}{r_{st}}\, ds\, dt}, \tag{7.141}$$

where the admissible functions should satisfy the edge condition (7.17) and ensure convergence of the integrals in (7.141). Principle (7.141) allows one to obtain some upper bounds for α_0.

Let us derive (7.141). Let $E_0' = E_0 e_3$ be the electric field in the half space $x_3 < 0$ and the aperture F is cut in the conducting plane $x_3 = 0$. Then the potential ϕ in the half space $x_3 > 0$ can be written as

$$\phi(x) = 2\int_F \phi(t)\frac{\partial G_0}{\partial x_3}dt, \quad x_3 > 0, \tag{7.142}$$

where

$$G_0(x,y) = \left(4\pi r_{xy}\right)^{-1}, \tag{7.143}$$

and

$$\phi(x) = -2\int_F \phi(t)\frac{\partial G}{\partial x_3}dt - E_0 x_3, \quad x_3 < 0. \tag{7.144}$$

The potential $\phi(x)$ and its derivatives are continuous when crossing the aperture F and

$$\phi|_S = 0. \tag{7.145}$$

Let $|x| \to \infty$, $x_3 > 0$. Then

$$\phi(x) \sim \frac{2\varepsilon_0 \int_F \phi(t)dt\, x_3}{4\pi\varepsilon_0 |x|^3} = \frac{(P,x)}{4\pi\varepsilon|x|^3}, \tag{7.146}$$

where

$$P = \alpha_0 \varepsilon_0 E_0 e_3, \tag{7.147}$$

and

$$\alpha_0 = \frac{2}{E_0}\int_F \phi(t)dt. \tag{7.148}$$

Let σ denote the charge density on S.

$$\sigma = -\varepsilon_0 \frac{\partial \phi}{\partial x_3}\bigg|_{x_3=+0}. \tag{7.149}$$

Green's formula implies

$$\phi(x) = \int_{x_3=0}\left(\phi(t)\frac{\partial G_0(x,t)}{\partial x_3} - G_0(x,t)\frac{\partial \phi}{\partial x_3}\right)dt. \tag{7.150}$$

From (7.145), (7.149), and (7.150) it follows that

$$\phi(x) = \int_F \frac{\partial G_0}{\partial x_3} dt + \frac{1}{\varepsilon_0} \int_S G_0(x,t)\sigma\, dt - \int_F G_0(x,t) \frac{\partial \phi}{\partial x_3} dt. \tag{7.151}$$

Let us show that

$$\left. \frac{\partial \phi}{\partial x_3} \right|_F = -\frac{E_0}{2}. \tag{7.152}$$

This follows from (7.142), (7.144), and the condition

$$\left. \left(\frac{\partial \phi}{\partial x_3} \right) \right|_{x_3=+0} = \left. \left(\frac{\partial \phi}{\partial x_3} \right) \right|_{x_3=-0} \quad \text{on } F. \tag{7.153}$$

Let us take $x \in S$ in (7.151) and take into account (7.152). This yields

$$\int_S \frac{\sigma(t)dt}{r_{st}} = -\frac{\varepsilon_0 E_0}{2} g(s), \quad s \in S, \tag{7.154}$$

where $g(s)$ is defined in (7.140). Let $x \to s \in F$, $x_3 > 0$ in (7.151). Then

$$\phi(s) = \frac{\phi(s)}{2} + \frac{1}{\varepsilon_0} \int_S G_0(s,t)\sigma\, dt + \frac{E_0}{8\pi} g(s), \tag{7.155}$$

which is equivalent to the equation

$$\phi(s) = \frac{1}{2\pi\varepsilon_0} \int_S \frac{\sigma(t)dt}{r_{st}} + \frac{E_0}{4\pi} g(s), \tag{7.156}$$

From (7.156) and (7.148) it follows that

$$\alpha_0 = \frac{1}{\pi \varepsilon_0 E_0} \int_S \sigma(t) g(t) dt + \alpha, \tag{7.157}$$

where α is defined in (7.140). This can be written as

$$\alpha - \alpha_0 = -\frac{1}{\pi \varepsilon_0 E_0} \int_S \sigma(t) g(t) dt. \tag{7.158}$$

From (7.154), (7.158), and Theorem 3.2 formula (7.141) follows. In the derivation of (7.141) we used some ideas from [25].

3. Consider the skin effect in thin wires. Let the axis of the wire be directed along the x_3-axis Γ be the boundary of the cross section D of the wire, D' be the plane domain extrior to D, a be the diameter of Γ, $ka \ll 1$. One can consider also wires the axes of which are curves with radius of curvature $R \gg a$. We assume that $\delta \ll a$ where δ is the skin depth defined

in Section 7.4. Let Γ be the length of Γ, J be the total current in the wire, and

$$Aj = \frac{1}{\pi} \int_\Gamma \frac{\partial}{\partial N_s} \ln \frac{1}{r_{st}} j(t)dt, \tag{7.159}$$

where N_s is the unit outward pointing normal to Γ at the point s.

Proposition 7.1 *Under the above assumptions the current density on the surface Γ can be found by the iterative process*

$$j_{n+1} = -Aj_n, \quad j_0 = \frac{J}{L}, \quad j = \lim_{n\to\infty} j_n(t). \tag{7.160}$$

Proof. It is sufficient to note that under the above assumptions the problem about the current distribution on Γ can be formulated as follows, Let $v(x_1, x_2)e_3$ be the vector potential of the static magnetic field corresponding to the current J. Then

$$\frac{\partial^2 v}{\partial x_1^2} + \frac{\partial^2 v}{\partial x_2^2} = 0 \text{ in } D', \quad v|_\Gamma = \text{const}, \tag{7.161}$$

$$v \sim \frac{\mu_0 J}{2\pi} \ln \frac{1}{r} \text{ as } r = \left(x_1^2 + x_2^2\right)^{1/2} \longrightarrow \infty, \tag{7.162}$$

$$-\frac{1}{\mu_0} \frac{\partial v}{\partial N}\bigg|_\Gamma = j(t), \quad J = \int_\Gamma j(t)dt. \tag{7.163}$$

If we look for the solution of the problem (7.161)–(7.163) of the form

$$v(x) = \frac{\mu_0}{2\pi} \int_\Gamma \ln \frac{1}{r_{xt}} j(t)dt, \tag{7.164}$$

then from (7.163) it follows that

$$j = -\frac{Aj - j}{2}$$

or

$$j = -Aj. \tag{7.165}$$

Proposition 7.1 follows now from Theorem 6.2. □

7.6 An Inverse Problem of Radiation Theory

1. Suppose that we are interested in measuring the electromagnetic field in the aperture of the mirror antenna. A possible method for making such measurements is as follows. Let us assume that the wavelength range is $\lambda \sim 3$ cm and let us place at some point x_0 in the aperture of the antenna a small probe of dimension a, $ka \ll 1$, $k = 2\pi\lambda^{-1}$. Let E_0, H_0 denote the electromagnetic field at the point x_0 and E, H denote the field scattered by the probe in the far-field zone. Note that for a small probe the far-field zone, which is defined by the condition $ka^2r^{-1} \ll 1$, is, in fact, close to the probe. For example, if $\lambda = 3$cm, $a = 0.3$ cm, then $ka^2 = 0.19$ cm. Therefore if $r = 2$ cm, then $ka^2r^{-1} \approx 0.1 \ll 1$. Let us assume for simplicity that the probe material is such that the magnetic dipole radiation from the probe is negligible. In this case the electric field scattered by the probe in the direction n can be calculated from the formula (7.95) as

$$E = \frac{k^2}{4\pi\varepsilon_0}\big[n, [P, n]\big], \tag{7.166}$$

where

$$P_i = \alpha_{ij}(\gamma)\varepsilon_0 V E_{0j}, \quad \gamma = \frac{\varepsilon' - \varepsilon_0}{\varepsilon' + \varepsilon_0}. \tag{7.167}$$

Here V is the volume of the probe, ε_0 is its dielectric constant, $\alpha_{ij}(\gamma)$ is its electric polarizability tensor, k is the wave number of the field in the aperture, E_0 is the electric field at the point x_0 where the probe was placed, n is the unit vector, and one sums up over the repeated indices. Let n_1 and n_2 be two non-collinear unit vectors, and E_j, $j = 1.2$, be the scattered fields corresponding to n_j. We will solve the following

Problem 7.1 Find E_0, H_0 from the measured E_j, $j = 1.2$.

We assume that the tensor $\alpha_{ij}(\gamma)$ is known. In Chapter 5 some explicit analytical approximate formulas for $\alpha_{ij}(\gamma)$ are given. From (7.166) it follows that

$$E_j = b\Big\{P - n_j\big(P, n_j\big)\Big\}, \quad b = \frac{k^2}{4\pi\varepsilon_0}, \quad j = 1, 2. \tag{7.168}$$

Therefore

$$bP = E_1 + bn_1\big(P, n_1\big) = E_2 + bn_2\big(P, n_2\big). \tag{7.169}$$

Let us choose for simplicity n_1 perpendicular to n_2. Then it follows from (7.169) that

$$b(P, n_2) = (E_1, n_2), \tag{7.170}$$

$$b(P, n_1) = (E_2, n_1). \tag{7.171}$$

Therefore

$$P = b^{-1}E_1 + b^{-1}n_1(E_2, n_1) = b^{-1}E_2 + b^{-1}n_2(E_1, n_2). \tag{7.172}$$

Thus, one can find vector P from the knowledge of E_1 and E_2. If P is known then E_0 can be found from the linear system

$$\alpha_{ij}(\gamma)\varepsilon_0 V E_{0j} = P_i, \quad 1 \leq i \leq 3. \tag{7.173}$$

The matrix of this system is positive definite because the tensor α_{ij} has this property (see Chapter 5. This follows also from the fact that $\frac{1}{2}\alpha_{ij}\varepsilon_0 V E_{0j} E_{0i}$ is the energy of the dipole P in the field E_0). Therefore the system (7.173) can be uniquely solved for E_{0j}, $1 \leq j \leq 3$. We proved that the above Problem has a unique solution and gave a simple algorithm for the solution of this problem. The key point in the above argument is the fact that the matrix $\alpha_{ij}(\gamma)$ is known explicitly (from Chapter 5).

2. In applications the problem of finding the distribution of particles according to their sizes is often of interest. Suppose that there is a medium consisting of many particles and the condition (7.85) is satisfied. We assume that the medium is rarefied, i.e., $d \gg a$, where a is the characteristic dimension of the particles. Let us assume for simplicity that the particles are spherical. Then the scattering amplitude for a single particle can be calculated from formulas (7.128) and (7.129). The scattering amplitude is the function $f(n, k, r)$ of the radius r of the particle. Suppose that $\phi(r)$ is the density of the distribution of the particles according to their sizes, so that $\phi(r)dr$ is the number of the particles per unit volume with the radius in the interval $(r, r + dr)$. Then the total scattered field in the direction n can be calculated from the formula

$$F(n, k) = \int_0^\infty \phi(r) f(n, k, r) dr. \tag{7.174}$$

Let us assume that we can measure $F(n, k)$ for a fixed k and all directions n. Then (7.174) can be considered as an integral equation of the first kind for an unknown function $\phi(r)$.

3. Suppose that we can measure the electric field scattered by a small particle ($ka \ll 1$) of an unknown shape. The initial field we denote by E_{0j}, the scattered field by f_j. Let us assume that the magnetic dipole radiation is negligible. The problem is to find the shape of the small particle.

First let us note that every small particle scatters electromagnetic wave like some ellipsoid. Indeed, the main term in the scattered field is the dipole scattering. We have seen above that the knowledge of the scattered field allows one to find the dipole moment P and that equation (7.167) holds. This equation allows one to find the tensor $\alpha_{ij}(\gamma)$ corresponding to the particle. This tensor is determined if one knows its diagonal form. Let α_1, α_2, α_3 be the eigenvalues of the tensor $\alpha_{ij}(\gamma)$. Then an ellipsoid with the semiaxes proportional to α_j scatters as the above body. Therefore one can identify the shape of the small scatterer by giving the three numbers $(\alpha_1, \alpha_2, \alpha_3)$. These numbers are the eigenvalues of the tensor $\alpha_{ij}(\gamma)$ which can be calculated from the known initial field E_{0j} and the measured scattered field f_i. For example, one can take $E_{0j} = \delta_{ij}$. Then $P_i = \alpha_{ij}(\gamma)V\varepsilon_0$. We assume that the particle is homogeneous and its dielectric constant ε is known, so that γ in (7.167) is known. For an ellipsoid the polarizability tensor in the diagonal form is $\alpha_{ij} = \alpha_j\delta_{ij}$, where $\alpha_j = (\varepsilon' - \varepsilon_0)(\varepsilon_0 + (\varepsilon' - \varepsilon_0)n^{(j)})^{-1}$, where ε' is the dielectric constant of the ellipsoid and $n^{(j)}$ are the depolarization coefficients. These coefficients are calculated explicitly with the help of the elliptic integrals, and they are tabulated in [58].

Chapter 8

Fredholm Alternative and a Characterization of Fredholm Operators

8.1 Fredholm Alternative and a Characterization of the Fredholm Operators

Let A be a linear bounded operator in a Hilbert space H, $N(A)$ and $R(A)$ its null-space and range, and A^* its adjoint. The operator A is called Fredholm with index zero iff $\dim N(A) = \dim N(A^*) := n < \infty$ and $R(A)$ and $R(A^*)$ are closed subspaces of H. Only the Fredholm operators with index zero are considered in this Chapter and are called Fredholm operators.

We give a simple and short proof of the following known (cf. [44]) result: a linear bounded operator A is Fredholm if and only if $A = B + F$, where B is an isomorphism and F is a finite-rank operator, that is an operator F with $\dim R(F) < \infty$, its rank is $\dim R(F)$. We call a linear bounded operator B on H an *isomorphism* if it is a bicontinuous injection of H onto H, that is, B^{-1} is defined on all of H and is bounded. Our proof of the Fredholm alternative consists of a reduction to a finite-dimensional linear algebraic system which is equivalent to the equation $Au = f$. For this linear algebraic system in a finite-dimensional space the Fredholm alternative is an elementary fact, easily proved and well-known. In Section 8.2 we give a characterization of unbounded Fredholm operators. This result appears to be new. In Section 8.3 the Fredholm alternative is established for operator-functions which depend on a parameter meromorphically, and the Laurent coefficients of their principal parts are finite-rank operators.

This chapter is based on the papers [88], [81] and [136].

8.1.1 *Introduction*

We prove the Fredholm alternative and give a characterization of the class of Fredholm operators by a reduction of the operator equation with a Fredholm operator to a linear algebraic system in a finite dimensional space.

The Fredholm alternative is a classical result whose proof for linear equations of the form $(I + T)u = f$, where T is a compact operator in a Banach space, can be found in most texts on functional analysis, of which we mention just [44]. A characterization of the set of Fredholm operators is given in [44], but it is missing in most texts on functional analysis. The proofs in [44] follow the classical Riesz argument used in the Riesz-Fredholm theory. Though beautiful, this theory is not very simple.

Our aim is to give a short and simple proof of the Fredholm alternative and of a characterization of the class of Fredholm operators. We give the argument for the case of Hilbert space, but the proof can be easily adjusted to the case of Banach space.

The idea is to reduce the problem to the one for linear algebraic systems in finite-dimensional case, for which the Fredholm alternative is a simple known result: in a finite-dimensional space R^N property (8.4) in the Definition 8.1 of Fredholm operators is a consequence of the closedness of any finite-dimensional linear subspace, since $R(A)$ is such a subspace in R^N, while property (8.18) is a consequence of the simple formulas $r(A) = r(A^*)$ and $n(A) = N - r(A)$, valid for matrices, where $r(A)$ is the rank of A and $n(A)$ is the dimension of the null-space of A.

If $\{e_j\}_{1\leq j\leq n}$, is an orthonormal basis of $R(F)$, then $Fu = \sum_{j=1}^{n}(Fu, e_j)e_j$, so

$$Fu = \sum_{j=1}^{n} (u, F^*e_j)e_j, \tag{8.1}$$

and

$$F^*u = \sum_{j=1}^{n} (u, e_j)F^*e_j, \tag{8.2}$$

where (u, v) is the inner product in H.

Definition 8.1 *An operator A is called Fredholm if and only if*

$$\dim N(A) = \dim N(A^*) := n < \infty, \tag{8.3}$$

and

$$R(A) = \overline{R(A)}, \quad R(A^*) = \overline{R(A^*)}, \tag{8.4}$$

where the overline stands for the closure.

Recall that

$$H = \overline{R(A)} \oplus N(A^*), \quad H = \overline{R(A^*)} \oplus N(A), \tag{8.5}$$

for any linear densely-defined (i.e., having a domain of definition dense in H) operator A, not necessarily bounded. For a Fredholm operator A one has:

$$H = R(A) \oplus N(A^*), \quad H = R(A^*) \oplus N(A). \tag{8.6}$$

Consider the equations:

$$Au = f, \tag{8.7}$$

$$Au_0 = 0, \tag{8.8}$$

$$A^*v = g, \tag{8.9}$$

$$A^*v_0 = 0. \tag{8.10}$$

Let us formulate the Fredholm alternative:

Theorem 8.1 *If B is an isomorphism and F is a finite-rank operator, then $A = B + F$ is Fredholm.*

For any Fredholm operator A the following (Fredholm) alternative holds:

(1) either (8.8) has only the trivial solution $u_0 = 0$, and then (8.10) has only the trivial solution, and equations (8.7) and (8.9) are uniquely solvable for any right-hand sides f and g,
or

(2) (8.8) has exactly $n > 0$ linearly independent solutions $\{\phi_j\}, 1 \le j \le n$, and then (8.10) has also exactly n linearly independent solutions $\{\psi_j\}, 1 \le j \le n$, equations (8.7) and (8.9) are solvable if and only if $(f, \psi_j) = 0$, $1 \le j \le n$, and, respectively, $(g, \phi_j) = 0$, $1 \le j \le n$. If they are solvable, their solutions are not unique and their general solutions are, respectively: $u = u_p + \sum_{j=1}^n a_j\phi_j$, and $v = v_p + \sum_{j=1}^n b_j\psi_j$, where a_j and b_j are arbitrary constants, and u_p and v_p are some particular solutions to (8.7) and (8.9), respectively.

Let us give a characterization of the class of Fredholm operators, that is, a necessary and sufficient condition for A to be Fredholm.

Theorem 8.2 *A linear bounded operator A is Fredholm if and only if $A = B + F$, where B is an isomorphism and F has finite rank.*

We prove these theorems in the next section.

8.1.2 *Proofs*

Let us first prove Theorem 8.2.

Proof of Theorem 8.2. From the proof of Theorem 8.1 that follows, we see that if $A = B + F$, where B is an isomorphism and F has finite rank, then A is Fredholm. To prove the converse, choose some orthonormal bases $\{\phi_j\}_{1\le j\le n}$ and $\{\psi_j\}_{1\le j\le n}$, in $N(A)$ and $N(A^*)$, respectively, using assumption (8.3). Define

$$Bu := Au - \sum_{j=1}^{n} (u, \phi_j)\psi_j := Au - Fu. \tag{8.11}$$

Clearly F has finite rank, and $A = B + F$. Let us prove that B is an isomorphism. If this is done, then Theorem 8.2 is proved.

We need to prove that $N(B) = \{0\}$ and $R(B) = H$. It is known (Banach's theorem), that if B is a linear bounded injection and $R(B) = H$, then B^{-1} is a bounded operator, so B is an isomorphism because B is bounded.

Suppose $Bu = 0$. Then $Au = 0$ (so that $u \in N(A)$), and $Fu = 0$ (because, according to (8.6), Au is orthogonal to Fu). Since $\{\psi_j\}, 1 \le j \le n$, is a linearly independent system, the equation $Fu = 0$ implies $(u, \phi_j) = 0$ for all $1 \le j \le n$, that is, u is orthogonal to $N(A)$. If $u \in N(A)$ and at the same time it is orthogonal to $N(A)$, then $u = 0$. So, $N(B) = \{0\}$.

Let us now prove that $R(B) = H$:

Take an arbitrary $f \in H$ and, using (8.6), represent it as $f = f_1 + f_2$, where $f_1 \in R(A)$ and $f_2 \in N(A^*)$ are orthogonal. Thus there is a $u_p \in H$ and some constants c_j such that $f = Au_p + \sum_1^n c_j\psi_j$. We choose u_p orthogonal to $N(A)$. This is clearly possible.

We claim that $Bu = f$, where $u := u_p - \sum_1^n c_j\phi_j$. Indeed, using the orthonormality of the system ϕ_j, $1 \le j \le n$, one gets $Bu = Au_p + \sum_1^n c_j\psi_j = f$.

Thus we have proved that $R(B) = H$. □

We now prove Theorem 8.1.

Proof of Theorem 8.1. If A is Fredholm, then the statements (1) and (2) of Theorem 8.1 are equivalent to (8.3) and (8.4), since (8.6) follows from (8.4).

Let us prove that if $A = B + F$, where B is an isomorphism and F has finite rank, then A is Fredholm. Both properties (8.3) and (8.4) are known for operators in finite-dimensional spaces. Therefore to prove that A is Fredholm it is sufficient to prove that equations (8.7) and (8.9) are equivalent to linear algebraic systems in a finite-dimensional space.

Let us prove this equivalence. We start with equation (8.7), denote $Bu := w$, and get an equation

$$w + Tw = f, \tag{8.12}$$

that is equivalent to (8.7). Here, $T := FB^{-1}$, is a finite-rank operator that has the same rank n as F because B is an isomorphism. Equation (8.11) is equivalent to (8.7): each solution to (8.7) is in one-to-one correspondence with a solution of (8.12) since B is an isomorphism. In particular, the dimensions of the null-spaces $N(A)$ and $N(I+T)$ are equal, $R(A) = R(I+T)$, and $R(I+T)$ is closed. The last claim is a consequence of the Fredholm alternative for finite-dimensional linear equations, but we give an independent proof of the closedness of $R(A)$ at the end of the Section.

Since T is a finite-rank operator, the dimension of $N(I+T)$ is finite and is not greater than the rank of T. Indeed, if $u = -Tu$ and T has finite rank n, then $Tu = \sum_{j=1}^{n}(Tu, e_j)e_j$, where $\{e_j\}_{1\leq j\leq n}$, is an orthonormal basis of $R(T)$, and $u = -\sum_{j=1}^{n}(u, T^*e_j)e_j$, so that u belongs to a subspace of dimension $n = r(T)$.

Since A and A^* enter symmetrically in the statement of Theorem 8.1, it is sufficient to prove (8.3) and (8.4) for A and check that the dimensions of $N(A)$ and $N(A^*)$ are equal.

To prove (8.3) and (8.4), let us reduce (8.9) to an equivalent equation of the form

$$v + T^*v = h, \tag{8.13}$$

where $T^* := B^{*-1}F^*$, is the adjoint to T, and

$$h := B^{*-1}g. \tag{8.14}$$

Since B is an isomorphism, $(B^{-1})^* = (B^*)^{-1}$. Applying B^{*-1} to equation (8.9), one gets an equivalent equation (8.13) and T^* is a finite-rank operator of the same rank n as T.

The last claim is easy to prove: if $\{e_j\}_{1\leq j\leq n}$ is a basis in $R(T)$, then $Tu = \sum_{j=1}^n (Tu, e_j)e_j$, and $T^*u = \sum_{j=1}^n (u, e_j)T^*e_j$, so $r(T^*) \leq r(T)$. By symmetry one has $r(T) \leq r(T^*)$, so $r(T) = r(T^*)$, and the claim is proved.

Writing explicitly the linear algebraic systems, equivalent to the equations (8.12) and (8.13), one sees that the matrices of these systems are adjoint. The system equivalent to equation (8.12) is:

$$c_i + \sum_1^n t_{ij}c_j = f_i, \tag{8.15}$$

where

$$t_{ij} := \big(e_j, T^*e_i\big),\ c_j := (w, T^*e_j),\ f_i := (f, T^*e_i),$$

and the one equivalent to (8.13) is:

$$\xi_i + \sum_1^n t_{ij}^*\xi_j = h_i, \tag{8.16}$$

where

$$t_{ij}^* = (T^*e_j, e_i),\ \xi_j := (v, e_j),\ h_i := (h, e_i),$$

and t_{ij}^* is the matrix adjoint to t_{ij}. For linear algebraic systems (8.15) and (8.16) the Fredholm alternative is a well-known elementary result. These systems are equivalent to equations (8.7) and (8.9), respectively. Therefore the Fredholm alternative holds for equations (8.7) and (8.9), so that properties (8.3) and (8.4) are proved. □

In conclusion let us explain in detail why equations (8.12) and (8.15) are equivalent in the following sense: every solution to (8.12) generates a solution to (8.15) and vice versa.

It is clear that (8.12) implies (8.15): just take the inner product of (8.12) with T^*e_j and get (8.15). So, each solution to (8.12) generates a solution to (8.15). We claim that each solution to (8.15) generates a solution to (8.12). Indeed, let c_j solve (8.15). Define $w := f - \sum_{j=1}^n c_je_j$. Then $Tw = Tf - \sum_{j=1}^n c_jTe_j = \sum_{i=1}^n[(Tf, e_i)e_i - \sum_{j=1}^n c_j(Te_j, e_i)e_i] = \sum_{i=1}^n c_ie_i = f - w$. Here we use (8.15) and take into account that $(Tf, e_i) = f_i$ and $(Te_j, e_i) = t_{ij}$. Thus, the element $w := f - \sum_1^n c_je_j$ solves (8.12), as claimed.

It is easy to check that if $\{w_1, \dots w_k\}$ are k linearly independent solutions to the homogeneous version of equation (8.12), then the corresponding k solutions $\{c_{1m}, \dots c_{nm}\}_{1 \le m \le k}$ of the homogeneous version of the system (8.15) are also linearly independent, and vice versa.

Let us give an independent proof of property (8.4):

$R(A)$ is closed if $A = B + F$, where B is an isomorphism and F is a finite-rank operator.

Since $A = (I + T)B$ and B is an isomorphism, it is sufficient to prove that $R(I + T)$ is closed if T has finite rank.

Let $u_j + Tu_j := f_j \to f$ as $j \to \infty$. Without loss of generality choose u_j orthogonal to $N(I + T)$. We want to prove that there exists a u such that $(I + T)u = f$. Suppose first that $\sup_{1 \le j < \infty} \|u_j\| < \infty$, where $\|\cdot\|$ denotes the norm in H. Since T is a finite-rank operator, Tu_j converges in H for some subsequence, which is denoted by u_j again. (Recall that in finite-dimensional spaces bounded sets are precompact). This implies that $u_j = f_j - Tu_j$ converges in H to an element u. Passing to the limit, one gets $(I + T)u = f$. To complete the proof, let us establish that $\sup_j \|u_j\| < \infty$. Assuming that this is false, one can choose a subsequence, denoted by u_j again, such that $\|u_j\| > j$. Let $z_j := u_j/\|u_j\|$. Then $\|z_j\| = 1$, z_j is orthogonal to $N(I+T)$, and $z_j + Tz_j = f_j/\|u_j\| \to 0$. As before, it follows that $z_j \to z$ in H, and passing to the limit in the equation for z_j one gets $z + Tz = 0$. Since z is orthogonal to $N(I + T)$, it follows that $z = 0$. This is a contradiction since $\|z\| = lim_{j\to\infty} \|z_j\| = 1$. This contradiction proves the desired estimate and the proof is completed. □

This proof is valid for any compact linear operator T. If T is a finite-rank operator, then the closedness of $R(I + T)$ follows also from a simple observation: finite-dimensional linear spaces are closed.

8.2 A Characterization of Unbounded Fredholm Operators

8.2.1 *Statement of the result*

This Section is a continuation of Section 8.1, where bounded Fredholm operators are studied.

We call a linear closed densely defined operator $A : X \to Y$ acting from a Banach space X into a Banach space Y a Fredholm operator, and write $A \in Fred(X, Y)$ if and only if

$$R(A) = \overline{R(A)} \tag{8.17}$$

and

$$n(A) = n(A^*) := n < \infty, \quad n(A) := \dim N(A), \tag{8.18}$$

where $N(A) := \{u : Au = 0, u \in D(A)\}$.

In the literature the Noether operators are sometimes called Fredholm operators. The Noether operators are operators for which (8.17) holds, $n(A) < \infty$, $n(A^*) < \infty$, but $n(A)$ may be not equal to $n(A^*)$. Thus $Fred(X, Y)$ is a proper subset of the Noether operators.

The Noether operators are called in honor of F. Noether, who was the first to study a class of singular integral equations with operators of this class in 1921 [77].

In Section 8.1 a proof of the Fredholm alternative and a characterization of Fredholm operators are given for bounded linear operators. Recall that a linear bounded operator F is called a finite-rank operator if $\dim R(F) < \infty$, where $R(F)$ is the range of F.

In this section these results are generalized to the case of closed unbounded linear operators. Namely, the following result is proved:

Theorem 8.3 *If A is a Fredholm operator, then*

$$A = B - F, \tag{8.19}$$

where B is a linear closed operator, $D(B) = D(A)$, $R(B) = Y$, $N(B) = \{0\}$, and F is a finite-rank operator. Conversely, if (8.19) *holds, where $B : X \to Y$ is a linear closed densely defined operator, $R(B) = Y$, $N(B) = \{0\}$, and F is a finite-rank operator, then A is closed, $D(A) = D(B)$, and* (8.17) *and* (8.18) *hold, so A is a Fredholm operator.*

Below a proof of Theorem 8.3 is given. In the literature a characterization of unbounded Fredholm operators is not discussed, as it seems. Theorem 8.3 is useful, for example, in the theory of elliptic boundary value problems, but we do not go into further detail (see, e.g., [44], [45]).

8.2.2 *Proof*

1. Assume that $A : X \to Y$ is linear, closed, densely defined operator, and (8.17) and (8.18) hold. Let us prove that then (8.19) holds, $D(B) = D(A), R(B) = Y, N(B) = \{0\}$, B is closed, and F is finite-rank operator.

Let $\{\varphi_j\}_{1 \le j \le n}$ be a basis of $N(A)$ and $\{\psi_j\}_{1 \le j \le n}$ be a basis of $N(A^*)$.

It is known that

$$R(A)^{\perp} = N(A^*), \tag{8.20}$$

where $R(A)^{\perp}$ is a set of linear functionals $\{\psi_j\}$ in Y^* such that $(\psi_j, Au) = 0$ $\forall u \in D(A)$, where (ψ_j, f) is the value of a linear functional $\psi_j \in Y^*$ on the element $f \in Y$. Clearly, $\psi_j \in N(A^*), 1 \leq j \leq n$.

Define

$$Bu := Au + \sum_{j=1}^{n} (h_j, u)\nu_j := (A + F)u, \quad \nu_j \in Y, \tag{8.21}$$

where F is a finite-rank operator, $\{\nu_j\}_{1\leq j\leq n}$ is the set of elements of Y, biorthogonal to the set $\{\psi_j\}_{1\leq j\leq n}$, $(\psi_j, \nu_m) = \delta_{jm} := \begin{cases} 0, & j \neq m \\ 1, & j = m \end{cases}$, and $\{h_j\}_{1\leq j\leq n}$ is the set of elements of X^*, biorthogonal to the set $\{\varphi_j\}_{1\leq j\leq n}$, $(h_j, \varphi_m) = \delta_{jm}$. Existence of sets biorthogonal to finitely many linearly independent elements of a Banach space follows from the Hahn-Banach theorem. An arbitrary element $u \in X$ can be uniquely represented as $u = u_1 + \sum_{j=1}^{n} c_j\varphi_j$, where $c_j = \text{const}$, and $(h_j, u_1) = 0$, $1 \leq j \leq n$.

Let us check that $N(B) = \{0\}$ and $R(B) = Y$. Assume $Bu = 0$, that is $Au + \sum_{j=1}^{n}(h_j, u)\nu_j = 0$. Apply ψ_m to this equation, use $(\psi_m, Au) = 0$, and get

$$0 = \sum_{j=1}^{n} (\psi_m, \nu_j)(h_j, u) = \sum_{j=1}^{n} \delta_{mj}(h_j, u) = (h_m, u), \quad 1 \leq m \leq n.$$

Therefore $Au = 0$. So $u \in N(A)$, and $u = \sum_{j=1}^{n} c_j\varphi_j$, $c_j = \text{const}$. Apply h_m to this equation and use $(h_m, \varphi_j) = \delta_{mj}$ to get $c_m = 0$, $1 \leq m \leq n$. Thus $u = 0$. We have proved that $N(B) = \{0\}$.

To prove $R(B) = Y$, take an arbitrary element $f \in Y$ and write $f = f_1 + f_2$, where $f_1 = Au_1$ belongs to $R(A)$, and $f_2 = \sum_{j=1}^{n} a_j\nu_j$, $a_j = \text{const}$. Note that

$$Y = R(A) \dotplus L_n, \tag{8.22}$$

where the sum is direct, L_n is spanned by the elements $\{\nu_j\}_{1\leq j\leq n}$, and $a_j = (\psi_j, f)$. Indeed,

$$(\psi_m, f) = (\psi_m, Au_1) + \sum_{j=1}^{n} a_j(\psi_m, \nu_j) = a_m, \quad 1 \leq m \leq n.$$

Given an arbitrary $f \in Y$, $f = Au_1 + \sum_{j=1}^{n}(\psi_j, f)\nu_j$, define $u = u_1 + \sum_{j=1}^{n}(\psi_j, f)\varphi_j$, where $(h_j, u_1) = 0$, $1 \le j \le n$. Then $Bu = f$. Indeed, using (8.21) one has:

$$B\left[u_1 + \sum_{j=1}^{n}(\psi_j, f)\varphi_j\right] = Au_1 + \sum_{j=1}^{n}(h_j, u_1)\nu_j + \sum_{j=1}^{n}\left(h_j, \sum_{m=1}^{n}(\psi_m, f)\varphi_m\right)\nu_j = f. \tag{8.23}$$

Here the relations $(h_j, \varphi_m) = \delta_{jm}$, and $(h_j, u_1) = 0$ are used. We have proved the relation $R(B) = Y$.

2. Let us now assume that $A = B - F$, where $B : X \to Y$ is a linear closed densely defined operator, $D(A) = D(B)$, $N(B) = \{0\}$, $R(B) = Y$, and F is a finite-rank operator. We wish to prove that (8.17) and (8.18) hold and A is closed.

Let us prove (8.17). Assume that $Au_n := f_n \to f$ and prove that $f \in R(A)$.

One has $Bu_n - Fu_n \to f$. Since $N(B) = \{0\}$, $R(B) = Y$, and B is closed, B^{-1} is bounded by Banach's theorem. Thus

$$u_n - B^{-1}Fu_n \longrightarrow B^{-1}f. \tag{8.24}$$

Since F is a finite-rank operator, $B^{-1}F$ is compact. Therefore, if $\sup_n \| u_n \| \le c$, where c is a constant, then a subsequence, denoted u_n again, can be found, such that $B^{-1}Fu_n$ converges in the norm of X. Consequently, (8.24) implies $u_n \to u$, $u - B^{-1}Fu = B^{-1}f$, so $u \in D(B)$ and $Bu - Fu = f$.

To finish the proof, let us establish the estimate $\sup_n \|u_n\| \le c$. Assuming $\|u_{n_k}\| \to \infty$ and denoting n_k by n and $B^{-1}F$ by T, define $v_n := \frac{u_n}{\|u_n\|}$, $\|v_n\| = 1$. Then $v_n - Tv_n \to 0$ as $n \to \infty$. One may assume that v_n is chosen in a direct complement of $N(I - T)$ in X. Arguing as above, one selects a convergent in X subsequence, denoted again by v_n, $v_n \to v$, and gets $v - Tv = 0$. Since v belongs to the direct complement of $N(I - T)$, it follows that $v = 0$. On the other hand, since $\|v\| = \lim_{n\to\infty} \|v_n\| = 1$, one gets a contradiction, which proves the desired estimate $\sup_n \|u_n\| \le c$. Property (8.17) is proved.

Let us prove that A is closed. If $Au_n \to f$ and $u_n \to u$, then $Bu_n - Fu_n \to f$, and the above argument shows that $Bu - Fu = f$ so $Au = f$. Thus A is closed.

Finally, let us prove (8.18).

Let $Au = 0$, i.e., $Bu - Fu = 0$. Applying the bounded linear injective operator B^{-1}, one gets an equivalent equation

$$u - Tu = 0, \quad T := B^{-1}F, \quad T : X \longrightarrow X, \tag{8.25}$$

with a finite-rank operator T. It is an elementary fact that $\dim N(I-T) := n < \infty$ if T is a finite-rank operator. Since $N(A) = N(I - T)$, one has $\dim N(A) = n < \infty$.

Now let $A^*v = 0$. Then

$$B^*v - F^*v = 0. \tag{8.26}$$

Since $(B^*)^{-1} = (B^{-1})^*$ is a bounded and injective linear operator, the elements v are in one-to-one correspondence with the elements $w := B^*v$, and (8.26) is equivalent to

$$w - T^*w = 0, \quad T^* = F^*(B^*)^{-1}, \tag{8.27}$$

so that T^* is the adjoint to operator $T := B^{-1}F$.

Since T is a finite-rank operator, it is an elementary fact that $\dim N(I - T^*) = \dim(I - T) = n < \infty$. Since $N(A^*) = N(I - T^*)$, property (8.18) is proved.

Theorem 8.3 is proved. □

An immediate consequence of this theorem is the Fredholm alternative for unbounded operators $A \in \text{Fred}\,(X, Y)$.

8.3 Fredholm Alternative for Analytic Operators

Let X and Y be Banach spaces and $A(k) : X \to Y$ be a linear bounded operator-function analytic in a connected domain Δ of a complex plane k. Assume that the range $R(A(k))$ is closed and $\dim N(A(k)) = \dim N(A^*(k)) = r < \infty$, so $A(k) \in \text{Fred}\,(X, Y)$ is Fredholm operator with index zero, and $\text{Fred}\,(X, Y)$ denotes the set of all such operators.

Theorem 8.4 *([136]) Under the above assumptions either $A^{-1}(k)$ does not exist $\forall k \in \Delta$, or $A^{-1}(k)$ exists for all $k \in \Delta$ except, possibly, for a discrete set $\{k_j\}$. The points k_j are poles of $A^{-1}(k)$, and the coefficients a_p in the expansion $A^{-1}(k) = \sum_{p=-m_j}^{\infty} a_p(k - k_j)^p$ are finite-rank operators. This conclusion remains valid if one assumes that $A(k)$ is a meromorphic operator-function of k in Δ, provided that $b_{-m_0} \in$ Fred(X, Y), where $A(k) = \sum_{p=-m_0}^{\infty} b_p(k - k_0)^p$ and $k_0 \in \Delta$ is a pole of $A(k)$.*

Proof. First we assume that $A(k)$ is analytic in Δ. Choose a finite-rank operator F such that $B(k) := A(k) + F$ is an isomorphism of X onto Y. This is possible, as was shown in Section 8.2. Equation $A(k)u = f$ is equivalent to $B(k)u = Fu + f$ and to $u = B^{-1}(k)Fu + B^{-1}(k)f$. Since $B(k)$ is analytic in Δ and the operator $B^{-1}(k)$ is bounded, it follows that $B^{-1}(k)$ is analytic in Δ.

Therefore

$$u = T(k)u + h(k), \quad T(k) := B^{-1}(k)F, \quad h(k) := B^{-1}(k)F,$$

where $h(k)$ is analytic in Δ and $T(k)$ is an analytic in Δ finite-rank operator-function. If $d(k) := \det(I - T(k)) \equiv 0$, then the operator $I - T(k)$ is not boundedly invertible for all $k \in \Delta$. Otherwise $d(k)$, which is an analytic function, may have only a discrete set of zeros in Δ. If $d(\kappa) = 0$, then $d(k) \neq 0$ for $|k - \kappa| < \delta$, where $\delta > 0$ is a sufficiently small number, and, by Kramer's formulas, one concludes that $u(k)$ is a meromorphic function in Δ. In a neighborhood of the point κ the Laurent expansion of the operator $A^{-1}(k) = (I - T(k))^{-1}B^{-1}(k)$ has coefficients which are finite-rank operators, because $T(k)$ is a finite-rank operator. This proves the first conclusion of Theorem 8.4 (under the assumptions of analyticity of $A(k)$ in Δ and of Fredholmness of $A(k)$: $A(k) \in \text{Fred}\,(X, Y)$).

Assume now that $A(k)$ is meromorphic in Δ, κ is a pole of order m. Then $(k - \kappa)^m A(k) := Q(k)$ is analytic in a neighborhood of κ. If $b_{-m_0} := \lim_{k \to \kappa}(k - \kappa)^m A(k)$ is a bounded Fredholm operator, then $Q(k)$ is an analytic bounded Fredholm operator for $|k - \kappa| < \delta$, for a sufficiently small $\delta > 0$. Thus the first conclusion of Theorem 8.4 applies and Theorem 8.4 is proved. □

Chapter 9

Boundary-Value Problems in Rough Domains

In this chapter boundary-value problems for the Laplace and Helmholtz operators are considered under weak assumptions on the smoothness of the domains. The theory we develop can be easily generalized to the case of uniformly elliptic formally self-adjoint differential operators with constant coefficients near infinity. We assume nothing about smoothness of the boundary S of a bounded domain D when the homogeneous Dirichlet boundary condition is imposed; we assume boundedness of the embedding $i_1 : H^1(D) \to L^2(D)$ when the Neumann boundary condition is imposed; we assume boundedness of the embeddings i_1 and of $i_2 : H^1(D) \to L^2(S)$ when the Robin boundary condition is imposed, and, if, in addition, i_1 and i_2 are compact, then the boundary-value problems with the spectral parameter are of Fredholm type. Here i_1 is the embedding of $H^1(D)$ (or $H^1(\tilde{D})$) into $L^2(D)$ ($L^2(\tilde{D})$), $D' := \mathbb{R}^n \setminus D$ is the exterior domain, and $\tilde{D} \subset D'$ is a bounded domain whose boundary consists of two components: $S := \partial D$ and $\tilde{S}$, where $\tilde{S}$ is a smooth compact manifold. The space $L^2(S)$ is the L^2 space on S with respect to Hausdorff $(n-1)$-dimensional measure on S.

Our theory is developed in such a way that the interior and exterior boundary-value problems are studied similarly in spite of the fact that the corresponding operators have discrete spectrum in the case of interior boundary-value problems and continuous spectrum in the case of exterior ones. The novel points consist of the usage of the limiting absorption principle, the relation between closed quadratic forms and selfadjoint operators, and the construction of the theory under weak assumptions about the boundaries of the domains, which can be much rougher than the Lipschitz domains. We give examples of admissible bounded domains whose boundaries have countably many connected components and admissible domains whose boundaries are not locally representable by a graph of a Lipschitz function . The results in this Chapter are based on the works [144], [107], [103], [104], [31], and the presentation follows closely [31].

9.1 Introduction

An essentially self-contained presentation of a method for a study of boundary-value problems for second-order elliptic equations in domains with non-smooth boundary is given in this Section. The novel points include the usage of the limiting absorption principle for the proof of the existence of the solution. The theory of boundary-value problems for interior and exterior domains are constructed similarly in spite of the fact that the Dirichlet Laplacian has a discrete spectrum in the former case and a continuous spectrum in the latter case. For brevity of the presentation we consider the boundary-value problems for Laplacian, and the three classical boundary conditions. We study interior and exterior boundary-value problems and obtain the existence results and the Fredholm property under weak assumptions on the smoothness of the boundary. The method we use is applicable for general second-order elliptic equations. Elliptic boundary-value problems were studied in numerous books and papers. We mention [29] and [57], where many references can be found. In [65] embedding theorems for a variety of non-smooth domains have been studied. In [104] the obstacle scattering problems were studied for non-smooth obstacles. In [107] the [107] the boundary-value problems and direct and inverse obstacle scattering problems have been studied. In [30] embedding theorems in some classes of non-smooth (rough) domains were studied.

Consider the boundary-value problems

$$-\Delta u = F \text{ in } D, \quad F \in L^2(D), \tag{9.1}$$

$$u = 0 \text{ on } S := \partial D. \tag{9.2}$$

The boundary conditions can be the Neumann one

$$u_N = 0 \text{ on } S, \tag{9.3}$$

where N is the outer unit normal to S, or the Robin one:

$$u_N + h(s)u = 0 \text{ on } S, \tag{9.4}$$

where $h(s) \geq 0$ is a bounded piecewise-continuous function on S.

We are interested in similar problems in the exterior domain $D' := R^n \setminus D$, and we consider the case $n = 3$. The case $n > 3$ can be treated similarly. If $n = 2$ some additional remarks are in order since the fundamental solution in this case changes sign and tends to infinity as $|x - y| := r_{xy} \to \infty$. If $n = 3$, then $g(x, y) := \frac{1}{4\pi r_{xy}}$, and if $n = 2$, then $g(x, y) = \frac{1}{2\pi} \ln \frac{1}{r_{xy}}$, $x, y \in \mathbb{R}^n$, $-\Delta g = \delta(x - y)$ in $\mathbb{R}^n$, and $\delta(x)$ is the delta-function.

Below $(\cdot, \cdot)$ denotes the inner product in $L^2(D) := H^0$, $L_0^2(D)$ is the set of $L^2(D)$ functions with compact support, $L_0^2(D')$ is the set of $L^2(D')$ functions which vanish near infinity, $\overset{\circ}{H}{}^1$ is the closure of $C_0^\infty(D)$ in the $H^1 := H^1(D)-$norm defined as $\|u\|_1 := (\int_D (|u|^2 + |\nabla u|^2) dx)^{1/2}$. We denote $\|u\| := (\int_D |u|^2 dx)^{1/2}$, and let $D_\varepsilon := \{x : x \in D, \text{dist}(x, S) < \varepsilon\}$, where $\varepsilon > 0$ is a small number, and $\text{dist}(x, S)$ is the distance from the point x to S, and D'_ε is defined similarly.

If the boundary conditions are non-homogeneous, e.g., $u = f$ on S, then we assume that there exists a function $v \in H^1(D) \bigcap H^2_{\text{loc}}(D)$, $\Delta v \in L^2(D)$, such that $v = f$ on S and consider $w := u - v$. The function w satisfies equation (9.1) with F replaced by $F + \Delta v$, and w satisfies (9.2). Similarly one treats inhomogeneous Neumann and Robin boundary conditions. In the case of inhomogeneous boundary conditions the smoothness assumptions on the boundary S are more restrictive than in the case of the homogeneous boundary conditions.

Let us reformulate problems (9.1)–(9.4) so that the assumptions on the smoothness of S are minimal.

In the case of the Dirichlet problem (9.1)–(9.2) we use the *weak formulation*:

u solves (9.1)–(9.2) iff $u \in \overset{\circ}{H}{}^1(D) := \overset{\circ}{H}{}^1$ and

$$[u, \phi] := (\nabla u, \nabla \phi) = (F, \phi) \quad \forall \phi \in \overset{\circ}{H}{}^1. \tag{9.5}$$

The weak formulation (9.5) of the Dirichlet problem (9.1)–(9.2) does not require any smoothness of S, and boundedness of D is the only restriction on D for the Dirichlet problem.

The weak formulation of the Neumann problem (9.1), (9.3) is:

$$[u, \phi] = (F, \phi) \quad \forall \phi \in H^1. \tag{9.6}$$

An obvious necessary condition on F for (9.6) to hold is

$$(F, 1) = 0. \tag{9.7}$$

Although the statement of the problem (9.6) does not require any smoothness assumption on S, one has to assume that S is smooth enough for the Poincare-type inequality to hold:

$$\inf_{m \in \mathbb{R}^1} \|u - m\| \leq c\|\nabla u\|, \quad c = \text{const} > 0, \tag{9.8}$$

see Remark 9.1 below.

The infimum in (9.8) is attained at $m_0 = \frac{1}{|D|}\int_D u dx$, $|D| := \text{meas}\, D$, and $(u - m_0, 1) = 0$. If $(u, 1) = 0$, then (9.8) implies $\|u\| \leq c\|\nabla u\|$. The role of this inequality will be clear from the proof of the existence of the solution to (9.6) (see Section 9.2).

Finally, for the Robin boundary condition the weak formulation of the boundary-value problem (9.1), (9.4) is:

$$[u, \phi] + \int_S hu\bar{\phi} ds = (F, \phi) \quad \forall \phi \in H^1. \tag{9.9}$$

For (9.9) to make sense, one has to be able to define u on S. For this reason we assume that the embedding $i_2 : H^1(D_\varepsilon) \to L^2(S)$ is bounded. We also assume the compactness of i_2, and this assumption is motivated in the proof of the existence and uniqueness of the solution to (9.9), it yields the Fredholm property of the boundary-value problem.

Let us formulate our results. We assume that $D \subset \mathbb{R}^n$, $n = 3$, is a bounded domain and $F \in L^2(D)$ is compactly supported. This assumption will be relaxed in Remark 9.5.

Theorem 9.1 *The solution* $u \in \overset{\circ}{H}{}^1(D)$ *of* (9.5) *exists and is unique.*

Theorem 9.2 *If* D *is such that* (9.8) *holds and* F *satisfies* (9.7), *then there exists a solution* u *to* (9.6), *and* $\{u + c\}$, $c = \text{const}$, *is the set of all solutions to* (9.6) *in* H^1.

Theorem 9.3 *If* D *is such that* $i_1 : H^1(D) \to L^2(D)$ *and* $i_2 : H^1(D) \to L^2(S)$ *are bounded,* $F \in L^2_0(D)$ *and* $h \geq 0$ *is a piecewise-continuous bounded function on* S, $h \not\equiv 0$, *then problem* (9.9) *has a solution in* $H^1(D)$ *and this solution is unique. If* i_1 *and* i_2 *are compact, then the problem*

$$[u, \phi] + \int_S hu\bar{\phi} ds - \lambda(u, \phi) = (F, \phi), \quad \lambda = \text{const} \in \mathbb{R}$$

is of Fredholm type.

Similar results are obtained in Section 9.3 for the boundary-value problems in the exterior domains (Theorem 9.4).

9.2 Proofs

Proof of Theorem 9.1. One has

$$|[u,\phi]| = |(F,\phi)| \leq \|F\|\|\phi\| \leq c\|F\|\|\varphi\|_1 \tag{9.10}$$

where we have used the inequality

$$\|\phi\| \leq c\|\phi\|_1, \quad \phi \in \overset{\circ}{H}{}^1, \tag{9.11}$$

which holds for any bounded domain, i.e., without any smoothness assumptions on D. Note that the norm $[u,u]^{1/2} := [u]$ is equivalent to H^1 norm on $\overset{\circ}{H}{}^1 : c_1\|u\|_1 \leq [u] \leq \|u\|_1$, $c_1 = \text{const} > 0$. Inequality (9.10) shows that (F,ϕ) is a bounded linear functional in $H^1(D)$ so, by the Riesz theorem about linear functionals in a Hilbert space, one has

$$[u,\phi] = [BF,\phi] \quad \forall\phi \in \overset{\circ}{H}{}^1,$$

where B is a bounded linear operator from $L^2(D)$ into $\overset{\circ}{H}{}^1$. Thus $u = BF$ is the unique solution to (9.5). □

Proof of Theorem 9.2. If $(F,1) = 0$, then one may assume that $(\phi,1) = 0$ because $(F,\phi) = (F,\phi - m)$ and the constant m can be chosen so that $(\phi - m, 1) = 0$ if D is bounded. If D is such that (9.8) holds, then

$$|[u,\phi]| = |(F,\phi - m)| \leq \|F\| \inf_m \|\phi - m\| \leq c\|F\|\|\nabla\phi\|. \tag{9.12}$$

Thus $(F,\phi) = [BF,\phi]$, where $B : L^2(D) \to H^1$ is a bounded linear operator. Thus $u = BF$ solves (9.6), for any constant m and $u + m$ also solves (9.6) because $[m,\phi] = 0$. If u and v solve (9.6), then $w := u - v$ solves the equation $[w,\phi] = 0 \quad \forall\phi \in H^1$. Take $\phi = w$ and get $[w,w] = \|\nabla w\|^2 = 0$. Thus $\nabla w = 0$ and $w = \text{const}$. Theorem 9.2 is proved. □

Remark 9.1 *Necessary and sufficient conditions on D for (9.8) to hold one can find in [65]. Inequality (9.8) is equivalent to the boundedness of the embedding in $i_1 : L^1_2(D) \to L^2(D)$. By L^1_2 the space of functions u such that $\|\nabla u\| < \infty$ is denoted.*

Remark 9.2 *If one wants to study the problem*

$$-\Delta u - \lambda u = F, \qquad u = 0 \text{ on } S \tag{9.13}$$

where $\lambda = \text{const}$, and a similar problem with the Neumann boundary condition (9.3) or with the Robin condition (9.4) to be of Fredholm type, then one *has to assume the operators B in the proofs of Theorem 9.1 and Theorem 9.2 to be compact in $\overset{\circ}{H}{}^1$ and in H^1 respectively. Originally the operators B were acting from $L^2(D)$ onto $\overset{\circ}{H}{}^1$ and H^1 (respectively in Theorem 9.1 and in Theorem 9.2). Thus, B are defined on $\overset{\circ}{H}{}^1 \subset L^2(D)$ and on $H^1 \subset L^2(D)$ respectively.*

Proof of Theorem 9.3. If the embedding $i_2 : H^1(D_\varepsilon) \to L^2(S)$ is bounded, then

$$|\int_S hu\bar{\phi}ds| \leq \sup_S |h| \|u\|_{L^2(S)} \|\phi\|_{L^2(S)} \leq c\|\phi\|_1. \tag{9.14}$$

By Riesz's theorem one gets

$$\int_S hu\bar{\phi}ds = (Tu, \phi)_1.$$

Equation (9.9) can be written as

$$(Au + Tu - BF, \phi)_1 = 0 \qquad \forall \phi \in H^1, \tag{9.15}$$

where $[u, \phi] = (Au, \phi)_1$, $(F, \phi) = (BF, \phi)_1$. Thus

$$Qu - BF := Au + Tu - BF = 0, \tag{9.16}$$

where A is a bounded linear operator in H^1, $\|A\| \leq 1$ because $[u, u] \leq (u, u)_1$, $||B|| \leq 1$ because $|(BF, \phi)_1| = |(F, \phi)| \leq \|F\| \|\phi\| \leq \|F\|_1 \|\phi\|_1$, and T is a bounded operator in H^1 if the embedding operator $i_2 : H^1(D_\varepsilon) \to L^2(S)$ is bounded. If i_2 is compact, then T is compact in H^1. The operator $Q := A + T$ is linear, defined on all of H^1, and bounded. The expression

$$N^2(u) := (Qu, u)_1 = [u, u] + \int_S h|u|^2 ds$$

defines a norm $N(u)$ equivalent to $\|u\|_1$.

Let us prove this equivalence.

By (9.14) one has $N^2(u) \leq c\|u\|_1^2$. Also

$$\|u\|_1^2 = [u, u] + (u, u) \leq N^2(u) + (u, u) \leq c_1 N^2(u)$$

because

$$\|u\| \leq cN(u),$$

where $c = \text{const} > 0$ stands for various constants independent of u.

Let us prove the inequality $\|u\| \leq cN(u)$.

Assuming that it fails, one finds a sequence $u_n \in H^1$, $\|u_n\| = 1$, such that $\|u_n\| \geq nN(u_n)$, so $N(u_n) \leq \frac{1}{n}$. Thus $\|\nabla u_n\| \to 0$ and $\int_S h|u_n|^2 ds \to 0$. Since $\|u_n\| = 1$ one may assume that $u_n \rightharpoonup v$, where $\rightharpoonup$ denotes weak convergence in $L^2(D)$. If $u_n \rightharpoonup v$ and $\nabla u_n \rightharpoonup 0$, then $\nabla v = 0$, so $v = C = \text{const}$, and

$$0 = \lim_{n\to\infty} \int_S h|u_n|^2 ds = C^2 \int_S h ds,$$

so $C = 0$. The inequality is proved.

Thus, the norms $N(u)$ and $\|u\|_1$ are equivalent, the operator Q is positive definite, selfadjoint as an operator in H^1, and therefore Q has a bounded inverse in H^1. Thus, equation (9.16) has a unique solution $u = Q^{-1}BF$ in H^1. The statement of Theorem 9.3 concerning Fredholm's type of problem (9.9) follows from Lemma 9.1 below.

Theorem 9.3 is proved. □

Remark 9.3 *As in Remark 9.2, if B is compact in H^1, then the problem*

$$[u, \phi] + \int_S hu\bar{\phi} ds = \lambda(u, \phi) + (F, \phi), \quad \lambda = \text{const} \qquad (9.17)$$

is of Fredholm type. This problem can be written as $Qu := Au + Tu = \lambda Bu + BF$, or

$$u = \lambda Q^{-1}Bu + Q^{-1}BF \qquad (9.18)$$

where the operator $Q^{-1}B$ is compact in H^1.

Lemma 9.1 *The operator B is compact in H^1 if and only if the embedding operator $i_1 : H^1(D) \to L^2(D_\varepsilon)$ is compact.*

Proof. Suppose that the embedding $i_1 : H^1(D) \to L^2(D)$ is compact. One has $\|u\|^2 = (Bu, u)_1 = (u, Bu)_1$, so B is a linear positive, symmetric, and bounded operator in H^1. Here the inner product $(Bu, u)_1$ is equivalent to the inner product $[Bu, u]$. One has $(u, \phi) = (Bu, \phi)_1$, so $\|Bu\|_1 \leq \|u\| \leq \|u\|_1$, so $\|B\|_{H^1 \to H^1} \leq 1$. A linear positive, symmetric, bounded operator B, defined on all of H^1, is selfadjoint. The operator $B^{1/2} > 0$ is well defined, B and $B^{1/2}$ are simultaneously compact, and $\|u\| = \|B^{1/2}u\|_1$. Thus, if i_1 is compact then the inequality $\|u_n\|_1 \leq 1$ implies the existence of a convergent in $L^2(D)$ subsequence, denoted again u_n, so that $B^{1/2}u_n$ converges in H^1. Thus, $B^{1/2}$ is compact in H^1 and so is B.

Conversely, if B is compact in H^1 so is $B^{1/2}$. Therefore, if u_n is a bounded in H^1 sequence, $\|u_n\|_1 \leq 1$, then $B^{1/2}u_{n_k}$ is a convergent in H^1

sequence. Denote the subsequence u_{n_k} again u_n. Then u_n is a convergent in $H^0 = L^2(D)$ sequence because $\|u_n\| = \|B^{1/2}u_n\|_1$. Therefore i_1 is compact.

Lemma 9.1 is proved. □

Remark 9.4 *We have used the assumptions $h \geq 0$ and $h \not\equiv 0$ in the proof of Theorem 9.3. If h changes sign on S but the embeddings $i_2 : H^1(D) \to L^2(S)$ and $i_1 : H^1(D) \to L^2(D)$ are compact, then problem (9.9) is still of Fredholm's type because T is compact in H^1 if i_2 is compact.*

9.3 Exterior Boundary-Value Problems

Consider boundary-value problems (9.1), (9.2), (9.3), (9.4) in the exterior domain $D' = \mathbb{R}^3 \setminus \overline{D}$. The closure of $H_0^1(D')$ in the norm $\|u\|_1 := \{\int_{D'}(|u|^2 + |\nabla u|^2)dx\}^{1/2}$ is denoted by $H^1 = H^1(D')$ and $H_0^1(D')$ is the set of functions vanishing near infinity and with finite norm $\|u\|_1 < \infty$. We assume that D is bounded. The weak formulation of the boundary-value problems is given similarly to (9.5), (9.6) and (9.9). The corresponding quadratic forms Dirichlet t_D, Neumann t_N and Robin t_R, where the forms

$$t_D[u,u] = (\nabla u, \nabla u), \quad u \in \overset{\circ}{H}{}^1(D'); \qquad t_N[u,u] = (\nabla u, \nabla u), \quad u \in H^1(D');$$

$$t_R[u,u] = (\nabla u, \nabla u) + \langle hu, u \rangle, \quad u \in H^1(D'), \quad \langle u, v \rangle := \int_S u\bar{v}ds, \quad 0 \leq h \leq c,$$

are nonnegative, symmetric and closable. Here and below, $c > 0$ stands for various constants. Nonnegativity and symmetry of the above forms are obvious.

Let us prove their closability.

By definition a quadratic form $t[u,u]$ bounded from below, i.e., $t[u,u] > -m(u,u)$, $m = \text{const}$, and densely defined in the Hilbert space $H = L^2(D')$, is closable if $t[u_n - u_m, u_n - u_m] \underset{n,m\to\infty}{\longrightarrow} 0$ and $u_n \underset{H}{\longrightarrow} 0$ imply $t[u_n, u_n] \underset{n\to\infty}{\longrightarrow} 0$. The closure of the domain $D[t]$ of the closable quadratic form in the norm $[u] := \{t[u,u] + (m+1)(u,u)\}^{1/2}$ is a Hilbert space $H_t \subset H$ densely embedded in H. The quadratic form $t[u,u]$ is defined on H_t and this form with the domain of definition H_t is closed.

To prove the closability, consider, for example, t_D, and assume $u_n \underset{H}{\longrightarrow} 0$, $(\nabla u_n - \nabla u_m, \nabla u_n - \nabla u_m) \to 0$ as $n, m \to \infty$. Then $\nabla u_n \underset{H}{\longrightarrow} f$, and

$$(f, \phi) = \lim_{n\to\infty} (\nabla u_n, \phi) = -\lim_{n\to\infty} (u_n, \nabla\phi) = 0, \qquad \forall \phi \in C_0^\infty. \tag{9.19}$$

Thus $f = 0$, so t_D is closable. Similarly one checks that t_N and t_R are closable. Let us denote by $H^1_{2,2}(D')$ the completion of $C(\overline{D'}) \bigcap C^\infty(D') \bigcap H^1_0(D')$ in the norm $\|u\| := (\|\nabla u\|^2_{L^2(D')} + \|\nabla u\|^2_{L^2(S)})^{1/2}$.

For an arbitrary open set $D \subset \mathbb{R}^3$ with a finite volume ($|D| < \infty$, where $|D| := \operatorname{meas} D$ is the volume of D) the inequality

$$\|u\|_{L^3(D)} \le c\big(\|\nabla u\|_{L^2(D)} + \|u\|_{L^2(S)}\big) \tag{9.20}$$

holds, and the embedding operator $i : H^1_{2,2}(D) \to L^q(D)$ is compact if $q < 3$ and $|D| < \infty$ ([65, p.258]).

If D is an extension domain, i.e., D has an extension property, then the inequality

$$\|u\|_{L^2(S)} \le c\|u\|_1, \tag{*}$$

where $c > 0$, may have no sense because the trace on the boundary may be not well defined, since the boundary may have Hausdorff dimension greater than $n - 1$, where n is the dimension of the space.

The extension property, $D \subset EV^\ell_p$, means that there exists a linear bounded extension operator $E : V^\ell_p(D) \to V^\ell_p(\mathbb{R}^3)$, $Eu = u$ in D.

The space $V^\ell_p(D) := \bigcap_{j=0}^{\ell} L^j_p(D)$ and $L^j_p(D)$ is the set of functions with the finite norm

$$\|u\|_{L^\ell_p(D)} = \left\| \left(\sum_{|\alpha|=\ell} |D^\alpha u|^2 \right)^{1/2} \right\|_{L^p(D)}, \quad p > 0.$$

If D satisfies cone property, then D is an extension domain. If $D \subset C^{0,1}$ is a Lipschitz domain, then D satisfies cone property and therefore D is an extension domain. Inequality (9.20) may hold for some domains which have no extension property. Estimate $(*)$ may fail for some domains for which (9.20) holds. If the Hausdorff 2-dimensional measure $|S| := s(S)$ of S is finite then a sufficient condition on D for $(*)$ to hold is given in [65, p.262].

Consider the closed symmetric forms t_D, t_N and t_R. Each of these forms define a unique selfadjoint operator A in $H = L^2(D')$, $D(A) \subset H^1(D') \subset H$, $(Au, v) = t[u, v]$, $u \in D(A)$, $v \in D[t]$, $A = A_D$, $A = A_N$, and $A = A_R$, respectively.

Let $L_{2,a} := L^2(D', (1 + |x|)^{-a})$, $a \in (1, 2)$, $\|u\|^2_{L_{2,a}} = \int_{D'} \frac{|u|^2 dx}{(1+|x|)^a}$ and L^2_0 be the set of $L^2(D')$ functions vanishing near infinity.

Recall that $\tilde{D} \subset D'$ is a bounded domain whose boundary consists of $S := \partial D$ and $\tilde{S}$, which is a smooth compact manifold. Assume that $i_1' : H^1(\tilde{D}) \longrightarrow L^2(\tilde{D})$ and $i_2' : H^1(\tilde{D}) \to L^2(S)$ are compact. Then the following theorem holds:

Theorem 9.4 *For any $F \in L_0^2$, each of the boundary-value problems:*

$$A_i u = F, \quad i = D, N \text{ or } R, \quad A_i u = -\Delta u, \tag{9.21}$$

has a solution $u = \lim_{\epsilon \downarrow 0}(A - i\epsilon)^{-1}F := (A - i0)^{-1}F$, $u \in H^2_{loc}(D')$, $u \in L_{2,a}$, $a \in (1, 2)$, and this solution is unique.

Remark Equation (9.21) is understood as the weak formulation of the exterior boundary- value problem with the Dirichlet or Neumann or Robin boundary condition.

Similar result holds for the operator $A - k^2$, where $k = const > 0$, in which case the solution u satisfies the radiation condition at infinity:

$$\lim_{r\to\infty} \int_{|s|=r} \left|\frac{\partial u}{\partial r} - iku\right|^2 ds = 0. \tag{9.22}$$

Proof of Theorem 9.4.

Uniqueness. If $k = 0$, then uniqueness follows from the maximum priciple. If $k > 0$, then uniqueness can be established with the help of the radiation condition. We give details for the case $k > 0$ at the end of this chapter. □

Existence. Since $A = A_i$ is selfadjoint, the equation

$$(A - i\varepsilon)u_\varepsilon = F, \qquad \varepsilon = \text{const} > 0, \qquad A = -\Delta, \tag{9.23}$$

has a unique solution $u_\varepsilon \in H = L^2(D')$. Let us prove that if $F \in L_{2,-a}$, then there exists the limit

$$u = \lim_{\varepsilon\to 0} u_\varepsilon, \qquad \lim_{\varepsilon\to 0} \|u - u_\varepsilon\|_{L_{2,a}} = 0, \tag{9.24}$$

and u solves (9.21). Thus the *limiting absorption principle* holds at $\lambda = 0$. Recall that the limiting absorption principle holds at a point λ if the limit $u := \lim_{\varepsilon\to 0}(A - \lambda - i\varepsilon)^{-1}F$ exists in some sense and solves the equation $(A - \lambda)u = F$.

To prove (9.24), assume first that

$$\sup_{1>\varepsilon>0} \|u_\varepsilon\|_{L_{2,a}} \le c, \tag{9.25}$$

where $c = \text{const}$ does not depend on ε. If (9.25) holds, then (9.24) holds, as we will prove. Finally, we prove (9.25).

Let us prove that (9.25) implies (9.24). Indeed, (9.25) implies

$$\|u_\varepsilon\|_{L^2(D'_R)} \le c, \tag{9.26}$$

where $D'_R := D' \bigcap B_R$, $B_R := \{x : |x| \le R\}$, and we choose $R > 0$ so that $\operatorname{supp} F \subset B'_R$. It follows from (9.26) that there exists a sequence $\varepsilon_n \to 0$ such that $u_n := u_{\varepsilon_n}$ converges weakly: $u_n \rightharpoonup u$ in $L^2(D'_R)$. From the relation

$$t[u_n, \phi] = (F, \phi), \tag{9.27}$$

where the form t corresponds to the operator A in (9.23) and the choice $\phi = u_n$ is possible, it follows that

$$t_i[u_n, u_n] = \|\nabla u_n\|_{L^2(D')} \le c, \qquad i = D, N \tag{9.28}$$

and

$$t_i[u_n, u_n] = \|\nabla u_n\|^2_{L^2(D')} + \int_S h|u_n|^2 ds \le c, \qquad i = R. \tag{9.29}$$

From (9.25) and (9.23) it follows that

$$\|\Delta u_n\|_{L^2(D'_R)} \le c. \tag{9.30}$$

By the known elliptic inequality:

$$\|u\|_{H^2(D_1)} \le c(D_1, D_2)\big(\|\Delta u\|_{L^2(D_2)} + \|u\|_{L^2(D_2)}\big), \qquad D_1 \Subset D_2, \tag{9.31}$$

where H^2 is the usual Sobolev space, it follows from (9.28) and (9.26) that

$$\|u_n\|_{H^2(D_1)} \le c, \tag{9.32}$$

where $D_1 \Subset D'$ is any bounded strictly inner subdomain of D'. By the embedding theorem, it follows that there exists a u such that

$$\lim_{n\to\infty} \|u_n - u\|_{H^1(D_1)} = 0, \qquad D_1 \Subset D'. \tag{9.33}$$

From (9.33) and (9.23) it follows that $\lim_{n\to\infty} \|\Delta u_n - \Delta u\| = 0$, and by (9.31) one concludes

$$\lim_{n\to\infty} \|u_n - u\|_{H^2(D_1)} = 0, \qquad D_1 \Subset D'. \tag{9.34}$$

Passing to the limit in (9.23) with $\varepsilon = \varepsilon_n$ one gets equation (9.21) for u in F'. From (9.28) or (9.29) it follows that

$$u_n \underset{H^1(D'_R)}{\longrightarrow} u \text{ for any } R > 0. \tag{9.35}$$

Outside the ball B_R one has the equation

$$-\Delta u_n - i\varepsilon_n u_n = 0 \text{ in } B'_R := \mathbb{R}^3 \setminus \overline{B}_R, \qquad u_n(\infty) = 0, \tag{9.36}$$

and, by Green's formula, one gets

$$u_n(x) = \int_{S_R} \left(g_n \frac{\partial u_n}{\partial N} - u_n \frac{\partial g_n}{\partial N} \right) ds, \quad x \in B'_R, \quad S_R := \{s : |s| = R\}, \tag{9.37}$$

where$N$ is the outer normal to S_R and $g_n = \frac{e^{\gamma\sqrt{\varepsilon_n}|x-y|}}{4\pi|x-y|}$, $\gamma := \frac{-1+i}{\sqrt{2}}$.

By (9.34) and the embedding theorem, one has

$$\lim_{n\to\infty} \left(\|u_n - u\|_{L^2(S_R)} + \left\| \frac{\partial u_n}{\partial N} - \frac{\partial u}{\partial N} \right\|_{L^2(S_R)} \right) = 0. \tag{9.38}$$

Passing to the limit in (9.37) one gets

$$u(x) = \int_{S_R} \left(g \frac{\partial u}{\partial N} - u \frac{\partial g}{\partial N} \right) ds, \quad x \in B'_R, \quad g := \frac{1}{4\pi|x-y|}. \tag{9.39}$$

Thus

$$|u(x)| \leq \frac{c}{|x|}, \quad x \in B'_R, \tag{9.40}$$

and $u_n(x)$ satisfies (9.40) with a constant c independent of n.

If the Dirichlet condition is imposed, then the embedding $i' : \overset{\circ}{H}{}^1(\tilde{D}) \to L^2(\tilde{D})$ is compact for any *bounded* domain $\tilde{D}$. If the Neumann condition is imposed, then the compactness of the embedding $i'_1 : H^1(\tilde{D}) \to L^2(\tilde{D})$ imposes some restriction on the smoothness of S (remember that $\tilde{S}$ is assumed smooth), and the above embedding operator is not compact for some open bounded sets $\tilde{D}$. However, this restriction on the smoothness of S is weak: it is satisfied for any extension domain. If the Robin condition is imposed, then we use compactness of the operator $i'_2 : H^1(\tilde{D}) \to L^2(S)$ for passing to the limit

$$\lim_{n\to\infty} [(\nabla u_n, \nabla u_n) + \langle h u_n, u_n \rangle] = (\nabla u, \nabla u) + \langle h u, u \rangle.$$

If the embedding operator $i'_1 : H^1(\tilde{D}) \to L^2(\tilde{D})$ is compact, then (9.28), (9.33) and (9.40) imply the following three conclusions:

$$\lim_{n\to\infty} \|u_n - u\|_{L^2(D'_R)} = 0, \quad \forall R < \infty, \tag{9.41}$$

$$u_n \rightharpoonup u \text{ in } H^1(D'_R), \tag{9.42}$$

$$\lim_{n\to\infty} \|u_n - u\|_{L_{2,a}} = 0. \tag{9.43}$$

Note that (9.43) follows from (9.41) and (9.40) if $a > 1$. Indeed,

$$\int_{|x|>R} \frac{|u_n - u|^2 dx}{(1+|x|)^{1+a}} \leq c \int_{|x|\geq R} \frac{dx}{(1+|x|)^{1+a}|x|^2} \leq \frac{c}{R^a}.$$

For an arbitrary small $\delta > 0$, one can choose R so that $\frac{c}{R^a} < \delta$ and fix such an R. For a fixed R one takes n sufficiently large and use (9.41) to get

$$\int_{D'_R} \frac{|u_n - u|^2 dx}{(1+|x|)^{1+a}} < \delta.$$

This implies (9.43).

The limit u solves problem (9.21). We have already proved uniqueness of its solution. therefore not only the subsequence u_n converges to u, but also $u_\varepsilon \to u$ as $\varepsilon \to 0$. We have proved that (9.25) implies (9.24).

To complete the proof of the existence of the solution to (9.21) one has to prove (9.25). Suppose (9.25) is wrong. Then there is a sequence $\varepsilon_n \to 0$ such that $\|u_{\varepsilon_N}\|_{L_{2,a}} := \|u_n\|_{L_{2,a}} \to \infty$. Let $v_n := \frac{u_n}{\|u_n\|_{2,a}}$. Then

$$\|v_n\|_{2,a} = 1 \tag{9.44}$$

$$Av_n - i\varepsilon_n v_n = \frac{F}{\|u_n\|_{2,a}}. \tag{9.45}$$

By the above argument, relation (9.44) implies the existence of $v \in L_{2,a}$ such that

$$\lim_{n\to\infty} \|v_n - v\|_{2,a} = 0, \tag{9.46}$$

and

$$Av = 0. \tag{9.47}$$

By the uniqueness result, established above, it follows that $v = 0$. Thus (9.46) implies $\lim_{n\to\infty} \|v_n\|_{2,a} = 0$. This contradicts to (9.44).

This contradiction proves Theorem 9.4. □

Remark 9.5 *The above argument is valid also for solving the problem*

$$A_i u - \lambda u = F, \qquad i = D, N, R, \qquad \lambda \in \mathbb{R}, \tag{9.48}$$

provided that problem (9.48) *with* $F = 0$ *has only the trivial solution.*

One may also weaken the assumption about F. If $F \in L_{2,-1}$, then (9.39) *should be replaced by*

$$u(x) = \int_{S_R} \left(g \frac{\partial u}{\partial N} - u \frac{\partial g}{\partial N} \right) ds - \int_{B'_R} g(x,y) F(y) dy. \tag{9.49}$$

If $a > 3$, then, using the Cauchy inequality, one gets:

$$\left| \int_{B'_R} g(x,y) F dy \right|^2 \leq c \int_{B'_R} \frac{dy}{|x-y|^2 (1+|y|)^a} \int_{B'_R} |F|^2 (1+|y|)^a dy \leq \frac{c}{|x|^2}, \tag{9.50}$$

for large $|x|$, so that (9.40) *holds if $F \in L_{2,-a}$, $a > 3$. The rest of the argument is unchanged.*

Remark 9.6 *We want to emphasize that the assumptions on the smoothness of the boundary S under which we have proved existence and uniqueness of the solutions to boundary-value problems are weaker than the usual assumptions for the Neumann and Robin boundary conditions. For the Dirichlet condition $u = 0$ on S no assumptions, except boundedness of D, are used. For the Neumann condition, $u_N = 0$ on S, only compactness of the embedding operator $i'_1 : H^1(\tilde{D}) \to L^2(\tilde{D})$ is used, and for the Robin boundary condition, $u_N + hu = 0$ on S, $0 \leq h \leq c$, compactness of both of the embedding operators i'_1 and $i'_2 : H^1(\tilde{D}) \to L^2(S)$ is used.*

Our arguments can be applied for a study of the boundary-value problems for second-order formally selfadjoint elliptic operators and for nonselfadjoint sectorial second-order elliptic operators. In [45] one finds the theory of sectorial operators and the corresponding sectorial sesquilinear forms.

In conclusion let us prove the uniqueness theorem mentioned below Theorem 9.4 in the case $k > 0$. Namely, if in (9.21) one has $F = 0$ and $A - k^2$ in place of $A = A_i$, where $k = const > 0$, then a weak solution to this homogeneous (9.21), which satisfies the radiation condition (9.22), must vanish. Let us prove this for the Robin boundary condition. Define $W := u_1 - u_2$. One has:

$$\int_{D'} \operatorname{grad} W \operatorname{grad} \bar{\phi} dx + \int_S \sigma W \bar{\phi} ds = 0, \tag{9.51}$$

for all $\phi \in H^1_{loc}$ vanishing near infinity, and W satisfies (9.22). From (9.22) one gets

$$\lim_{r\to\infty}\int_{|x|=r}\left(|W_r|^2+k^2|W|^2\right)ds+\lim_{r\to\infty} ik\int_{|x|=r}\left(W_r\bar{W}-\bar{W}_r W\right)ds=0\,. \tag{9.52}$$

The second integral vanishes because of the radiation condition. Thus,

$$\lim_{r\to\infty}\int_{|x|=r}\left(|W_r|^2+k^2|W|^2\right)ds=0. \tag{9.53}$$

This and the known lemma (see e.g., [133], p. 25) imply that $W=0$ near infinity. By the unique continuation property for the solution to homogeneous Helmholtz equation, $W=0$ in D'.

See also [104] for more details.

9.4 Quasiisometrical Mappings

The main purpose of this section is to study boundary behavior of quasiisometrical homeomorphisms.

9.4.1 *Definitions and main properties*

Let us start with some definitions.

Definition 9.1 *(Quasiisometrical homeomorphisms)* Let A and B be two subsets of R^n. A homeomorphism $\varphi : A \to B$ is $Q-$quasiisometrical if for any point $x\in A$ there exists such a ball $B(x,r)$ that

$$Q^{-1}|y-z|\le|\varphi(y)-\varphi(z)|\le Q|y-z| \tag{9.54}$$

for any $y,z\in B(x,r)\cap A$. Here the constant $Q>0$ does not depends on the choice of $x\in A$, but the radius r may depend on x.

Obviously the inverse homeomorphism $\varphi^{-1} : A \to B$ is also Q-quasiisometrical. A homeomorphism $\varphi : A\to B$ is a quasiisometrical homeomorphism if it is a Q-quaiisometrical one for some Q. Sets A and B are quasiisometrically equivalent if there exists a quasiisometrical homeomorphism $\varphi : A \to B$.

Definition 9.2 *(Lipschitz Manifolds)* A set $M\subset R^n$ is an m-dimensional Q-lipschitz manifold if for any point $a\in M$ there exists a Q-quasiisometrical homeomorphism $\varphi_a : B(0,1)\to R^n$ such that $\varphi(0)=a$ and $\varphi(B(0,1)\cap R^m)\subset M$. Here $R^m := \{x\in R^n : x_{m+1}=...=x_n=0\}$.

We are interested in compact lipschitz manifolds that are boundaries of domains in R^n and/or in $(n-1)$-dimensional lipschitz manifolds that are dense subsets of boundaries in the sense of $(n-1)$-dimensional Hausdorff measure H^{n-1}.

Definition 9.3 *(Class L)* We call a bounded domain $U \subset R^n$ a domain of class L if:

1. There exist a bounded smooth domain $V \subset R^n$ and a quasiisometrical homeomorphism $\varphi : V \to U$;
2. The boundary ∂U of U is a $(n-1)$-dimensional lipschitz manifold.

The following proposition is well known and will be useful for a study of domains of class L and boundary behavior of quasisisometrical homeomorphisms.

Proposition 9.1 *Let A and B be two subsets of R^n. A homeomorphism $\varphi : A \to B$ is $Q-$quasiisometrical if and only if for any point $a \in A$ the following inequality holds:*

$$Q^{-1} \leq \liminf_{x \to a, x \in A} \frac{|\varphi(x) - \varphi(a)|}{|x-a|} \leq \limsup_{x \to a, x \in A} \frac{|\varphi(x) - \varphi(a)|}{|x-a|} \leq Q.$$

Here the constant $Q > 0$ does not depend on the choice of $a \in U$.

This proposition is a motivation for the following definition.

Definition 9.4 *(Quasilipschitz mappings)* Let A be a set in R^n. A mapping $\varphi : A \to R^m$ is $Q-$quasilipshitz if for any $a \in A$ one has:

$$\limsup_{x \to a, x \in A} \frac{|\varphi(x) - \varphi(a)|}{|x-a|} \leq Q$$

Here the constant $Q > 0$ does not depend on the choice of $a \in A$.

A mapping is quasilipschitz if it is Q-quasilipschitz for some Q.

A homeomorphism $\varphi : A \to B$ is a quasiisometrical homeomorphism iff φ and φ^{-1} are quasilipschitz.

By definition any quasilipschitz mapping is a locally lipschitz one. A restriction of a Q-quasilipshitz mapping on any subset $B \subset A$ is a quasilipschitz mapping also.

9.4.2 *Interior metric and boundary metrics*

Suppose A is a linearly connected set in R^n. An interior metric μ_A on A can be defined by the following way:

Definition 9.5 For any $x, y \subset A$

$$\mu_A(x,y) = \inf_{\gamma_{x,y}} l(\gamma_{x,y}),$$

where $\gamma_{x,y} : [0,1] \to A$, $\gamma_{x,y}(0) = x$, $\gamma_{x,y}(1) = y$ is a rectifiable curve and $l(\gamma_{x,y})$ is length of $\gamma_{x,y}$.

As follows from Definition 9.4 a Q-quasilipschitz mapping can change the length of a rectifiable curve by a factor Q at most. Hence a Q-quasilipschitz mapping $\varphi : A \to B$ of a linearly connected set A onto a linearly connected set B is a lipschitz mapping of the metric space (A, μ_A) onto the metric space (B, μ_B). Any Q-quasiisometrical homeomorphism $\varphi : A \to B$ is a bilipschitz homeomorphism of the metric space (A, μ_A) onto the metric space (B, μ_B).

Because any domain U of the class L is quasiisometrically equivalent to a smooth bounded domain and for a smooth bounded domain the interior metric is equivalent to the Euclidian metric, the interior metric μ_U is equivalent to the Euclidian metric for the domain U also. It means that for any domain $U \in L$

$$K^{-1}|x-y| \le \mu_U(x,y) \le K|x-y|$$

for any $x, y \in U$. Here a positive constant K does not depend on the choice of the points x, y. Therefore for any bounded domain $U \in L$ any quasilipshitz mapping $\varphi : U \to V$ is a lipschitz mapping $\varphi : (U, \mu_U) \to R^m$ for the interior metric.

We will use the following definition of locally connected domain $U \in R^n$ that is equivalent to the standard one.

Definition 9.6 Suppose $(x_k \in U), (y_k \in U)$ are two arbitrary convergent sequences such that $\liminf_{k\to\infty} \mu_U(x_k, y_k) > 0$. Call a domain $U \in R^n$ locally connected if for any such sequences one has $\lim_{k\to\infty} x_k \ne \lim_{k\to\infty} y_k$

If a boundary of a bounded domain is a topological manifold then this domain is locally connected. Therefore, domains of the class L are locally connected domains because their boundaries are compact lipschitz manifolds.

Definition 9.7 Let A be a closed linearly connected subset of R^n and $H^{n-1}(A) > 0$. Call the interior metric μ_A a quasieuclidean metric almost everwhere if there exists a closed set $Q \subset A$ with $H^{n-1}(Q) = 0$, such that for any point $x \in A \setminus Q$ the following condition holds:

There exists such ball $B(x, r)$ that for any $y, z \in B(x, r)$

$$\frac{1}{K}|y - z| \leq \mu_A(y, z) \leq K|y - z|,$$

where $K = const > 0$ does not depends on choice y, z and x.

By definition of lipschitz manifolds any domain of the class L is quasieuclidean at any boundary point.

Definition 9.8 Suppose U is a domain in R^n and $x_0, y_0 \in \partial U$. Let us call the following quantity

$$\widetilde{\mu}_{\partial U}(x_0, y_0) := \lim_{\varepsilon \to 0} \inf_{|x - x_0| < \varepsilon, |y - y_0| < \varepsilon} \mu_U(x, y)$$

relative interior boundary metric.

Because boundary of any domain U of the class L is a compact lipschitz manifold, the relative interior boundary metric on ∂U is equivalent to the interior boundary metric on ∂U for such domains. This motivates the following definition:

Definition 9.9 A bounded domain $U \subset R^n$ has an almost quasiisometrical boundary if $H^{n-1}(\partial U) < \infty$ and there exists a closed set $A \in \partial U$ with $H^{n-1}(A) = 0$ such that for any point $x_0 \in \partial U \setminus A$ the following condition holds:

There exists a ball $B(x_0, r)$, $B(x_0, r) \cap A = \emptyset$, such that for any $x, y \in \partial U \cap B(x_0, r)$ one has:

$$\frac{1}{K}\mu_{\partial U}(x, y) \leq \widetilde{\mu_{\partial U}}(x, y) \leq K\mu_{\partial U}(x, y),$$

where $K = const > 0$ does not depend on the choice of x_0, x and y.

We will use for the two-sided inequalities similar to the above one the following short notation $\widetilde{\mu}_{\partial U}(x, y) \sim \mu_{\partial U}(x, y)$.

If a domain U has an almost quasiisometric boundary ∂U and this boundary is locally almost quasieuclidian then $\mu_{\partial U}(x, y) \sim |x - y|$ for any $x, y \in \partial U$.

Definition 9.10 We call a bounded domain $U \subset R^n$ an almost quasiisometrical domain if $H^{n-1}(\partial U) < \infty$ and there exists such a closed set $A \in \partial U$, with $H^{n-1}(A) = 0$, that the following condition holds:

There exists a ball $B(x_0, r) \cap A = \emptyset$ such that for any $x, y \in \partial U \cap B(x_0, r)$ one has:

$$\mu_{\partial U}(x, y) \sim \widetilde{\mu}_{\partial U}(x, y) \sim |x - y|\,.$$

By the extension theorem for lipschitz mappings any quasiisometrical homeomorphism φ of a smooth bounded domain in R^n onto a domain V in R^n has a lipschitz extension $\widetilde{\psi}$ onto R^n. Denote by ψ the restriction of a lipschitz extension $\widetilde{\psi}$ on ∂U. By continuity, the extension ψ is unique.

Definition 9.11 Let U be a smooth domain in R^n and V be a domain in R^n such that $H^{n-1}(\partial V) < \infty$. A quasiisometrical homeomorphism $\varphi : U \to V$ has N^{-1}-property on the boundary if for any $A \in \partial V$ with $H^{n-1}(A) = 0$ one has $H^{n-1}(\psi^{-1}(A)) = 0$.

The definition makes sense because the extension ψ of a quasiisometrical homeomorphism φ on ∂U is unique.

Definition 9.12 *(Class QI)* Let us call a bounded domain V a domain of class QI if:

1) There exists a quasisiometrical homeomorphism $\varphi : U \to V$ of a smooth bounded domain U onto the domain V that has the N^{-1}-property on the boundary.

2) there exists such a closed set $A \in \partial V$, $H^{n-1}(A) = 0$, that $\partial V \setminus A$ is a Q-lipschitz manifold for some Q;

3) V is a locally connected almost quasiisometrical domain.

Remark 9.7 *The class L is a subclass of the class QI.*

9.4.3 *Boundary behavior of quasiisometrical homeomorphisms*

Proposition 9.2 *Suppose a Q-quasiisometrical homeomorphism $\varphi : U \to V$ maps a smooth bounded domain U onto a locally connected domain V. Then there exists an extension ψ of φ on ∂U such that $\psi(\partial U) = \partial V$ and the mapping $\psi|_{\partial U}$ is a lipshitz mapping of multiplicity one.*

Proof. Because U is a smooth domain, φ is a lipschitz mapping. By the extension theorem for lipschitz mappings there exists a Q-lipschitz extension $\widetilde{\psi} : R^n \to R^n$ of φ. This extension is not necessarily a quasiisometrical homeomorphism. By continuity of $\widetilde{\psi}$ and because $\varphi : U \to V$ is a homeomorphism we have $\widetilde{\psi}(\partial U) = \partial V$.

Suppose $\psi := \widetilde{\varphi}|_{\partial U}$ has multiplicity more than one. Then there exist two different points $x_0, y_0 \in \partial U$, $x_0 \neq y_0$ such that $\psi(x_0) = \psi(y_0)$. Choose two sequences: $x_k \in U$ and $y_k \in U$ such that $\lim_{k\to\infty} x_k = x_0, \lim_{k\to\infty} y_k = y_0$. Because U is a smooth bounded domain the interior metric μ_U is equivalent to the Euclidean metric, i.e. there exists a positive constant Q such that

$\mu_U(x_k, y_k) \geq Q^{-1}|x_k - y_k| \geq Q^{-1}|x_0 - y_0| > 0$ for all sufficiently large k. The homeomorphism $\varphi : (U, \mu_U) \to (V, \mu_V)$ is a bi-lipschitz homeomorphism. Therefore $\liminf_{k\to\infty} \mu_V(\varphi(x_k), \varphi(y_k)) > 0$. Because U is a locally connected domain $\psi(x_0) = \lim_{k\to\infty} \varphi(x_k) \neq \lim_{k\to\infty} \varphi(y_k) = \psi(y_0)$. This contradiction proves the Proposition. □

For any lipschitz m-dimensional compact manifold $M \subset R^n$ and for any lipschitz mapping $\varphi : M \to R^n$ the set $\varphi(M)$ is H^m-measurable for the m-dimensional Hausdorff measure H^m and $H^m(\varphi(M)) < \infty$.

The next theorem, dealing with area formulas, is a particular case of the result proved in [3] and used for domains of the class QI.

Let us start with an abstract version of this theorem.

Definition 9.13 Call a metric space X a H^k-rectifiable metric space if there exists such finite or countable set of lipschitz mappings $\alpha_i : A_i \to X$ of mesurable sets $A_i \subset R^k$ into X that $H^k(X \setminus \bigcup_j \alpha_i(A_i)) = 0$.

By the definition of the class QI a boundary ∂U of any domain $U \in QI$ is a H^{n-1}-rectifiable metric space.

Our next definition represents an abstract version of Jacobian for H^k-rectifiable metric spaces.

Definition 9.14 Let X and Y be H^k-rectifiable complete metric spaces and $F : X \to Y$ be a lipschitz mapping. Call the quantity

$$J(x, F) := \lim_{r\to 0} \frac{H^k(F(B(x,r))}{H^k(B(x,r))}$$

a formal Jacobian of F at a point x.

Theorem 9.5 *Suppose X and Y are H^k-rectifiable complete metric spaces and $F : X \to Y$ is a lipschitz mapping of multiplicity one.*

Then

1. *Formal Jacobian $J(x, F)$ exists H^k–almost everywhere;*
2. *The following area formula holds:*

$$\int_X J(x, F)dH^k = \int_{F(X)} dH^k;$$

3. *For any $u \in L^1(Y)$*

$$\int_X u(F(x))J(x, F)dH^k = \int_{F(X)} u(y)dH^K.$$

Corollary 9.1 *If a domain V belongs to the class QI, and $\varphi : V \to U$ is a Q-quasiisometrical homeomorphism, then $H^{n-1}(\partial U) < \infty$.*

Proof. Any Q-quasiisometrical homeomorphism $\varphi : V \to U$ of a smooth domain $V \in R^n$ onto a domain U of the class QI has a lipschitz extension $\widetilde{\psi} : R^n \to R^n$. By definition of the class QI the domain V is a locally connected domain. Hence by Proposition 9.2 the Q-lipschitz mapping $\psi := \widetilde{\psi} \setminus \partial V$ has multiplicity one and $\psi(\partial V) = \partial U$. By Theorem 9.5 one gets $H^{n-1}(\partial U) < \infty$. □

9.5 Quasiisometrical Homeomorphisms and Embedding Operators

By Corollary 9.1, $H^{n-1}(\partial V) < \infty$ for any domain $V \in QI$. Therefore we can define Banach space $L^2(\partial V)$ using the Hausdorff measure H^{n-1}.

Proposition 9.3 *Let U be a smooth domain and $V \in QI$. Any Q-quasiisometrical homeomorphism $\varphi : U \to V$ that has N^{-1}property on the boundary induces a bounded composition operator $\psi^* : L^2(\partial V) \to L^2(\partial U)$ by the rule $\psi^*(u) = u \circ \psi$.*

Proof. Denote by m the $(n-1)$-dimensional Lebesgue measure on ∂U and by ψ the extension of ϕ onto ∂U. By Theorem 9.5 for any $u \in L^2(\partial V)$

$$\int_{\partial U} |u(\psi(x))|^2 J(x, \psi) dm = \int_{\partial V} |u(y)|^2 dH^m.$$

Suppose that there exists such a constant $K > 0$ that $J(x, \psi) \geq K^{-1}$ for almost all $x \in \partial U$. Denote by $A \in \partial U$, with $H^{n-1}(A) = 0$, a set of all points for which the previous inequality does not hold. Then

$$\begin{aligned}
\|\psi^* u\|^2_{L^2(\partial U)} &= \int_{\partial U \setminus A} |u(\psi(x))|^2 dH^{n-1} \\
&= \int_{\partial U \setminus A} |u(\psi(x))|^2 \frac{J(x, \psi)}{J(x, \psi)} dH^{n-1} \\
&\leq K \int_{\partial V \setminus \psi^{-1}(A)} |u(\psi(x))|^2 J(x, \psi) dm = K \|u\|^2_{L^2(\partial V)} .
\end{aligned}$$

The last equality is valid because φ has the N^{-1}-property on the boundary, i.e. $m(\psi^{-1}(A)) = 0$. Therefore $\psi^* : L^2(\partial V) \to L^2(\partial U)$ is a bounded operator and $\|\psi^*\| \leq K$.

To finish the proof we have to demonstrate that $J(x,\psi) \geq K^{-1}$. Remember that any domain of the class QI has an almost quasiisometric boundary.

It means that we can choose such a closed subset $A \subset \partial V$, with $H^{n-1}(A) = 0$, that the following property holds:

For any $x_0 \in \partial V \setminus A$ there exists such a ball $B(x_0, r) \cap A = \emptyset$ that:

1) $B(x_0, r) \cap A = \emptyset$ and $\widetilde{\mu}_{\partial V}(x,y) \sim \mu_{\partial V}(x,y) \sim |x-y|$ for any $x, y \in \partial V \cap B(x_0, r)$,

2) $B(x_0, r) \cap \partial V$ is a lipschitz manifold.

Let $B := \psi^{-1}(A)$. Choose a point $z_0 \in \partial U \setminus B$ and such a ball $B(z_0, R)$ that relations $\widetilde{\mu}_{\partial U}(x,y) \sim \mu_{\partial U}(x,y) \sim |x-y|$ hold for any $x, y \in \psi(B(z_0, R)) \cap \partial U$. This is possible because U is a smooth domain.

Because φ is a Q-quasiisometric, the length $|\gamma|$ of any curve $\gamma \subset V$ satisfies the estimate:

$$\frac{1}{Q}|\varphi(\gamma)| \leq |\gamma| \leq Q|\varphi(\gamma)|$$

where $|\varphi(\gamma)|$ is the length of the curve $\varphi(\gamma) \in U$. In terms of the relative interior metric $\widetilde{\mu}_{\partial U}$ it means that

$$B_{\widetilde{\mu}_{\partial U}}(x_0, \frac{1}{Q}R) \subset \psi(B(z_0, R) \cap \partial V \subset B_{\widetilde{\mu}_{\partial U}}(x_0, QR)$$

where $x_0 = \psi(z_0)$. Without loss of generality we can suppose that $\widetilde{\mu}_{\partial U}(x,y) \sim \mu_{\partial U}(x,y) \sim |x-y|$ for any $x, y \in B_{\widetilde{\mu}_{\partial U}}(x_0, QR)$. Finally we obtain

$$B(x_0, \frac{1}{K}R) \subset \psi(B(z_0, R) \cap \partial V \subset B(x_0, KR) \tag{9.55}$$

for some constant K that depends only on Q and constants in relations $\widetilde{\mu}_{\partial U}(x,y) \sim \mu_{\partial U}(x,y) \sim |x-y|$.

We have proved the inequality $J_{\psi}(x) \geq K^{-1}$ almost everywhere on ∂U.

□

9.5.1 *Compact embedding operators for rough domains*

It is well known that the embedding operator $H^1(\Omega) \to L^2(\partial\Omega)$ is compact for bounded smooth domains.

We will prove compactness of the embedding operator for the class QI. Then we extend the embedding theorem to the domains that are finite unions of the QI-domains. Our proof is based on the following result: a quasiisometrical homeomorphism $\varphi : U \to V$ induces a bounded composition operator $\varphi^* : H^1(V) \to H^1(U)$ by the rule $\varphi^*(u) = u \circ \varphi$ (see, for example [32] or [165]).

Definition 9.15 *A domain U is a domain of class Q if it is a finite union of elementary domains of class QI.*

Let us use the following result:

Theorem 9.6 *(see for example [32] or [165]) Let U and V be domains in R^n. A quasiisometrical homeomorphism $\varphi : U \to V$ induces a bounded composition operator $\varphi^* : H^1(V) \Rightarrow H^1(U)$ by the rule $\varphi^*(u) = u \circ \varphi$.*

Combining this Theorem with Theorem 9.3, one gets:

Theorem 9.7 *If U is a domain of the class QI, then the embedding operator $i_U : H^1(U) \to L^2(\partial U)$ is compact.*

Proof. By definition of the class QI there exist a smooth bounded domain V and a quasiisometrical homemorphism $\varphi : V \to U$. By Proposition 9.3 φ induces a bounded composition operator $\psi^* : L^2(\partial U) \to L^2(\partial V)$ by the rule $\psi^*(u) = u \circ \psi$. Because the embedding operator $i_U : H^1(U) \to L^2(\partial U)$ is compact and the composition operator$(\varphi^{-1})^* : H^1(V) \to H^1(U)$, induced by quasiisometrical homeomorphism φ, is bounded the embedding operator $i_V : H^1(V) \to L^2(\partial V)$, $i_U = (\varphi^{-1})^* \circ i_V \circ (\overline{\varphi})^*$ is compact. □

To apply this result for domains of the class Q we need the following lemma:

Lemma 9.2 *If U and V are domains of the class QI, then the embedding operator $H^1(U \cup V) \to L^2(\partial(U \cup V))$ is compact.*

Proof. By previous proposition operators $i_U : H^1(U) \to L^2(\partial U)$ and $i_V : H^1(V) \to L^2(\partial V)$ are compact. Choose a sequence $\{w_n\} \subset H^1(U \cup V)$, $\|w_n\|_{H^1(U \cup V)} \leq 1$ for all n. Let $u_n := w_n|_{\partial U}$ and $v_n := w_n|_{\partial V}$. Then $u_n \in L^2(\partial U)$, $v_n \in L^2(\partial V)$, $\|u_n\|_{L^2(\partial U)} \leq \|i_U\|$, $\|v_n\|_{L^2(\partial V)} \leq \|i_V\|$.

Because the embedding operator $H^1(U) \to L^2(\partial U)$ is compact, we can choose a subsequence $\{u_{n_k}\}$ of the sequence $\{u_n\}$ which converges in $L^2(\partial U)$ to a function $u_0 \in L^2(\partial U)$. Because the embedding operator $H^1(V) \to L^2(\partial V)$ is also compact we can choose a subsequence $\{v_{n_{k_m}}\}$ of the sequence $\{v_{n_k}\}$ which converges in $L^2(\partial V)$ to a function $v_0 \in L^2(\partial V)$. One has: $u_0 = v_0$ almost everywhere in $\partial U \cap \partial V$ and the function $w_0(x)$ which is defined as $w_0(x) := u_0(x)$ on $\partial U \cap \partial(U \cup V)$ and $w_0(x) := v_0(x)$ on $\partial V \cap \partial(U \cup V)$ belongs to $L^2(\partial(U \cup V))$.

Hence

$$\|w_{n_{k_m}} - w_0\|_{L^2(\partial(U\cup V))} \le \|u_{n_{k_m}} - u_0\|_{L^2(\partial U)} + \|v_{n_{k_m}} - v_0\|_{L^2(\partial V)}.$$

Therefore $\|w_{n_{k_m}} - w_0\|_{L^2(\partial(U\cup V))} \to 0$ for $m \to \infty$. □

From Theorem 9.7 and Lemma 9.2 the main result of this section follows immediately:

Theorem 9.8 *If a domain Ω belongs to class Q then the embedding operator $H^1(\Omega) \to L^2(\partial\Omega)$ is compact.*

Proof. Let U be an elementary domain of class Q. By Theorem 9.7 the embedding operator $i_U : H^1(U) \to L^2(\partial U)$ is compact.

Because any domain V of class Q is a finite union of domains of class QI the result follows from Lemma 9.2. □

9.5.2 *Examples*

Example 5.7 shows that a domain of the class Q can have unfinite number of connected boundary components.

Example 9.1 Take two domains:

1. Let domain U is a union of rectangles $P_k := \{(x_1, x_2) : |x_1 - 2^{-k}| < 2^{-k-2}; 0 \le x_2 < 2^{-k-2}\}$, $k = 1, 2, ...$ and the square $S := (0,1) \times (-1, 0)$;
2. $V := \{(x_1, x_2) : 0 < x_1 < 1; 10^{-1}x_1 \le x_2 < 1\}$.

In the book [65] it was proved that U is a domain of the class L. It is obvious that V is also a domain of the class L. Therefore $\Omega = U \cup V$ is a domain of class Q. By Theorem 9.8 the embedding operator $H^1(\Omega) \Rightarrow L^2(\partial\Omega)$ is compact.

The boundary $\partial\Omega$ of the plane domain Ω contains countably many connected components that are boundaries of domains

$$S_k := \{(x_1, x_2) : |x_1 - 2^{-k}| < 2^{-k-2}; 10^{-1}x_1 \le x_2 < 2^{-k-2}\}.$$

The boundary of the rectangle $S_0 := \{(x_1, x_2) : 0 < x_1 < 1; -1 \le x_2 < 1\}$ is also a large connected component of $\partial\Omega$.

Any neighboorhood of the point $\{0, 0\}$ contains countably many connected components of $\partial\Omega$ and therefore can not be represented as a graph of any continuous function.

Higher-dimensional examples can be constructed using the rotation of the plane domain Ω around x_1-axis.

Next, we show that the class QI contains simply-connected domains with non-trivial singularities.

Let us describe first a construction of a new quasiisometrical homeomorphism using a given one. Suppose that $S_k(x) = kx$ is a similarity transformation (which is called below a similarity) of R^n with the similarity coefficient $k > 0$, $S_{k_1}(x) = k_1 x$ is another similarity and $\varphi : U \to V$ is a $Q-$quasiisometrical homeomorphism. Then a composition $\psi := S_k \circ \varphi \circ S_{k_1}$ is a $k_1 k Q$ -quasiisometrical homeomorphism.

This remark was used in [30] for construction of an example of a domain with "spiral" boundary which is quasiisometrically equivalent to a cube. At "the spiral vertex" the boundary of the "spiral" domain is not a graph of any lipschitz function. Here we will show that the "spiral" domain belongs to the class QI. Let us recall the example from [30].

Example 9.2 We can start with the triangle $T := \{(s,t) : 0 < s < 1, s < t < 2s\}$ because T is quasiisometrically equivalent to the unit square $Q_2 = (0,1) \times (0,1)$. Hence we need to construct only a quasiisometrical homeomorphism φ_0 from T into R^2.

Let (ρ, θ) be polar coordinates in the plane. Define first a mapping $\varphi : R^2_+ \to R^2$ as follows: $\varphi(s,t) = (\rho(s,t), \theta(s,t))$, $\rho(s,t) = s$, $\theta(s,t) = 2\pi \ln \frac{t}{s^2}$. Here $R^2_+ := \{(s,t) | 0 < s < \infty, 0 < t < \infty\}$. An inverse mapping can be calculated easily: $\varphi^{-1}(\rho,\theta) = (s(\rho,\theta), t(\rho,\theta))$, $s(\rho,\theta) = \rho$, $t(\rho,\theta)) = \rho^2 e^{\frac{\theta}{2\pi}}$. Therefore φ and $\varphi_0 := \varphi | T$ are diffeomorphisms.

The image of the ray $t = ks$, $s > 0, k > 0$, is the logarithmic spiral $\rho = k \exp(-\frac{\theta}{2\pi})$. Hence the image $S := \varphi(T) = \varphi_0(T)$ is an "elementary spiral" plane domain, because ∂T is a union of two logarithmic spirals $\rho = \exp(-\frac{\theta}{2\pi})$, $\rho = 2\exp(-\frac{\theta}{2\pi})$ and the segment of the circle $\rho = 1$.

The domain T is a union of countably many subdomains $T_n := \{(s,t) : e^{-(n+1)} < s < e^{-(n-1)}, s < t < 2s\}$, $n = 1, 2, \ldots$. On the first domain T_1 the diffeomorphism $\varphi_1 := \varphi | T_1$ is $Q-$quasiisometrical, because φ_1 is the restriction on T_1 of a diffeomorphism φ defined in R^2_+ and $\overline{T_1} \subset R^2_+$. We do not calculate the number Q.

In [30] it was proven that any diffeomorphism $\varphi_n := \varphi | T_n$ that is the composition $\varphi_n = S_{e^{-(n-1)}} \circ \varphi_1 \circ S_{e^{n-1}}$ of similarities $S_{e^{-(n-1)}}$, $S_{e^{n-1}}$ and the $Q-$quasiisometrical diffeomorphism φ_1 is $Q-$quasiisometrical. Therefore the diffeomorphism φ_0 is also $Q-$quasiisometrical, and the "elementary spiral" domain $U = \varphi_0(T)$ is quasiisometrically equivalent to the unit square.

By construction, the boundary of the domain $U := \varphi(T)$ is smooth at any point except the point $\{0\}$. This domain is a locally connected domain. The quasiisometrical homeomorphism φ has N^{-1} property because all the homeomorphisms φ_n have this property. Except the point $\{0\}$ the boundary ∂U is a Q-lipschitz manifold. All other properties of QI-domains are subject of simple direct calculations. Therefore the domain T is a QI-domain.

9.6 Conclusions

In this section we combine the results about elliptic boundary-value problems with these about embedding operators.

The first result is a formulation of Theorem 9.3 for a large concrete class of rough domains. This result follows immediately from Theorem 9.3, Theorem 3.11 from [30] and Theorem 9.8.

Theorem 9.9 *If D is a domain of the class Q, $F \in L_0^2(D)$, and $h \geq 0$ is a piecewise-continuous bounded function on ∂D, $h \not\equiv 0$, then problem (9.9) has a solution in $H^1(D)$, this solution is unique, and the problem*

$$[u, \phi] + \int_{\partial D} hu\bar{\phi}ds - \lambda(u, \phi) = (F, \phi), \qquad \lambda = \text{const} \in \mathbb{R}$$

is of Fredholm type.

The next result is a formulation of Theorem 9.4 for a large class of rough exterior domains D'.

Fix a bounded domain $\widetilde{D} \subset D'$ whose boundary consists of two parts ∂D and a smooth compact manifold S. Assume that $\widetilde{D}$ belongs to the class Q. By the definition of the class Q, this assumption holds for any choice of $\widetilde{D}$ because for the smooth component S the conditions defining the class Q hold.

Theorem 9.4, Theorem 3.11 from [30], and Theorem 9.8 imply the following result:

Theorem 9.10 *For any $F \in L_0^2$, each of the boundary-value problems:*

$$A_i u = F, \qquad i = D, N \text{ or } R, \qquad A_i u = -\Delta u, \tag{9.56}$$

has a solution $u = \lim_{\epsilon \downarrow 0}(A - i\epsilon)^{-1}F := (A - i0)^{-1}F$, $u \in H^2_{loc}(D')$, $u \in L_{2,a}$, $a \in (1, 2)$, and this solution is unique.

Chapter 10

Low Frequency Asymptotics

10.1 Introduction

In this chapter the exterior domain $D' = D_e$ is denoted by Ω. The material presented in this chapter is taken mainly from [90], [120], and the presentation follows closely [120].

Let us consider the behavior of the solution to the problems

$$(\nabla^2 + k^2)u = 0 \text{ in } \Omega \subset \mathbb{R}^n, \ n \geq 2, \quad k = \text{const} > 0 \tag{10.1}$$

$$u = f \text{ on } \Gamma, \quad f \in H^1(\Gamma) \tag{10.2}$$

where u for $k > 0$ always satisfies the radiation condition, D is a bounded domain with Liapunov's boundary (this means that $\Gamma \in C^{1,\lambda}$, $\lambda > 0$). We are also interested in the boundary conditions

$$u_N = f \text{ on } \Gamma \tag{10.3}$$

and

$$u_N + \eta(s)u = f \text{ on } \Gamma \tag{10.4}$$

where N is the unit normal on Γ pointing into Ω. The case when $Lu = \partial_i[a_{ij}(x)\partial_j u]$ stands in place of ∇^2 can be treated as well, provided that $a_{ij} \in C^1$, $a_{ij}(x) = \delta_{ij}$ for $|x|$ sufficiently large, and the matrix $a_{ij}(x)$ satisfies the ellipticity condition $c_1 t_i \bar{t}_i \leq a_{ij} t_i \bar{t}_j \leq c_2 t_i \bar{t}_i$, c_1 and c_2 are positive constants, the bar stands for complex conjugate and over repeated indices one sums up. The function $\eta(s) \in C(\Gamma)$ is assumed to satisfy conditions

$$\operatorname{Im} \eta(s) \geq 0; \text{ if } \operatorname{Im} \eta = 0 \text{ then } \eta \leq 0 \tag{10.5}$$

Assumption (10.5) implies uniqueness of the solution to problem (10.1), (10.4). If equation (10.1) would be nonhomogeneous, say

$$(\nabla^2 + k^2)u = F \tag{10.6}$$

then one can consider

$$v = u + \int_{\mathbb{R}^n} g(x, y, k)F(y)dy \tag{10.7}$$

where

$$(\nabla^2 + k^2)g = -\delta(x - y) \text{ in } \mathbb{R}^n, \ g \text{ satisfies the radiation condition} \tag{10.8}$$

and for v one obtains the above problems with the homogeneous equation (10.1).

We want to describe some methods to study the behavior as $k \to 0$ of the solutions to equation (10.1) satisfying one of the conditions (10.2), (10.3) or (10.4). Note that the limit $u(x, k)$ as $k \to 0$ does not always exist. This will be clear from our results. To make it transparent without going into detail, let us consider problem (10.1), (10.3) in $\mathbb{R}^2$.

Suppose that the solution of this problem has a limit in $H^2_{\text{loc}}(\Omega)$,

$$u(x, k) \underset{H^2_{\text{loc}}(\Omega)}{\longrightarrow} u_0(x) \text{ as } k \to 0. \tag{10.9}$$

Then

$$\nabla^2 u_0 = 0 \text{ in } \Omega, \quad u_{0N} = f \text{ on } \Gamma. \tag{10.10}$$

By Green's formula one has

$$u(x, k) = \int_\Gamma (ug_N - gf)ds = \int_\Gamma (u_0 g_{0N} - g_0 f)ds + \alpha(k) \int_\Gamma f ds + o(1) \text{ as } k \to 0, \tag{10.11}$$

where $\alpha(k) = [\ln(\frac{2}{k}) - \gamma]/2\pi + i/4$ and $\gamma = 0.5572\cdots$ is the Euler's constant,

$$\begin{aligned} g(x, y, k) &= \alpha(k) + g_0(x, y) + O\big(k^2 \ln k\big) \text{ as } k \to 0, \\ g_0(x, t) &= (2\pi)^{-1} \ln r_{xy}^{-1}, \quad r_{xy} = |x - y|. \end{aligned} \tag{10.12}$$

The estimate $O(k^2 \ln k)$ holds uniformly in x, y in the region $0 < c_1 \leq |x - y| \leq c_2$, where c_j, $j = 1, 2$, are constants. Therefore (10.9) cannot hold unless

$$\int_\Gamma f\, ds = 0. \tag{10.13}$$

It is also clear from (10.11) that if (10.13) holds then (10.9) holds and u_0 solves (10.10) and satisfies the condition

$$u_0(x) = O(|x|^{-1}) \text{ as } |x| \to \infty. \tag{10.14}$$

Indeed, if (10.13) holds then

$$u_0(x) = \int_\Gamma \left[u_0(s)g_{0N}(x,s) - g_0(x,s)f(s)\right]ds = O\big(|x|^{-1}\big) \text{ as } |x| \to \infty \tag{10.15}$$

since $g_{0N}(x,s) = O(|x|^{-1})$ and

$$\int_\Gamma g_0(x,s)f(s) = \frac{1}{2\pi}\ln|x|^{-1}\int_\Gamma f(s) + O(|x|^{-1}) = O(|x|^{-1}). \tag{10.16}$$

We give several approaches to the problem of low frequency asymptotics of the solutions to the exterior problem. The first approach is based on integral equations of the first kind. It provides a detailed information and allows one to obtain asymptotic expansion of the solution as $k \to 0$. Its drawback is that it works efficiently for the equation with constant coefficients for which the behavior of the Green function as $k \to 0$ is known in detail. The second approach is rather general. It gives a convenient necessary and sufficient condition for the limit (10.9) to exist, but it does not give (at least without extra work) the rate of convergence. The third approach is based on *a priori* estimates and uses the fact that for sufficiently small k equation (10.1) in a bounded domain satisfies the maximum principle. The first two approaches belong to the author [90], [133], [120], the third one is due to [160]. Section 10.7 is based on the works [137], [138], and the presentation follows [138].

Let us describe these approaches. The results obtained by the integral equation method give necessary and sufficient conditions for the existence of the limit (10.9) and the asymptotics of the solution as $k \to 0$.

A discussion of low-frequency scattering is given in [19].

10.2 Integral Equation Method for the Dirichlet Problem

Let us look for the solution to (10.11), (10.12) of the form

$$u = \int_\Gamma g(x,s,k)\sigma(s)ds := Q(k)\sigma. \tag{10.17}$$

Consider the case $n \geq 3$ first. The method is valid in $\mathbb{R}^n$ for $n \geq 2$, but the results for $n = 2$ are sometimes different because the Green's function $g(x,k) = \frac{i}{4}H_0^{(1)}(k|x|)$ does not have a finite limit as $k \to 0$.

Let us take $n = 3$. The case $n > 3$ is treated similarly. If $n = 3$ then $g = (4\pi|x|)^{-1}\exp(ik|x|)$ and (10.17) solves (10.1). To satisfy (10.2) choose σ as the solution to the equation

$$Q(k)\sigma = f. \tag{10.18}$$

It is known [133, p. 199], that $Q : H^0(\Gamma) \to H^1(\Gamma)$ is an isomorphism if $k > 0$ is sufficiently small (so that k^2 is not a Dirichlet eigenvalue of $-\nabla^2$ in D). Therefore

$$\sigma = Q^{-1}(k)f. \tag{10.19}$$

The operator $Q(k)$ depends on k analytically and $Q^{-1}(k)$ is analytic in k in a sufficiently small neighborhood of $k = 0$. Indeed, $Q(k) := Q_0 + B(k)$, $Q_0 := Q(0)$, $\|B(k)\|_{H^0(\Gamma)\to H^1(\Gamma)} \leq c|k|$, Q_0 is an isomorphism of $H^0(\Gamma)$ onto $H^1(\Gamma)$. Therefore

$$Q^{-1}(k) = \left[Q_0 + B(k)\right]^{-1} = \left[I + Q_0^{-1}B(k)\right]^{-1}Q_0^{-1}. \tag{10.20}$$

The operator $Q_0^{-1}B(k)$ is analytic in k as an operator in $H^0(\Gamma)$ and $\|Q_0^{-1}B(k)\|_{H^0(\Gamma)\to H^1(\Gamma)} \to 0$ as $|k| \to 0$. Therefore $Q^{-1}(k)$ is an isomorphism of $H^1(\Gamma)$ onto $H^0(\Gamma)$ which is analytic in k in a sufficiently small neighborhood of $k = 0$. This implies that σ defined by (10.19) is analytic in k, in particular:

$$\sigma(s,k) \underset{H}{\longrightarrow} \sigma_0(s) \text{ as } k \to 0, \quad \sigma_0 = Q_0^{-1}f \in H^0(\Gamma). \tag{10.21}$$

Thus we have

Theorem 10.1 *For any $f \in H^1(\Gamma)$ the limit (10.9) exists and*

$$u_0(x) = \int_\Gamma g_0(x,s)\sigma_0(s)ds, \quad \sigma_0(s) = Q_0^{-1}f \tag{10.22}$$

solves the limiting problem

$$\nabla^2 u_0 = 0 \textit{ in } \Omega, \quad u_0 = f \textit{ on } \Gamma, \quad u(\infty) = 0. \tag{10.23}$$

One has

$$u(x,k) = u_0(x) + u_1(x,k), \quad \left|u_1(x,k)\right| = O(k) \textit{ as } k \to 0 \tag{10.24}$$

and the term $u_1(x,k)$ *can be calculated:*

$$u_1(x,k) = ik \int_\Gamma (g_1\sigma_0 + g_0\sigma_1)\, ds + O(k^2) \text{ as } k \to 0, \tag{10.25}$$

where g_1 *and* σ_1 *are defined by the formula*

$$\begin{aligned} \sigma(s,k) &= \sigma_0(x) + ik\sigma_1(s) + o(k), \\ g(x,k) &= g_0(x) + ikg_1(x) + o(k), \quad k \to \infty, \end{aligned} \tag{10.26}$$

so that

$$g_1(x) = \frac{1}{4\pi}, \quad \sigma_1(s) = -\frac{1}{4\pi} Q_0^{-1} \int_\Gamma Q_0^{-1} f\, ds. \tag{10.27}$$

Proof. We have already proved all but the second formula (10.27). To prove this formula, one calculates $B(k)$ explicitly:

$$B(k)f = [Q(k) - Q(0)]f = \frac{ik}{4\pi} \int_\Gamma f\, ds + O(k^2) \text{ as } k \to 0. \tag{10.28}$$

From (10.20) and (10.19) one gets

$$\sigma = Q_0^{-1} f - Q_0^{-1} B(k) Q_0^{-1} f + \cdots. \tag{10.29}$$

From (10.28) and (10.29) the second formula (10.26) follows. Theorem 10.1 is proved. □

Consider now the case $n = 2$. There are some new features in this case: Green's function $g = \frac{i}{4} H_0^{(1)}(kr_{xy})$ does not have a finite limit as $k \to 0$, the operator Q_0 may have for some domains a nontrivial null-space $N(Q_0)$ so that Q_0^{-1} does not exist for these domains. By $N(A) = \{u : Au = 0\}$ we denote the null space of an operator A.

Lemma 10.1 *There exist* $\Omega \subset \mathbb{R}^2$ *such that* $N(Q_0) \neq \{0\}$. *If* $N(Q_0) \neq \{0\}$ *then* $\dim N(Q_0) = 1$ *and one can choose* $\phi \in N(Q_0)$ *so that* $\phi \geq 0$.

Proof. Let us prove that a disc D of a suitable radius a will have a nontrivial null-space $N(Q_0) \neq \{0\}$. The integral equation $Q_0\phi = 0$ for the disc of radius a can be written as

$$-\frac{1}{2\pi} \int_0^{2\pi} \ln \left[2a^2 - 2a^2 \cos(\alpha - \beta)\right]^{1/2} \phi(\beta) d\beta = 0$$

or

$$-\frac{1}{4\pi} \ln(2a^2) \int_0^{2\pi} \phi(\beta) d\beta - \frac{1}{4\pi} \int_0^{2\pi} \ln\left(\sin^2 \frac{\alpha - \beta}{2}\right) \phi(\beta) d\beta = 0$$

or

$$\begin{aligned} B\phi := \nu(a)\int_0^{2\pi}\phi(\beta)d\beta - \frac{1}{2\pi}\int_0^{2\pi}\left|\sin\frac{\alpha-\beta}{2}\right|\phi(\beta)d\beta = 0,\\ \nu(a) := -\frac{1}{2\pi}\ln(2a). \end{aligned} \tag{10.30}$$

Note that $\nu(a) \to +\infty$ as $a \to 0$, $\nu(a) \to -\infty$ as $a \to \infty$. Since B is a compact in $L^2(0, 2\pi)$ selfadjoint operator its spectrum is discrete. From the variational definition of the eigenvalues $\lambda_j(B)$ of B it follows that there exists an a such that $\lambda_j(B) = 0$ for some j. More explicitly, take $\phi(\beta) = 1$, then (10.30) reduces to

$$-\ln(2a) - \frac{1}{2\pi}\int_0^{2\pi}\left|\sin\frac{\alpha-\beta}{2}\right| d\beta = 0. \tag{10.31}$$

One has $\int_0^{2\pi}|\sin\frac{\alpha-\beta}{2}|d\beta = 2\int_{\frac{-\alpha}{2}}^{\pi-\frac{\alpha}{2}}|\sin\gamma|d\gamma = 4$, so that (10.31) becomes $-\ln(2a) = 2/\pi$. This equation holds for $a = a_0 = \frac{1}{2}\exp(-2/\pi)$. Thus, if Ω is the disc of radius a_0 then $N(Q_0) \neq \{0\}$, $1 \in N(Q_0)$.

Let us prove the second statement of Lemma 10.1. The proof is valid also in the case of Γ which has m connected components. We assume $m = 1$ and leave the case $m > 1$ to the reader. We claim that if $\sigma \in N(Q_0)$, $\sigma \not\equiv 0$, then $\int_\Gamma \sigma dt \neq 0$. Indeed, otherwise $Q_0\sigma = 0$ and $\int_\Gamma \sigma dt = 0$ imply $\sigma \equiv 0$. To prove this, let $w(x) := \int_\Gamma g_0(x,s)\sigma(s)\,ds$, $g_0(x,s) = \frac{1}{2\pi}\ln|x-s|^{-1}$. One has

$$w(x) = \frac{1}{2\pi}\left(\ln|x|^{-1}\right)\int_\Gamma \sigma dt + w_1(x) \tag{10.32}$$

where

$$\nabla^2 w_1 = 0 \text{ in } \Omega,\ w_1 = O(|x|^{-1}),\ w_1 = 0 \text{ on } \Gamma. \tag{10.33}$$

Thus $w_1 = 0$ in Ω. If $\int_\Gamma \sigma dt = 0$ then $w_1 = w$. Thus $w = 0$ in Ω. Also $\nabla^2 w = 0$ in D, $w = 0$ on Γ, implies $w = 0$ in D. By the jump relation $\sigma = w_N^+ - w_N^- = 0$ where w_N^+ (w_N^-) denotes the limiting value of the normal derivative of w on Γ from D (Ω). The claim is proved.

Suppose now that $\sigma_j \in N(Q_0)$, $j = 1, 2$, $\sigma_j \not\equiv 0$. By the claim, $\int_\Gamma \sigma_j dt \neq 0$, $j = 1, 2$. One can find a constant c such that $\int_\Gamma(\sigma_1 - c\sigma_2)dt = 0$. Since $\sigma_1 - c\sigma_2 \in N(Q_0)$ one concludes that $\sigma_1 = c\sigma_2$. Thus, $\dim N(Q_0) = 1$. Let us finally prove that $\sigma \in N(Q_0)$ can be chosen so that $\sigma \geq 0$. Choose $\sigma \not\equiv 0$, $\sigma \in N(Q_0)$, such that $\int_\Gamma \sigma dt > 0$. This is possible (take $-\sigma$ if $\int_\Gamma \sigma dt < 0$). We claim that if $\sigma \in N(Q_0)$ and $\int_\Gamma \sigma dt > 0$ then $\sigma \geq 0$. Indeed, formulas

(10.32) and (10.33) show that $w \to -\infty$ as $|x| \to \infty$. By the maximum principle $w(x) < 0$ in Ω. Therefore $w_N \leq 0$. Since $w(x) = 0$ in D one has $w_N^+ = 0$. Thus $\sigma = w_N^+ - w_N^- \geq 0$. In fact, in the case $m = 1$, the strong maximum principle implies $\sigma(s) > 0$. Lemma 10.1 is proved. □

Let $n = 2$. Then the following result holds.

Theorem 10.2 *For the solution to problem* (10.1), (10.2) *the limit* (10.9) *exists and* $u_0(x)$ *solves the limiting problem*

$$\nabla^2 u_0 = 0 \text{ in } \Omega, \ u_0 = f \text{ on } \Gamma, \ |u(\infty)| = O(1). \tag{10.34}$$

Moreover

$$u_0(x) = \beta + \int_\Gamma g_0(x,t)\sigma_0(t)dt, \ \beta = \text{const}, \ g_0 = \frac{1}{2\pi} \ln \frac{1}{r_{xt}} \tag{10.35}$$

where

$$\sigma_0 = Q_0^{-1}(d)f - \beta Q_0^{-1}(d)1, \quad d = \text{const} > \operatorname{diam} D, \tag{10.36}$$

and

$$Q_0(d)f := \frac{1}{2\pi} \int_\Gamma \ln \frac{d}{r_{st}} f(t)dt. \tag{10.37}$$

One has

$$u(x,k) = u_0(x) + O(|\ln k|^{-1}) \ \text{as } k \to 0. \tag{10.38}$$

Proof. In contrast to the case $n = 3$, one can have domains in $\mathbb{R}^2$ such that Q_0 is not invertible, where $Q_0 f = \int_\Gamma g_0(s,t)f(t)dt$. To avoid this complication, choose $d = \text{const} > \operatorname{diam} D$ and define $Q_0(d)$ by formula (10.37). Look for the solution to (10.1), (10.2) of the form

$$u = \int_\Gamma g(x,t)\sigma(t)dt = \alpha_1(k)(\sigma,1) + \frac{1}{2\pi} \int_\Gamma \ln \frac{d}{r_{xt}} \sigma(t)dt + \epsilon(k)\sigma \tag{10.39}$$

where

$$\alpha_1(k) := \alpha(k) - \frac{1}{2\pi} \ln d, \quad \|\epsilon(k)\|_{H^0(\Gamma)\to H^1(\Gamma)} \leq c|k| \text{ as } k \to 0 \tag{10.40}$$

and $\alpha(k)$ is defined below (10.11).

Note that $Q_0(d)$ is an isomorphism of $H^0(\Gamma)$ onto $H^1(\Gamma)$ if $d > \operatorname{diam} D$. Indeed, it is injective: if $Q_0(d)\sigma = 0$ then $Q_0\sigma = \frac{-\ln}{2\pi}(\sigma,1)$. Since Q_0 is symmetric, this equation is solvable only if the orthogonality condition holds: $-\frac{\ln d}{2\pi}(\sigma,1)(\sigma_0,1) = 0$, where $Q_0\sigma_0 = 0$, $(\sigma_0,1) \neq 0$. Therefore

$(\sigma, 1)\ln d = 0$. If $\ln d \neq 0$ then $(\sigma, 1) = 0$, $Q_0\sigma = 0$, and this implies $\sigma = 0$ by Lemma 10.1. If $\ln d = 0$ then $Q_0\sigma = 0$ and $\ln \frac{1}{r_{st}} > 0$, $s, t \in \Gamma$. This implies $\sigma = 0$. Indeed, by Lemma 10.1 it is sufficient to prove that $(\sigma, 1) = 0$. Suppose $(\sigma, 1) > 0$. Without loss of generality assume $\sigma = 1 + \sigma_1$, $(\sigma, 1) = 0$. One has, if $\ln \frac{1}{r_{st}} > 0$, $Q_0\sigma_1 = -Q_0 1 < 0$. Again, using the orthogonality condition necessary for the solvability of the last equation, one obtains $(Q_0 1, \sigma) = 0$. Since $\sigma \geq 0$ and $Q_0 1 > 0$ it follows that $\sigma = 0$. The inequality $\sigma \geq 0$ is established as in Lemma 10.1. Let us now prove that $Q_0(d)$ is surjective as an operator from $H^0(\Gamma)$ into $H^1(\Gamma)$. Take an arbitrary $f \in H^1(\Gamma)$. Consider the equation $Q_0(d)\sigma = f$. The operator Q_0 with the kernel $-\frac{1}{2\pi}\ln|s - t|$ is an elliptic selfadjoint pseudodifferential operator in $H^0(\Gamma)$ of order -1 with index 0, so that $Q_0(d)$, which differs from Q_0 by a rank-one operator, has index zero as well. Since $N[Q_0(d)] = \{0\}$ one concludes that $Q_0(d) : H^0(\Gamma) \to H^1(\Gamma)$ is surjective. It is now easy to finish the proof of Theorem 10.2. Consider the equation (cf. (10.39)):

$$\alpha_1(k)(\sigma, 1) + Q_0(d)\sigma + \epsilon(k)\sigma = f. \tag{10.41}$$

Write it as

$$\sigma + \alpha_1(k)(\sigma, 1)Q_0^{-1}(d)1 + Q_0^{-1}(d)\epsilon(k)\sigma = Q_0^{-1}(d)f. \tag{10.42}$$

It follows from (10.42) that

$$(\sigma, 1) + \alpha_1(k)(\sigma, 1)(Q_0^{-1}(d)1, 1) + (Q_0^{-1}(d)\epsilon(k)\sigma, 1) = (Q_0^{-1}(d)f, 1). \tag{10.43}$$

Thus

$$(\sigma, 1) = \frac{(Q_0^{-1}(d)f, 1) - (Q_0^{-1}(d)\epsilon(k)\sigma, 1)}{1 + \alpha_1(k)(Q_0^{-1}(d)1, 1)}. \tag{10.44}$$

We have proved earlier that $c_0 := (Q_0^{-1}(d)1, 1) > 0$. This and (10.40) imply that

$$\alpha_1(k)(\sigma, 1) = \beta + O(|\alpha_1^{-1}(k)|) \text{ as } k \to 0, \quad \beta := \frac{(Q_0^{-1}(d)f, 1)}{(Q_0^{-1}(d)1, 1)}. \tag{10.45}$$

The equation

$$T\sigma = h, \quad T\sigma := \sigma + \alpha_1(\sigma, 1)p, \quad p := Q_0^{-1}(d)1 \tag{10.46}$$

can be solved explicitly:

$$\sigma = h - \frac{\alpha_1 p(h, 1)}{1 + \alpha_1(p, 1)}, \quad (p, 1) = c_0 > 0. \tag{10.47}$$

Thus

$$T^{-1} = I - \frac{\alpha_1 p(\cdot,1)}{1+\alpha_1 c_0} = I - \frac{p}{c_0}(\cdot,1) + O(|\alpha_1^{-1}|), \quad k \to 0. \tag{10.48}$$

It follows from (10.42) that

$$\left[I + T^{-1}Q_0^{-1}(d)\epsilon(k)\right]\sigma = T^{-1}Q_0^{-1}(d)f. \tag{10.49}$$

From (10.40), (10.48) and (10.49) it follows that

$$\sigma(k) = \sigma_0 + O(|\ln k|^{-1}) \text{ as } k \to 0 \tag{10.50}$$

where

$$\sigma_0 = Q_0^{-1}(d)f - \beta Q_0^{-1}(d)1, \tag{10.51}$$

and

$$(\sigma, 1) = \frac{\beta}{\alpha_1} + O(|\alpha_1|^{-2}) \text{ as } k \to 0. \tag{10.52}$$

Therefore by (10.39), (10.50)–(10.52),

$$u(x,k) = \beta + \frac{1}{2\pi}\int_\Gamma \ln\frac{d}{r_{xt}}\sigma_0(t)dt + O(|\alpha_1|^{-1}) \text{ as } k \to 0, \tag{10.53}$$

where σ_0 is given by (10.51). It follows from (10.52) that

$$(\sigma_0, 1) = 0. \tag{10.54}$$

From (10.53), (10.54) and (10.51) formulas (10.35) and (10.36) follow. Theorem 10.2 is proved. □

Remark 10.1 *Equation (10.49) can be solved by iterations since the norm of the operator $T^{-1}Q_0^{-1}(d)\epsilon(k)$ in $H^1(\Gamma)$ goes to zero as $k \to 0$. Therefore this equation can be used for obtaining full asymptotic expansion of $\sigma(s,k)$ and, using formula (10.39), one can obtain asymptotics of $u(x,k)$ as $k \to 0$.*

10.3 Integral Equation Method for the Neumann Problem

The basic result of this section is:

Theorem 10.3 *The limit* (10.9) *holds for the solution to problem* (10.1), (10.3) *in* $\mathbb{R}^3$ *for any* $f \in L^2(\Gamma)$. *It holds in* $\mathbb{R}^2$ *iff*

$$\int_\Gamma f dt = 0. \tag{10.55}$$

If (10.55) *holds then the limit* $u_0(x)$ *solves the problem*

$$\begin{aligned} &\nabla^2 u_0 = 0 \text{ in } \Omega, \quad u_{0N} = f \text{ on } \Gamma, \quad \Omega \subset \mathbb{R}^2, \\ &u_0(\infty) = 0 \end{aligned} \tag{10.56}$$

Proof. First assume $n = 3$. Then Green's formula yields for the solution to (10.1), (10.3):

$$\begin{aligned} u(x) &= \int_\Gamma (u g_N - g f) dt, \\ g(x,t) &= \frac{\exp(ik|x-t|)}{4\pi|x-t|}. \end{aligned} \tag{10.57}$$

Taking $x \to s \in \Gamma$, $x \in \Omega$, and using the well known formulas for the limiting values of the potential of double layer on Γ, one gets

$$\begin{aligned} \sigma &= \frac{A'\sigma + \sigma}{2} - \int_\Gamma g(s,t) f(t) dt, \quad \sigma := u\Big|_\Gamma, \\ A'\sigma &:= A'(k)\sigma := 2\int_\Gamma \frac{\partial g(s,t)}{\partial N_t}\sigma(t)dt, \ s \in \Gamma \end{aligned} \tag{10.58}$$

or

$$\sigma = A'\sigma - 2\int_\Gamma g(s,t) f(t) dt.$$

The operator $I - A'(0)$ has a bounded inverse in $L^2(\Gamma)$ and $\|A'(k) - A'(0)\|_{L^2(\Gamma)} \to 0$ as $k \to 0$. Therefore $\|(I - A'(k))^{-1} - (I - A'(0))^{-1}\|_{L^2(\Gamma)} \to 0$ as $k \to 0$. This and (10.58) imply that

$$\|\sigma(s,k) - \sigma_0(s)\| := \delta(k) \to 0 \text{ as } k \to 0. \tag{10.59}$$

In fact $\delta(k) \le ck$, $c = \text{const} > 0$. The function $\sigma_0(s)$ in (10.59) can be calculated from (10.58):

$$\sigma_0 = 2(I - A'(0))^{-1}\int_\Gamma g(s,t) f(t) dt. \tag{10.60}$$

From (10.59) and (10.57) one obtains(10.59). It is easy to check that $u_0(x) = \lim_{k\to 0} u(x,k)$ solves (10.56). Theorem 10.3 is proved in the case $n = 3$.

If $n = 2$ the proof is basically the same. The role of condition (10.55) is explained below formula (10.11), and, in fact, $u_0(x) = O(|x|^{-1})$ if (10.55) holds, a refinement of (10.56). Theorem 10.3 is proved. □

10.4 Integral Equation Method for the Robin Problem

Consider now problem (10.1), (10.4), in $\Omega \subset \mathbb{R}^2$. This problem is uniquely solvable under the assumptions (10.5) [133, p. 37]. Look for its solution of the form

$$u = \int_\Gamma g(x,t)\sigma(t)dt, \quad g = \frac{i}{4} H_0^{(1)}(k|x-t|). \tag{10.61}$$

The boundary condition (10.4) yields

$$\frac{A(k)\sigma - \sigma}{2} + \eta(s)Q(k)\sigma = f, \tag{10.62}$$

where

$$A(k)\sigma := 2\int_\Gamma \partial g(s,t)N_s\sigma(t)dt, \quad Q(k)\sigma = \int_\Gamma g(s,t)\sigma(t)dt. \tag{10.63}$$

Define $A_0 = A(0)$, $Q_0 = Q(0)$,

$$B = \frac{A_0 - I}{2} + \eta(s)Q_0. \tag{10.64}$$

Equation (10.62) can be written as

$$B\sigma + \alpha(k)\eta(s)(\sigma, 1) + \epsilon(k)\sigma = f \tag{10.65}$$

where $\alpha(k)$ is defined below (10.11) and $\|\epsilon(k)\|_{L^2(\Gamma)\to L^2(\Gamma)} \to 0$ as $k \to 0$.

Theorem 10.4 *If $N(B) = \{0\}$ and $(B^{-1}\eta, 1) \neq 0$ then for the solution to (10.1), (10.4) equation (10.9) holds. If $N(B) \neq \{0\}$ and (10.5) holds then $\dim N(B) = 1$. Let (10.5) hold, $h \in N(B)$, $h \not\equiv 0$, then $(\eta Q_0 h, 1) \neq 0$, $B'Q_0 h = 0$, and for the solution to (10.1), (10.4) equation (10.9) holds. Conversely, if for the solution (10.1), (10.4) equation (10.9) holds, then either $N(B) = \{0\}$ and $(B^{-1}\eta, 1) \neq 0$, or $\dim N(B) = 1$ and the equations $(\eta Q_0 h, 1) \neq 0$, $B'Q_0 h = 0$ hold. If $\dim N(B) > 1$, then there exists an $f \in L^2(\Gamma)$ for which (10.9) fails.*

The proof of this theorem is given in a series of lemmas.

Lemma 10.2 *If $N(B) = \{0\}$ then B is invertible and the condition*

$$(B^{-1}\eta, 1) \neq 0 \tag{10.66}$$

is necessary and sufficient for (10.9) *to hold for the solution to* (10.1), (10.4).

Proof. The operator B is of Fredholm type. Therefore $N(B) = \{0\}$ implies that B^{-1} is bounded and defined on all of the space $L^2(\Gamma)$. Equation (10.65) can be written as

$$\sigma + \alpha(k)c_\sigma B^{-1}\eta + B^{-1}\epsilon(k)\sigma = B^{-1}f, \quad c_\sigma := (\sigma, 1). \tag{10.67}$$

Integrate (10.67) over Γ to get

$$c_\sigma[1 + \alpha(k)(B^{-1}\eta, 1)] + (B^{-1}\epsilon(k)\sigma, 1) = (B^{-1}f, 1). \tag{10.68}$$

If (10.66) holds then one obtains from (10.68), as in the proof of Theorem 10.2, that there exists

$$\lim_{k\to 0} \alpha(k)c_\sigma = \frac{(B^{-1}f, 1)}{(B^{-1}\eta, 1)} := \beta. \tag{10.69}$$

If (10.69) holds then (10.67) implies that

$$\sigma(s, k) = \sigma_0(s) + O(|\ln k|^{-1}) \text{ as } k \to 0 \tag{10.70}$$

where

$$\sigma_0(s) = B^{-1}f - \frac{(B^{-1}f, 1)}{(B^{-1}\eta, 1)}B^{-1}\eta, \quad (\sigma_0, 1) = 0. \tag{10.71}$$

Therefore $u(x, k) = \int_\Gamma g(x, t)\sigma(t, k)dt$ satisfies (10.9) with

$$u_0(x) = \beta + \int_\Gamma g_0(x, t)\sigma_0(t)dt, \quad g_0(x, t) = \frac{1}{2\pi} \ln \frac{1}{r_{xt}} \tag{10.72}$$

and

$$u(x, k) = u_0(x) + O(|\ln k|^{-1}) \text{ as } k \to 0 \tag{10.73}$$

where $u_0(x)$ solves the limiting problem and is bounded at infinity.

Conversely, if $N(B) = \{0\}$ and $(B^{-1}\eta, 1) = 0$, then (10.9) does not hold for some f. Indeed, one can find $f \in L^2(\Gamma)$ such that $(B^{-1}f, 1) \neq 0$. For this f equation (10.9) does not hold. To prove this, note that equations

(10.67) and (10.68) and the condition $(B^{-1}f, 1) \neq 0$ imply that $\lim_{k\to 0} c_\sigma \neq 0$, and

$$\sigma = B^{-1}f - B^{-1}\epsilon(k)\sigma - \alpha(k)B^{-1}\eta[(B^{-1}f, 1) - (B^{-1}\epsilon(k)\sigma, 1)]. \quad (10.74)$$

It follows from (10.74) and (10.40) that

$$\sigma = -\alpha(k)B^{-1}\eta(B^{-1}f, 1) + B^{-1}f + o(|\alpha(k)|^{-1}) \text{ as } k \to 0. \quad (10.75)$$

Therefore $(\sigma, 1) = (B^{-1}f, 1) + o(|\alpha(k)|^{-1})$ and

$$u(x,k) = \int_\Gamma g(x,t)\sigma dt = \alpha(k)(B^{-1}f, 1) + \int_\Gamma g_0(x,t)\sigma dt + o(1) \text{ as } k \to 0. \quad (10.76)$$

It is clear from (10.76) that $|u(x,k)| \to \infty$ as $k \to 0$ at least at some points x. Lemma 10.2 is proved. □

Lemma 10.3 *If* (10.5) *holds and* $N(B) \neq \{0\}$ *then* $\dim N(B) = 1$.

Proof. Suppose $Bh = 0$, $h \not\equiv 0$. Then $(h, 1) \neq 0$. Indeed, suppose $(h, 1) = 0$. Define

$$w(x) = \frac{1}{2\pi}\int_\Gamma \ln \frac{1}{r_{xt}} h dt. \quad (10.77)$$

Then

$$\begin{aligned} \nabla^2 w &= 0 \text{ in } \Omega \cup D, \\ w_N^- + \eta w &= 0 \text{ on } \Gamma \\ w &= O(|x|^{-1}) \text{ as } |x| \to \infty \end{aligned} \quad (10.78)$$

Assumptions (10.5) imply that $w = 0$ in Ω, so $w = 0$ on Γ, $\nabla^2 w = 0$ in D and $w = 0$ in D. Therefore $h = w_N^+ - w_N^- = 0$.

It is now easy to see that $\dim N(B) = 1$. Indeed, let $h_j \in N(B)$, $(h_j, 1) \neq 0$, $j = 1, 2$. Then there exists a $c = \text{const}$ such that $(h_1 - ch_2, 1) = 0$. Since $h_1 - ch_2 \in N(B)$, it follows that $h_1 = ch_2$. Lemma 10.3 is proved. □

Lemma 10.4 *If* $h \in N(B)$, $h \not\equiv 0$, *then*

$$(\eta Q_0 h, 1) \neq 0 \quad (10.79)$$

and

$$B'Q_0 h = 0. \quad (10.80)$$

Here B' *is the transpose of* B.

Proof. Assume that

$$Bh := \frac{A_0 h - h}{2} + \eta Q_0 h = 0. \tag{10.81}$$

Since $A_0^* = A_0'$, one has $(A_0 h, 1) = (h, A_0' 1) = -(h, 1)$. This and (10.80) imply

$$0 = \frac{(A_0 h, 1) - (h, 1)}{2} + (\eta Q_0 h, 1) = -(h, 1) + (\eta Q_0 h, 1),$$

or

$$(h, 1) = (\eta Q_0 h, 1). \tag{10.82}$$

Since $(h, 1) \neq 0$ equation (10.79) follows.

Note that

$$Q_0 A_0 = A_0' Q_0. \tag{10.83}$$

Indeed, by Green's formula

$$\int_\Gamma ds' g(s, s') \partial g(s', t) N_{s'} = \int_\Gamma ds' \partial g(s, s') N_{s'} g(s', t) \tag{10.84}$$

which implies (10.83). Apply Q_0 to (10.81) and use (10.83) to get

$$B' p := \frac{A_0' p - p}{2} + Q_0 \eta p = 0, \quad p := Q_0 h, \tag{10.85}$$

so that (10.80) is proved. Lemma 10.4 is proved. □

Lemma 10.5 *If* (10.5) *holds and* $N(B) \neq \{0\}$ *then* (10.9) *holds for the solution to* (10.1), (10.4). *Conversely, if* (10.9) *holds for any* $f \in L^2(\Gamma)$ *for the solution to* (10.1), (10.4), *then* $\dim N(B) \leq 1$. *In the case* $\dim N(B) = 0$, *that is* $N(B) = \{0\}$, *condition* (10.66) *holds. In the case* $\dim N(B) = 1$ *conditions* (10.79) *and* (10.80) *hold.*

Proof. The last statement is a part of Lemma 10.5. Let us prove that (10.5) and $N(B) \neq \{0\}$ imply (10.9) for the solution to (10.1), (10.4). By Lemma 10.3, $\dim N(B) = 1$ and by Lemma 10.4, equations (10.79), (10.80) hold. To prove (10.9) it is sufficient to prove existence of the finite limits

$$\begin{aligned} \lim_{k \to 0} \sigma(s, k) &= \sigma_0(s), \\ \lim_{k \to 0} \alpha(k)(\sigma, 1) &= \beta. \end{aligned} \tag{10.86}$$

Let us prove (10.86). Define

$$B_1 = B + P, \quad P = (\cdot, h)\eta, \quad h \in N(B), \quad h \not\equiv 0. \tag{10.87}$$

The operator B_1 has a bounded inverse since it is injective and of Fredholm type. Its injectivity is easy to prove: if $B_1\sigma = 0$ then $B\sigma = -(\sigma, h)\eta$, and, by the necessary condition for solvability of the last equation, $(\sigma, h)(p, \bar{\eta}) = 0$, where $p \in N(B')$. Since, by (10.79), $(p, \bar{\eta}) = (\eta p, 1) \neq 0$, it follows that $(\sigma, h) = 0$ and therefore $B\sigma = 0$. Since $\dim N(B) = 1$, one concludes that $\sigma = ch$, $c(h, h) = 0$, so that $c = 0$ and $\sigma = 0$.

Write equation (10.65) as

$$\begin{aligned} \sigma + \alpha(k)c_\sigma B_1^{-1}(k)\eta - B_1^{-1}(k)P\sigma &= B_1^{-1}(k)f, \\ c\sigma &= (\sigma, 1), \\ B_1(k) &= B_1 + \epsilon(k). \end{aligned} \tag{10.88}$$

Let $\sigma_h := (\sigma, h)$ and $c_\sigma = (\sigma, 1)$. Then it follows from (10.88) that

$$\begin{aligned} \sigma_h + \alpha(k)c_\sigma(B_1^{-1}(k)\eta, h) - \sigma_h(B_1^{-1}(k)\eta, h) &= (B_1^{-1}(k)f, h), \\ c_\sigma + \alpha(k)c_\sigma(B_1^{-1}(k)\eta, h) - \sigma_h(B_1^{-1}(k)\eta, 1) &= (B_1^{-1}(k)f, 1). \end{aligned} \tag{10.89}$$

The matrix of this system for σ_h and c_σ is

$$T := \begin{bmatrix} 1 - (B_1^{-1}(k)\eta, h) & \alpha(k)(B_1^{-1}(k)\eta, h) \\ -(B_1^{-1}(k)\eta, 1) & 1 + \alpha(k)(B_1^{-1}(k)\eta, 1) \end{bmatrix}. \tag{10.90}$$

Note that

$$\|B_1^{-1}(k) - B_1^{-1}\| \to 0 \text{ as } k \to 0 \tag{10.91}$$

and

$$\det T = 1 + \alpha(B_1^{-1}(k)\eta, 1) - (B_1^{-1}(k)\eta, h) := 1 + \alpha b - a. \tag{10.92}$$

Let us prove that $b := (B_1^{-1}(k)\eta, 1) \neq 0$ as $k \to 0$. Indeed, by (10.90) one has $b \to (B_\eta^{-1}, 1)$. Denote $B_1^{-1}\eta := q$. We want to prove that $(q, 1) \neq 0$. One has $\eta = B_1 q = Bq + (q, h)\eta$. We will prove that $(q, h) = 1$, so $\eta = Bq + \eta$. Thus $Bq = 0$, $q = ch$, $q \not\equiv 0$ so $c \neq 0$. Thus $(q, 1) = c(h, 1) \neq 0$. Let us prove that $(q, h) = 1$. One has $\eta = Bq + (q, h)\eta$, so $(\eta, \bar{p}) = (q, h)(\eta, \bar{p})$. Note that $B'p = 0$, so $(Bq, \bar{p}) = (q, \bar{B'p}) = 0$. Therefore $(\eta, \bar{p}) = (q, h)(\eta, \bar{p})$. Since $(\eta, \bar{p}) = (\eta p, 1) \neq 0$ by (10.79), it follows that $(q, h) = 1$. Therefore, for all sufficiently small k,

$$\det T \neq 0 \tag{10.93}$$

and the system (10.89) is uniquely solvable for σ_h and c_σ.

Let us solve (10.89) for c_σ and σ_n. One has

$$c_\sigma = (\det T)^{-1}\Big\{\big[1-\big(B_1^{-1}(k)\eta, h\big)\big]\big(B_1^{-1}(k)f, 1\big) + \big(B_1^{-1}(k)\eta, 1\big)\big(B_1^{-1}(k)f, h\big)\Big\}.$$

Using (10.91) and taking $k \to 0$ one gets

$$\lim_{k\to 0} \alpha(k)c_\sigma = \beta := (B_1^{-1}f, h) \tag{10.94}$$

where the formula

$$\lim_{k\to 0} (B_1^{-1}(k)\eta, h) = 1 \tag{10.95}$$

was used. Similarly,

$$\lim_{k\to 0} \sigma_h = \beta - \frac{B_1^{-1}f, 1)}{(B_1^{-1}\eta, 1)} := \gamma. \tag{10.96}$$

So, formula (10.86) is proved.

From (10.94) and (10.61) it follows that

$$\begin{aligned} u(x,k) &= u_0(x) + O\big(|\ln k|^{-1}\big) \\ &:= \beta + \int_\Gamma g_0(x,s)\sigma_0(s)\,ds + O\big(|\ln k|^{-1}\big) \text{ as } k \to 0 \end{aligned} \tag{10.97}$$

where $\sigma_0(s)$ is defined by (10.86) and $u_0(x)$ solves the limiting problem. Existence of the limit (10.86) follows immediately from (10.88), (10.91), (10.94) and (10.96), and

$$\sigma_0(s) = B_1^{-1}f - \beta B_1^{-1}\eta + \gamma B_1^{-1}\eta. \tag{10.98}$$

We have proved that $N(B) \neq \{0\}$ and (10.5) imply (10.9) for the solution to (10.1), (10.4).

Let us assume now that (10.9) holds for all $f \in L^2(\Gamma)$, and prove that this implies either that $N(B) = \{0\}$ and (10.6) holds, or that $\dim N(B) = 1$ and (10.79), (10.80) hold. If (10.9) holds then the limits

$$\lim_{k\to 0} \alpha(k)(\sigma, 1) = \beta, \quad \lim_{k\to 0} \sigma(s,k) = \sigma_0(s) \tag{10.99}$$

exist. Indeed, if the limit (10.99) exists then the existence of (10.99) follows from the formula

$$u(x,k) = \alpha(k)(\sigma, 1) + \int_\Gamma g_0(x,t)\sigma(t,k)dt + o(1), \quad k \to 0 \qquad (10.100)$$

and the assumed existence of the limit

$$\lim_{k\to 0} u(x,k) = u_0(x). \qquad (10.101)$$

Let us prove existence of the limit (10.99) assuming (10.101). Assume first that B is injective. Then Lemma 10.5 yields (10.66) and the existence of (10.99). If $\dim N(B) = 1$, then Lemma 10.4 yields (10.79) and (10.80), and the existence of the limits (10.99) can be established as follows. First assume that $N(Q_0) = \{0\}$ and $(Q_0^{-1}1, 1) \neq 0$. Later we will drop these extra assumptions. Write (10.99), with $x = s \in \Gamma$ and $c_\sigma := (\sigma, 1)$, as

$$u(x,k) = \alpha(k)c_\sigma + (Q_0 + 3)\sigma, \quad \epsilon\sigma := (Q - Q_0)\sigma. \qquad (10.102)$$

For $k \to 0$ the operator $Q_0 + \epsilon$ is invertible, so

$$\sigma + \alpha(k)c_\sigma(Q_\sigma + \epsilon)^{-1}1 = (Q_0 + \epsilon)^{-1}u. \qquad (10.103)$$

The right-hand side of (10.103) has a finite limit as $k \to 0$ because of (10.101) and the equation $\|\epsilon(k)\|_{H^1(\Gamma)\to H^0(\Gamma)} \to 0$ as $k \to 0$. Integrate (10.103) over Γ to get

$$c_\sigma[1 + \alpha(k)(Q_0 + \epsilon)^{-1}1, 1)] = (Q_0 + \epsilon)^{-1}u, 1). \qquad (10.104)$$

Since $(Q_0^{-1}1, 1) \neq 0$ it follows that

$$c_\sigma = O(|\alpha(k)|^{-1}), \quad \lim_{k\to 0} \alpha(k)c_\sigma = \left(Q_0^{-1}u, 1\right)/\left(Q_0^{-1}1, 1\right) := \beta. \qquad (10.105)$$

Here $u = u|_\Gamma$. If at least one of the two extra assumptions ($N(Q_0) = \{0\}$ and $(Q^{-1}1, 1) \neq 0$) is not satisfied then find a constant $c > 0$ such that $Q_1 := Q_0 + c(\cdot, 1)$ is invertible and $(Q_1^{-1}1, 1) \neq 0$. This is possible as follows from the argument given in the proof of Theorem 10.2. Equation (10.102) can be written as

$$u = \alpha_1(k)c_\sigma + (Q_1 + \epsilon)\sigma, \quad \alpha_1(k) := \alpha(k) - c. \qquad (10.106)$$

Now one can repeat the argument given below formula (10.102) and obtain (10.105) with $\alpha_1(k)$ in place of $\alpha(k)$ and Q_1 in place of Q_0. This proves (10.99). Equations (10.103), (10.105) and (10.101) imply (10.99).

To complete the proof of Lemma 10.5 let us prove that if $\dim N(B) > 1$ there exists an $f \in L^2(\Gamma)$ for which (10.9) does not hold for the solution to (10.1), (10.4). Note that $\dim N(B) > 1$ implies that (10.5) does not hold, since we have proved in Lemma 10.3 that if (10.5) holds then $\dim N(B) \leq 1$. Suppose to the contrary that (10.9) holds for all $f \in L^2(\Gamma)$. Then the limits (10.99) exist. Therefore the limiting form of the equation (10.65)

$$B\sigma - \beta\eta = f, \quad \beta = \beta(\sigma) = \text{const} \tag{10.107}$$

is solvable for all $f \in L^2(\Gamma)$. Since B is a Fredholm operator, the necessary condition for the solvability of (10.107) is

$$(f - \beta\eta, p) = 0 \quad \forall p \in N(B). \tag{10.108}$$

If $\dim N(B) \geq 2$, condition (10.108) cannot be satisfied for all $f \in L^2(\Gamma)$ since one has only one parameter β to satisfy two or more conditions (10.108). Lemma 10.5 is proved.

Theorem 10.4 follows from Lemmas 10.2–10.5. □

10.5 The Method based on the Fredholm Property

The basic result of this section is: *a necessary and sufficient condition for the existence of the limit* (10.9) *is, roughly speaking, uniqueness of the solution of the limiting problem.* The approach we take is this: suppose that the problem at hand, for example, (10.1), (10.2), or (10.1), (10.3), or (10.1), (10.4), is of Fredholm type in the appropriate spaces. Then uniqueness of the solution to the limiting problem implies, by the Fredholm property, boundedness of the inverse operator.

The operator (10.1) depends continuously on k at $k = 0$, so its inverse has the same property if it exists and is bounded. These ideas are used in this section. The outlined approach allows one to handle operators with variable coefficients since it does not use the detailed information about the fundamental solution to equation (10.1).

Consider the problem

$$Lu + k^2 u = F \text{ in } \Omega \subset \mathbb{R}^n, \quad k > 0, \; n \geq 2 \tag{10.109}$$

with one of the boundary conditions (10.2), (10.3) or (10.4) and the radiation condition at infinity. Here

$$Lu = \partial_i(a_{ij}\partial_j u) - q(x)u, \tag{10.110}$$

over the repeated indices one sums up, $a_{ij}(x) \in C^1(\mathbb{R}^2)$ is a strongly elliptic real-valued matrix,

$$a_{ij} = \delta_{ij} \text{ for } |x| > R$$

where $R > 0$ is an arbitrary large fixed number, $q \in Q(\beta)$. Consider the limiting problem

$$Lu_0 = F \text{ in } \Omega \tag{10.111}$$

$$u_0 \text{ satisfies (10.2) or (10.3) or (10.4)} \tag{10.112}$$

$$u_0(x) = O(1) \text{ as } |x| \to \infty \text{ if } n = 2 \ (a); \quad u_0(\infty) = 0 \text{ if } n > 2 \ (b) \tag{10.113}$$

Let us assume that

$$\int_\Omega |F|^2(1+|x|)^s dx < \infty, \quad s > 1, \quad f \in H^{\frac{3}{2}}(\Gamma) \tag{10.114}$$

where f is the boundary function in (10.2), (10.3) or (10.4).

The basic assumption is:

Problem (10.111)–(10.113) *has at most one solution.* (10.115)

Let us introduce the space $L^2_s(\Omega) := L^2_s$ of functions with finite norm $\|u\| = (\int_\Omega |u|^2(1+|x|)^s dx)^{\frac{1}{2}}$. Our argument is valid with obvious modifications in $\mathbb{R}^n$, $n > 2$. All Fredholm operators in this Chapter are assumed to have index zero.

Lemma 10.6 *Consider a Fredholm operator $A(k) : X_1 \to X_2$ from a Banach space X_1 into a Banach space X_2, $k \in [0, b]$ is a parameter. Assume that $\|A(k) - A(k')\| \to 0$ as $k \to k'$, $k, k' \in [0, b]$, and $N(A(0)) = \{0\}$. Then $A^{-1}(k)$ is an isomorphism of X_1 onto X_2 for all $k \in [0, \delta]$ provided that $\delta > 0$ is sufficiently small and*

$$\left\|A^{-1}(k) - A^{-1}(0)\right\| \to 0 \text{ as } k \to 0. \tag{10.116}$$

Proof. By definition of a Fredholm operator, the range $R(A(k))$ of $A(k)$ is closed and its index is zero, that is, $\dim N(A(k)) = \operatorname{codim} R(A(k))$. In particular, $N(A(0)) = \{0\}$ implies that $R(A(0)) = X_2$, so that $A(0)$ is an isomorphism of X_1 onto X_2. We claim that, for sufficiently small $\delta > 0$, the operator $A(k)$ is an isomorphism of X_1 onto X_2 for all $k \in [0, \delta]$. Indeed, $A(k) = A(0) + A(k) - A(0) = A(0)[I + A^{-1}(0)B(k)]$ where $B(k) := A(k) - A(0)$, $\|B(k)\| \to 0$ as $k \to 0$ by the assumption. Therefore the operator $I + A^{-1}(0)B(k)$ is an isomorphism of X_1 onto X_1 and $A(0)$ is an isomorphism of X_1 onto X_2, and the claim follows. The conclusion (10.116) of Lemma 10.6 follows from the formula

$$A^{-1}(k) - A^{-1}(0) = -A^{-1}(k)[A(k) - A(0)]A^{-1}(0) \tag{10.117}$$

if one takes into account that

$$\sup_{0 \le k \le \delta} \|A^{-1}(k)\| < c. \tag{10.118}$$

To prove (10.118), assume the contrary. Then there is a sequence $k_n \in [0, \delta]$ such that $\|A^{-1}(k_n)\| \ge n$. One can assume that $k_n \to k_0 \in [0, \delta]$, and use the identity

$$A^{-1}(k_n) = \Big\{I + A^{-1}(k_0)\Big[A(k_n) - A(k_0)\Big]\Big\}A^{-1}(k_0). \tag{10.119}$$

As $k_n \to k_0$, it follows from (10.119) that $\|A^{-1}(k_n) - A^{-1}(k_0)\| \to 0$, so that $\|A^{-1}(k_0)\| < c$. Therefore (10.118) is proved. This completes the proof of Lemma 10.6. $\square$

Define the operator $\mathcal{L} : \mathcal{H}^2_{-s} \to \mathcal{H}_s$, where $\mathcal{H}^2_s = \{u : \|u\|_{L^2_s} + \|\partial u\|_{L^2_s} + \|\partial^2 u\|_{L^2_s}\} < \infty$, $s > 1$ and $\mathcal{H}_s = L^2_s \oplus H^0(\Gamma)$, by the formula

$$\mathcal{L}u := \mathcal{L}(k)u = \begin{pmatrix} F \\ f \end{pmatrix}, \quad (L + k^2)u = F, \quad \gamma u = f \tag{10.120}$$

where γu is the boundary operator (10.2), (10.3), or (10.4) and L is defined in (10.110). For example, in the case (10.4),

$$\gamma u = u_N + \eta(s)u. \tag{10.121}$$

The domain of definition of $\mathcal{L}$ belongs to $H^2_{\text{loc}}(\Omega) \cap L^2_s(\Omega)$. The operator $\mathcal{L}$ satisfies the estimate which follows from elliptic theory

$$\|u\|_{\mathcal{H}^2_{-s}} \le c\big(\|F\|_{L^2_s} + \|f\|_{H^\alpha(\Gamma)}\big), \quad 0 < c_1 < \Re k < c_2, 0 \le \operatorname{Im} k \le 1, \tag{10.122}$$

where c does not depend on u, $\alpha = \frac{3}{2}$ for the Dirichlet boundary condition and $\alpha = \frac{1}{2}$ for the conditions (10.3) or (10.4). Let us consider first $n \ge 3$. In this case using assumption (10.115) and the limiting absorption principle, one can prove that the fundamental solution to the equation

$$Lg + k^2 g = -\delta(x - y) \text{ in } \mathbb{R}^n, \tag{10.123}$$

where g satisfies the radiation condition at infinity, exists and is continuous in k at $k=0$. One can look for the solution to (10.9), (10.2) of the form

$$\begin{aligned} u &= \int_\Omega gF\,dy + v, \quad (L+k^2)v = 0 \text{ in } \Omega \\ v &= \int_\Gamma g(x,s,k)\sigma(s)ds, \quad Q\sigma = f - \int_\Omega gF\,dy\big|_\Gamma \end{aligned} \tag{10.124}$$

where $Q\sigma := v|_\Gamma$. The operator $Q(k): H^0(\Gamma) \to H^1(\Gamma)$ is norm continuous in k in the interval $0 \le k \le k_0$, where $k_0 > 0$ is a sufficiently small number, and $Q^{-1}(k): H^1(\Gamma) \to H^0(\Gamma)$ is bounded for $k \in [0, k_0]$ by (10.115). By lemma 10.6, it follows that $\|Q^{-1}(k) - Q^{-1}(0)\| \to 0$ as $k \to 0$. We have proved

Theorem 10.5 *If $n \ge 3$ and (10.115) holds, then $|u(x,k) - u_0(x)| \to 0$ as $k \to 0$ uniformly on compacts in Ω.*

If $n = 2$ then the result and the basic idea of the argument are the same, but some modifications are needed to take into account that g is no longer continuous as $k \to 0$. For example, if $L = \nabla^2$ then $g = \frac{i}{4}H_0^{(1)}(kr)$, $r := |x-y|$, so that (10.120) holds. However, due to the fact that the operator $Q(k): H^0(\Gamma) \to H^1(\Gamma)$ acts as a differentiation, the unbounded component of the operator $Q(k)\sigma$ which, for $L = \nabla^2$, is the constant $\alpha(k)(\sigma, 1)$, does not bring difficulties and one has $\|Q(k) - Q(k')\|_{H^0(\Gamma)\to H^1(\Gamma)} \to 0$ as $k \to k'$, $k, k' \in [0, k_0)$. As above, assumption (10.115) implies that $Q^{-1}(k)$ exists and, therefore, is bounded from $H^1(\Gamma)$ onto $H^0(\Gamma)$, since $Q(k): H^0(\Gamma) \to H^1(\Gamma)$ is Fredholm-type operator. The second point which needs a discussion is the possible unboundedness of $\int_\Omega gFdy|_\Gamma$ as $k \to 0$. Indeed, $\int_\Omega gFdy = \alpha(k)\int_\Omega Fdy + \int_\Omega g_0 Fdy + o(1)$ as $k \to 0$, and if $\int_\Omega Fdy \ne 0$, this expression is $O(|\ln k|)$ as $k \to 0$. The argument similar to the given in the proof of Theorem 10.2 is applicable now. Indeed, equation (10.124) can be written, for $L = \nabla^2$, as

$$\alpha(k)c_\sigma + Q_0\sigma + \epsilon(k)\sigma = f - \alpha(k)c - h(k), \quad c_\sigma := (\sigma, 1) \tag{10.125}$$

where Q_0 and $\epsilon(k)$ are as in the proof of Theorem 2.2, $c = \int_\Omega Fdy$, and $\|h(k) - h(0)\|_{C(\Gamma)} \to 0$ as $k \to 0$. Assume that $N(Q_0) = \{0\}$. Then (10.125) can be written as

$$\sigma + Q_0^{-1}\epsilon(k)\sigma + \alpha(k)c_\sigma Q_0^{-1}1 = Q_0^{-1}f - \alpha(k)cQ_0^{-1} - Q_0^{-1}h(k). \tag{10.126}$$

Integrate (10.126) over Γ to get

$$c_\sigma\Big[1+\alpha(k)(Q_0^{-1},1)\Big]=\Big(Q_0^{-1}[f-h(k)-\epsilon(k)\sigma],1\Big)-\alpha(k)c(Q_0^{-1},1). \tag{10.127}$$

As in Section 10.2, $(Q_0^{-1},1)\neq 0$, so that (10.127) implies

$$c_\sigma=-c+O(|\alpha|^{-1}) \text{ as } k\to 0. \tag{10.128}$$

From (10.126)–(10.128) one gets

$$\begin{aligned}&\sigma=Q_0^{-1}[f-h(0)]-bQ_0^{-1}+O\big(|\alpha|^{-1}\big),\quad b=\lim_{k\to 0}\big[(c\sigma+c)\alpha(k)\big]\\ &b=(Q_0^{-1},1)^{-1}\Big\{(Q_0^{-1}(f-h(0)),1)+c\Big\}.\end{aligned} \tag{10.129}$$

Therefore $\sigma(s,k)=\sigma_0(s)+O(|\alpha|^{-1})$ and

$$\begin{aligned}u(x,k)&=\int_\Gamma g(x,s)\sigma\,ds+\int_\Omega g(x,y)F(y)dy\\ &=\alpha\big(c_\sigma+c\big)+u_0+O\big(|\alpha|^{-1}\big)\\ &=b+\phi_0(x)+O\big(|\alpha|^{-1}\big):=u_0(x)+O\big(|\alpha|^{-1}\big)\end{aligned} \tag{10.130}$$

where b is defined in (10.129), $\phi_0(x) := \int_\Gamma g_0(x,s)\sigma_0(s)\,ds + \int_\Omega g_0(x,y)F(y)dy$, and $u_0(x)$ solves the limit problem

$$Lu_0=F \text{ in } \Omega,\quad u_0=f \text{ on } \Gamma,\quad \big|u_0(\infty)\big|<\infty. \tag{10.131}$$

The last condition (10.131) is a consequence of the equation

$$\phi_0(x)=\frac{1}{2\pi}\ln\frac{1}{|x|}\left[\int_\Gamma\sigma_0\,ds+\int_\Omega F(y)dy\right]+o(1) \text{ as } |x|\to\infty \tag{10.132}$$

which implies $\phi_0(x)=o(1)$ as $|x|\to\infty$, since $\int_\Gamma\sigma_0\,ds+\int_\Gamma F dy=0$.

In this argument we used essentially the special form of L, namely $L=\nabla^2$, since the behavior of g as $k\to 0$ has been used essentially. In the general case of L given by (10.2) one needs a different method of reducing the problem to a Fredholm-type equation, a method which does not use the properties of the fundamental solution g in a neighborhood of $k=0$.

This method is as follows. Assume that

$$\sup_{0\le k\le k_0}\|u(x,k)\|<c \tag{10.133}$$

where $u(x,k)$ is the solution to (10.109), (10.2) or (10.109), (10.3) or (10.109), (10.4), and the norm in (10.133) is $L^2_{-s}(\Omega)$, $s>1$. Then using the elliptic regularity estimate one concludes that there is a sequence $u(x, k_n)$, $k_n \to 0$, which converges to a limit $u_0(x)$ in $H^2_{\rm loc}(\Omega)$ and in $L^2_{-s}(\Omega)$. Passing to the limit $k_n \to 0$ in (10.109) and (10.2) yields (10.111)–(10.113) for $u_0(x)$. Therefore, the desired result (10.9) follows from (10.133).

Lemma 10.7 *Inequality* (10.133) *follows from* (10.115).

Proof. If (10.133) is false then there is a sequence $k_n \to 0$ such that $\|u(x, k_n)\| \geq n$. Define $v_n := u(x, k_n)/\|u(x, k_n)\|$. Then

$$Lv_n + k_n^2 v_n = F_n := F\|u(x, k_n)\|^{-1}, \quad v_n \Big|_{\Gamma} = f_n := f\|u(x, k_n)\|^{-1}$$
$$\|v_n\| = 1 \tag{10.134}$$

As above, one derives from (10.134) that

$$\begin{aligned} &v_n \to v_0 \text{ in } H^2_{\rm loc}(\Omega), \quad \|v_n - v_0\| \to 0 \text{ as } n \to \infty \\ &Lv_0 = 0 \text{ in } \Omega, \quad v_0 = 0 \text{ on } \Gamma, \quad v_0 \text{ satisfies (10.113)} \end{aligned} \tag{10.135}$$

By the assumption (10.115), equation (10.135) imply $v_0 = 0$, and (10.135) implies $\|v_n\| \to 0$. This contradicts to (10.29), and the contradiction proves (10.133). □

Let us summarize the result.

Theorem 10.6 *If* (10.115) *holds then* (10.9) *holds for the solution to* (10.109), (10.2) *or* (10.109), (10.3) *or* (10.109), (10.4). *Conversely, if* (10.9) *holds for the solution to* (10.109), (10.2) *or* (10.109), (10.3) *or* (10.109), (10.4) *for any* $F \in L^2_0(\Omega)$ *and* $f = 0$, *or for* $F = 0$ *and any* $f \in H^{3/2}(\Gamma)$, *then* (10.115) *holds.*

Proof. The first part of the conclusion of Theorem 10.6 has been proved above.

The second part follows from the solvability for all F and f of the limiting problem (10.111)–(10.113) in the spaces $\mathcal{H}^2_{-s} \to \mathcal{H}_s$, $s > 2$. Indeed, a necessary condition for the solvability of (10.111)–(10.113) for some $f \in H^{\frac{3}{2}}(\Gamma)$ and $F \in L^2_s(\Omega)$ is the orthogonality of F to all solutions w of the homogeneous adjoint problem. Since this condition $\int_\Omega Fw\,dx = 0$ holds for all $F \in L^2_0(\Omega)$, it follows that $w = 0$. If $F = 0$ then the orthogonality condition takes the form $\int_\Gamma f w_N\,ds = 0$, $\forall f \in H^{\frac{3}{2}}(\Gamma)$. This implies that $w_N = 0$ on Γ. Since $w = 0$ on Γ by the assumption, one applies uniqueness theorem for the solution of the Cauchy problem for elliptic equation $Lw = 0$

in Ω and concludes that $w = 0$. Thus, (10.115) holds. Theorem 10.6 is proved. □

10.6 The Method based on the Maximum Principle

Consider the problem (10.1), (10.2) in $\Omega \subset \mathbb{R}^n$, $n \geq 3$. Let $R = k^{-1}$. Define $c_1(k) := \sup |u(x,k)|$ where the supremum is taken over $x \in \Omega$, $|x| \leq 2\,\mathrm{diam}\, D := 2r_0$. From the elliptic estimates it follows that $|\nabla u|_{|x|=r_0} \leq cc_1(k)$. We will use Green's formula

$$u(x) = \int_{|s|=r_0} (ug_N - gu_N)\, ds, \quad x \in B'_{r_0}. \tag{10.136}$$

Here g satisfies the radiation condition $B'_{r_0} = \mathbb{R}^n \backslash B_{r_0}$, $B_{r_0} = \{x : |x| \leq r_0\}$,

$$(\nabla^2 + k^2)g = -\delta(x-y) \text{ in } \mathbb{R}^n,\ g(x) = \frac{i}{4}\frac{H_\nu^{(1)}(k|x|)}{(2\pi k|x|)^\nu}k^{2\nu},\ \nu := \frac{n-2}{2} \tag{10.137}$$

where $H_\nu^{(1)}$ is the Hankel function. One has, uniformly in $0 < r < r_0$,

$$|\phi(r)| \leq c\begin{cases} r^{-2\nu}, & n > 2 \\ |\ln r| + 1, & n = 2 \end{cases}, \quad \phi(r) := \frac{i}{4}H_\nu^{(1)}(\rho)(2\pi\rho)^\nu. \tag{10.138}$$

By c we denote various positive constants independent of k. From (10.137)–(10.138), one obtains

$$|g| \leq c|x-y|^{-n+2}, \quad |\nabla g| \leq c|x-y|^{-n+1} \text{ for } |x-y| < 2k^{-1}. \tag{10.139}$$

Choose an arbitrary $\theta \in [0, 2\pi)$ and define

$$v(x,k,\theta) = \Re\Big\{\exp(i\theta)\big[u(x,k) - u_0(x)\big]\Big\} \tag{10.140}$$

where $u_0(x)$ is the solution to problem (10.1), (10.2) with $k = 0$. Note that

$$\Delta v = -k^2\Re\{\exp(i\theta)u\}, \quad v = 0 \text{ on } \Gamma \tag{10.141}$$

$$v|_{|x|=k^{-1}} = \Re\Big\{\exp(i\theta)\big[u(x,k) - u_0(x)\big]\Big\}\Big|_{|x|=k^{-1}}. \tag{10.142}$$

If $v_+ := \alpha + \beta r^{-n+4}$, $r = |x|$, α and β are positive constants, then

$$\Delta v_+ = 2\beta(-N+4)r^{-n+2} \leq \Delta v \text{ for } \beta = \beta_1 c_1(k)k^2 \tag{10.143}$$

where $c_1(k)$ was defined earlier and $\beta_1 > 0$ is a sufficiently large fixed constant. Note that (10.136) and (10.139) imply that

$$|u(x,k)| \leq cc_1(k)|x|^{-n+2} \text{ for } |x| < k^{-1}. \tag{10.144}$$

Note also that

$$v_+\big|_{|x|=k^{-1}} \geq v\big|_{|x|=k^{-1}} \text{ for } \alpha = \alpha_1 c_1(k) k^{n-2} \tag{10.145}$$

where $\alpha_1 > 0$ is a sufficiently large constant. By the maximum principle, $v(x,k) \leq v_+(x,k)$ for $|x| \leq k^{-1}$. Choosing $\theta = \theta(x,k)$, which was arbitrary, in a suitable way, one concludes that

$$\big|u(x,k) - u_0(x,k)\big| \leq v_+(x) \text{ for } \quad |x| \leq k^{-1}. \tag{10.146}$$

One has, if $n > 4$ and $k \to 0$,

$$v_+ \leq \alpha + \beta \max_{0<d\leq r\leq k^{-1}} r^{-n+4} \leq \alpha_1 c_1(k) k^{n-2} + \beta_1 c_1(k) k^{n-2} \leq cc_1(k)k^2 \tag{10.147}$$

where $d > 0$ is a constant. From (10.146) and (10.147) it follows that

$$\max_{x\in\Omega,\, |x|\leq k^{-1}} \big|u(x,k) - u_0(x)\big| \leq cc_1(k)k^2, \quad n > 4. \tag{10.148}$$

By the definition of $c_1(k)$ and from the triangle inequality one gets

$$c_1(k) \leq \max_{x\in\Omega,\, |x|\leq k^{-1}} |u_0(x)| + cc_1(k)k^2. \tag{10.149}$$

This implies

$$c_1(k) \leq c_0 := \max_{x\in\Omega} |u_0(x)|. \tag{10.150}$$

Indeed, for $n \geq 3$, by the maximum principle, for example, one has

$$\max_{x\in\Omega} |u_0(x)| = c_0 < \infty. \tag{10.151}$$

We have assumed $n > 4$ so far. If $n = 4$ then one uses $v_+ = \alpha + \beta \ln r$, $\Delta v_+ = 2\beta r^{-2}$, $\beta < 0$ and $|\beta_1|$ is sufficiently large (see (10.143)), $\alpha = \alpha_1 c_1(k) k^2 |\ln k|$. Equation (10.148) becomes

$$\max_{x\in\Omega,\, |x|\leq k^{-1}} \big|u(x,k) - u_0(x)\big| \leq cc_1(k)k^2|\ln k| \leq cc_0 k^2 |\ln k|, \text{ for } n = 4. \tag{10.152}$$

If $n = 3$ then $v_+ = \alpha + \beta r$, $\Delta v_+ = 2\beta r^{-1}$, $\beta_1 < 0$, $|\beta_1|$ is sufficiently large, $\alpha = \alpha_1 k$ and (10.148) becomes

$$\max_{x\in\Omega,\,|x|\le k^{-1}} \big|u(x,k) - u_0(x)\big| \le c_0 k, \qquad \text{for } n = 3. \tag{10.153}$$

We leave to the reader to check that the estimates (10.148), (10.152), (10.153) are sharp: the rates given in the right-hand sides of these equalities occur for the function $H_\nu^{(1)}(k|x|)|x|^{-\nu}[H_\nu^{(1)}(k)]^{-1}$. Let us formulate the results.

Theorem 10.7 *If $n \ge 3$ then estimates* (10.148), (10.152) *and* (10.153) *hold.*

The case $n = 2$ can also be treated similarly. In this case one has

Theorem 10.8 *If $n = 2$ then*

$$\sup_{x\in\Omega,\,|x|\le k^{-1}} \big|u(x,k) - u_0(x)\big| \le c|\ln|x||\,|\ln k|^{-1}. \tag{10.154}$$

Proof. Choose $v_+ := \alpha \ln|x| + \beta s(k|x|)$, where $\alpha = \alpha_1|\ln k|^{-1}$ $\alpha_1 > 0$ is sufficiently large, $\beta = \beta_1|\ln k|^{-1}$, $\beta_1 < 0$, $|\beta_1|$ is sufficiently large, $s(r) := \frac{1}{4}r^2(|\ln r| + 2)$, $r \in (0,1)$. Using the argument similar to the one in the proof of Theorem 10.7, one gets (10.154). □

The case of the data f depending on k and the operator (10.110) with $q \ge 0$ so that for this operator the maximum principle holds, can be treated similarly. The arguments in this section are taken from [160].

Another idea based on coercivity estimates is developed in [134] and [160] and applied to a study of the low-frequency behavior of the solutions to dissipative Maxwell's equations.

10.7 Continuity of Solutions to Operator Equations with Respect to a Parameter

Let $A(k)u(k) = f(k)$ be an operator equation in a Banach space X, $k \in \Delta \subset \mathbb{C}$ is a parameter, $A(k) : X \to Y$ is a map, possibly nonlinear, $\Delta \subset \mathbb{C}$ is a domain. Sufficient conditions are given in this Section for the continuity of $u(k)$ with respect to k.

10.7.1 *Introduction*

Let X and Y be Banach spaces, $k \in \Delta \subset \mathbb{C}$ be a parameter, Δ be an open bounded set on a complex plane $\mathbb{C}$, $A(k) : X \to Y$ be a map, possibly nonlinear, $f := f(k) \in Y$ be a function.

Consider an equation

$$A(k)u(k) = f(k). \tag{10.155}$$

We are interested in conditions, sufficient for the continuity of $u(k)$ with respect to $k \in \Delta$. The novel points in our presentation include necessary and sufficient conditions for continuity of the solution to equation (10.155) and sufficient conditions for its continuity when the operator $A(k)$ is nonlinear.

Consider separately the cases when $A(k)$ is a linear map and when $A(k)$ is a nonlinear map.

Assumptions 1. $A(k) : X \to Y$ is a linear bounded operator, and

(a) equation (10.155) is uniquely solvable for any $k \in \Delta_0 := \{k : |k - k_0| \leq r\}$, $k_0 \in \Delta$, $\Delta_0 \subset \Delta$,
(b) $f(k)$ is continuous with respect to $k \in \Delta_0$, $\sup_{k\in\Delta_0} \|f(k)\| \leq c_0$;
(c) $\lim_{h\to 0} \sup_{\substack{k\in\Delta_0 \\ v\in M}} \|[A(k+h) - A(k)]v\| = 0$, where $M \subset X$ is an arbitrary bounded set,
d) $\sup_{\substack{k\in\Delta_0 \\ f\in N}} \|A^{-1}(k)f\| \leq c_1$, where $N \subset Y$ is an arbitrary bounded set, and c_1 may depend on N.

Theorem 10.9 *If* **Assumptions 1** *hold, then*

$$\lim_{h\to 0} \Big\| u(k+h) - u(k) \Big\| = 0. \tag{10.156}$$

Proof. One has

$$\begin{aligned} u(k+h) - u(k) &= A^{-1}(k+h)f(k+h) - A^{-1}(k)f(k) \\ &= A^{-1}(k+h)f(k+h) - A^{-1}(k)f(k+h) \\ &\quad + A^{-1}(k)f(k+h) - A^{-1}(k)f(k). \end{aligned} \tag{10.157}$$

$$\Big\| A^{-1}(k)\big[f(k+h) - f(k)\big] \Big\| \leq c_1 \Big\| f(k+h) - f(k) \Big\| \to 0 \text{ as } h \to 0. \tag{10.158}$$

$$\begin{aligned} \|A^{-1}(k+h) - A^{-1}(k)\| &= \Big\| A^{-1}(k+h)\big[A(h+k) - A(k)\big]A^{-1}(k) \Big\| \\ &\leq c_1^2 \Big\| A(k+h) - A(k) \Big\| \to 0 \text{ as } h \to 0. \end{aligned} \tag{10.159}$$

From (10.157)–(10.159) and **Assumptions 1** the conclusion of Theorem 10.9 follows. □

Remark 10.2 **Assumptions 1** *are not only sufficient for the continuity of the solution to* (10.155), *but also necessary if one requires the continuity of* $u(k)$ *uniform with respect to* f *running through arbitrary bounded sets. Indeed, the necessity of the assumption a) is clear; that of the assumption b) follows from the case* $A(k) = I$, *where* I *is the identity operator; that of the assumption c) follows from the case* $A(k) = I$, $A(k+h) = 2I$, $\forall h \neq 0$, $f(k) = g \neq 0\ \forall k \in \Delta_0$. *Indeed, in this case assumption c) fails and one has* $u(k) = g$, $u(k+h) = \frac{g}{2}$, *so* $||u(k+h) - u(k)|| = \frac{||g||}{2}$ *does not tend to zero as* $h \to 0$.

To prove the necessity of the assumption (d), assume that $\sup_{k\in\Delta_0} \|A^{-1}(k)\| = \infty$. *Then, by the Banach-Steinhaus theorem, there is an element* f *such that* $\sup_{k\in\Delta_0} \|A^{-1}(k)f\| = \infty$, *so that* $\lim_{j\to\infty} \|A^{-1}(k_j)f\| = \infty$, $k_j \to k \in \Delta_0$. *Then* $\|u_j\| := \|u(k_j)\| = \|A^{-1}(k_j)f\| \to \infty$, *so* u_j *does not converge to* $u := u(k) = A^{-1}(k)f$, *although* $k_j \to k$.

Assumptions 2. $A(k) : X \to Y$ is a nonlinear map, and (a), (b), (c) and (d) of **Assumptions 1** hold, and the following assumption holds:

(e) $A^{-1}(k)$ *is a homeomorphism of* X *onto* Y *for each* $k \in \Delta_0$.

Remark 10.3 *Assumption (e) is included in (d) in the case of a linear operator* $A(k)$ *because if* $\|A(k)\| \leq c_2$ *and* $\|A^{-1}(k)\| \leq c_1$ *then* $A(k)$, $k \in \Delta_0$, *is an isomorphism of* X *onto* Y.

Theorem 10.10 *If* **Assumptions 2** *hold, then* (10.156) *holds.*

Let us make the following **Assumption** A_d:

Assumption A_d: Assumptions 2 hold and

(f) $\dot{f}(k) := \frac{df(k)}{dk}$ is continuous in Δ_0,
(g) $\dot{A}(u,k) := \frac{\partial A(u,k)}{\partial k}$ is continuous with respect to (wrt) k in Δ_0 and wrt $u \in X$,
(j) $\sup_{k\in\Delta_0} \|[A'(u,k)]^{-1}\| \leq c_3$, where $A'(u,k)$ is the Fréchet derivative of $A(u,k)$ and $[A'(u,k)]^{-1}$ is continuous with respect to u and k. $\dot{f}(k) := \frac{df(k)}{dk}$ is continuous in Δ_0.

Remark 10.4 *If* **Assumption** A_d *holds, then*

$$\lim_{h\to 0} \|\dot{u}(k+h) - \dot{u}(k)\| = 0. \tag{10.160}$$

Remark 10.5 *If* **Assumptions 1** *hold except one: $A(k)$ is not necessarily a bounded linear operator, $A(k)$ may be unbounded, closed, densely defined operator-function, then the conclusion of Theorem 10.10 still holds and its proof is the same. For example, let $A(k) = L + B(k)$, where $B(k)$ is a bounded linear operator continuous with respect to $k \in \Delta_0$, and L is a closed, linear, densely defined operator from $D(L) \subset X$ into Y. Then*

$$\|A(k+h) - A(k)\| = \|B(k+h) - B(k)\| \to 0 \quad \textit{as} \quad h \to 0,$$

although $A(k)$ and $A(k+h)$ are unbounded.

In Section 10.7.2 proofs of Theorem 10.10 and of Remark 10.4 are given.

10.7.2 *Proofs*

Proof of Theorem 10.10. One has:

$$A(k+h)u(k+h) - A(k)u(k) = f(k+h) - f(k) = o(1) \quad \text{as} \quad h \to 0.$$

Thus

$$A(k)u(k+h) - A(k)u(k) = o(1) - [A(k+h)u(k+h) - A(k)u(k+h)].$$

Since $\sup_{\{u(k+h):\|u(k+h)\| \leq c\}} \|A(k+h)u(k+h) - A(k)u(k+h)\| \underset{h \to 0}{\to} 0$, one gets

$$A(k)u(k+h) \to A(k)u(k) \text{ as } h \to 0. \tag{10.161}$$

By the **Assumptions 2**, item (e), the operator $A(k)$ is a homeomorphism. Thus (10.161) implies (10.156).

Theorem 10.10 is proved. □

Proof of Remark 10.4. First, assume that $A(k)$ is linear. Then

$$\frac{d}{dk}A^{-1}(k) = -A^{-1}(k)\dot{A}(k)A^{-1}(k), \quad \dot{A} := \frac{dA}{dk}. \tag{10.162}$$

Indeed, differentiate the identity $A^{-1}(k)A(k) = I$ and get $\frac{dA^{-1}(k)}{dk}A(k) + A^{-1}(k)\dot{A}(k) = 0$. This implies (10.162). This argument proves also the existence of the deriviative $\frac{dA^{-1}(k)}{dk}$. Formula $u(k) = A^{-1}(k)f(k)$ and the continuity of $\dot{f}$ and of $\frac{dA^{-1}(k)}{dk}$ yield the existence and continuity of $\dot{u}(k)$. Remark 10.4 is proved for linear operators $A(k)$.

Assume now that $A(k)$ is nonlinear, $A(k)u := A(k,u)$. Then one can differentiate (10.155) with respect to k and get

$$\dot{A}(k,u) + A'(k,u)\dot{u} = \dot{f}, \tag{10.163}$$

where A' is the Fréchet derivative of $A(k,u)$ with respect to u. Formally one assumes that $\dot{u}$ exists, when one writes (10.163), but in fact (10.163) proves the existence of $\dot{u}$, because $\dot{f}$ and $\dot{A}(k,u) := \frac{\partial A(k,u)}{\partial k}$ exist by the **Assumption** A_d and $[A'(k,u)]^{-1}$ exists and is an isomorphism by the **Assumption** A_d, item (j). Thus, (10.163) implies

$$\dot{u} = [A'(k,u)]^{-1}\dot{f} - [A'(k,u)]^{-1}\dot{A}(k,u). \tag{10.164}$$

Formula (10.164) and **Assumption** A_d imply (10.160).

Remark 10.4 is proved. □

Consider some application of the above results to Fredholm equations depending on a parameter.

Let

$$Au := u - \int_D b(x,y,k)u(y)dy := [I - B(k)]u = f(k), \tag{10.165}$$

where $D \subset R^n$ is a bounded domain, $b(x,y,k)$ is a function on $D \times D \times \Delta_0$, $\Delta_0 := \{|k - k_0| < r\}$, $k_0 > 0$, $r > 0$ is a sufficiently small number. Assume that $A(k_0)$ is an isomorphism of $H := L^2(D)$ onto H, for example, $\int_D \int_D |b(x,y,k_0)|^2 dxdy < \infty$ and $N(I - B(k_0)) = \{0\}$, where $N(A)$ is the null-space of A. Then, $A(k_0)$ is an isomprohism of H onto H by the Fredholm alternative, and **Assumptions 1** hold if $f(k)$ is continuous with respect to $k \in \Delta_0$ and

$$\lim_{h\to 0} \int_D \int_D |b(x,y,k+h) - b(x,y,k)|^2 dx\, dy = 0 \quad k \in \Delta_0. \tag{10.166}$$

Condition (10.166) implies that if $A(k_0)$ is an isomorphism of H onto H, then so is $A(k)$ for all $k \in \Delta_0$ if $|k - k_0|$ is sufficiently small.

Remark 10.4 implies to (10.165) if $\dot{f}$ is continuous with respect to $k \in \Delta_0$, and $\dot{b} := \frac{\partial b}{\partial k}$ is continuous with respect to $k \in \Delta_0$ as an element of $L^2(D \times D)$. Indeed, under these assumptions $\dot{u} = [I - B(k)]^{-1}(\dot{f} - \dot{B}(k)u)$ and the right-hand side of this formula is continuous in Δ_0.

Chapter 11

Finding Small Inhomogeneities from Scattering Data

A new method for finding small inhomogeneities from surface scattering data is proposed and mathematically justified in this chapter. The presentation follows [107] and is based on [105]. The method allows one to find small holes and cracks in metallic and other objects from the observation of the acoustic field scattered by the objects.

11.1 Introduction

In many applications one is interested in finding small inhomogeneities in a medium from the observation of the scattered field, acoustic or electromagnetic, on the surface of the medium.

We have two typical examples of such problems in mind. The first one is in the area of material science and technology. Suppose that a piece of metal or other material is given and one wants to examine if it has small cavities (holes or cracks) inside. One irradiates the metal by acoustic waves and observes on the surface of the metal the scattered field. From these data one wants to determine:

(1) are there small cavities inside the metal?

(2) if there are cavities, then where are they located and what are their sizes?

Similar questions can be posed concerning localization not only of the cavities, but any small in comparison with the wavelength inhomogeneities. Our methods allow one to answer such questions.

As a second example, we mention the mammography problem. Currently x-ray mammography is widely used as a method of early diagnistics of breast cancer in women. However, it is believed that the probability for a woman to get a new cancer cell in her breast as a result of an x-ray mammography test is rather high. Therefore it is quite important to introduce ultrasound mammography tests. This is being done currently. A new cancer cells can be considered as small inhomogeneities in the healthy breast tissue. The problem is to localize them from the observation on the surface of the breast of the scattered acoustic field.

The purpose of this Section is to describe a new idea of solving the problem of finding inhomogeneities, small in comparison with the wavelength, from the observation of the scattered acoustic or electromagnetic waves on the surface of the medium.

For simplicity we present the basic ideas in the case of acoustic wave scattering. These ideas are based on the earlier results on wave scattering theory by small bodies presented in Chapter 7. Our objective in solving the inverse scattering problem of finding small inhomogeneities from surface scattering data are:

(1) to develop a computationally simple and stable method for a partial solution of the above inverse scattering problem. The exact inversion procedures (see [120], [107], and references therein) are computationally difficult and unstable. In practice it is often quite important, and sometimes sufficient for practical purposes, to get a "partial inversion", that is, to answer questions of the type we asked above: given the scattering data, can one determine if these data correspond to some small inhomogeneities inside the body? If yes, where are these inhomogeneities located? What are their intensities? We define the notion of intensity v_m of an inhomogeneity below formula (11.1).

Some theoretical and numerical results based on a version of the proposed approach one can find in [105], [36].

11.2 Basic Equations

Let the governing equation be

$$\left[\nabla^2 + k^2 + k^2 v(x)\right]u = -\delta(x-y) \quad \text{in } \mathbb{R}^3, \tag{11.1}$$

where u satisfies the radiation condition, $k = \text{const} > 0$, and $v(x)$ is the inhomogeneity in the velocity profile.

Assume that $\sup_{x\in\mathbb{R}^3} |v(x)| \leq c_0$, $\operatorname{supp} v = U_{m=1}^{M} B_m(\tilde{z}_m, \rho_m) \subset \mathbb{R}^3_- = \{x \mid x_{(3)} < 0\}$, where $x_{(3)}$ denotes the third component of vector x in Cartesian coordinates, $B_m(\tilde{z}_m, \rho_m)$ is a ball, centered at $\tilde{z}_m$ with radius ρ_m, $k\rho_m \ll 1$.

Denote

$$\overline{v}_m := \int_{B_m} v(x)dx.$$

Problem 11.1 (**Inverse Problem (IP):**) Given $u(x, y, k)$ for all $x, y \in P$, $P = \{x \mid x_{(3)} = 0\}$ and a fixed $k > 0$, find $\{\tilde{z}_m, \overline{v}_m\}$, $1 \le m \le M$.

In this section we propose a numerical method for solving the (IP). To describe this method let us introduce the following notations:

$$P := \{x \mid x_{(3)} = 0\} \tag{11.2}$$

$$\{x_j, y_j\} := \xi_j, \quad 1 \le j \le J,\ x_j, y_j \in P \tag{11.3}$$

are the points at which the data $u(x_j, y_j, k)$ are collected

$$k > 0 \text{ is fixed} \tag{11.4}$$

$$g(x, y, k) := \frac{\exp(ik|x - y|)}{4\pi|x - y|} \tag{11.5}$$

$$G_j(z) := G(\xi_j, z) := g(x_j, z, k)g(y_j, z, k) \tag{11.6}$$

$$f_j := \frac{u(x_j, y_j, k) - g(x_j, y_j, k)}{k^2} \tag{11.7}$$

$$\Phi(z_1, \dots, z_M, v_1, \dots, v_M) := \sum_{j=1}^{J} \left| f_j - \sum_{m=1}^{M} G_j(z_m)v_m \right|^2. \tag{11.8}$$

The proposed method for solving the (IP) consists in finding the global minimizer of function (11.8). This minimizer $(\tilde{z}_1, \dots, \tilde{z}_M, \tilde{v}_1, \dots, \tilde{v}_M)$ gives the estimates of the positions $\tilde{z}_m$ of the small inhomogeneities and their intensities $\overline{v}_m$. This is explained in more detail below formula (11.14). Numerical realization of the proposed method, including a numerical procedure for estimating the number M of small inhomogeneities from the surface scattering data is described in [36].

Our approach with a suitable modification is valid in the situation when the Born approximation fails, for example, in the case of scattering by delta-type inhomogeneities [2], [28], [107].

In this case the basic condition

$$Mk^2c_0\rho^2 \ll 1, \tag{$*$}$$

which guarantees the applicability of the Born approximation, is violated. Here $\rho := \max_{1\le m\le M}\rho_m$ and c_0 was defined below formula (11.1). We assume throughout that M is not very large, between 1 and 15.

In the scattering by a delta-type inhomogeneity the assumption is $c_0\rho^3 = \text{const} := V$ as $\rho \to 0$, so that for any fixed $k > 0$ one has $k^2c_0\rho^2 = k^2V\rho^{-1} \to \infty$ as $\rho \to 0$, and clearly condition $(*)$ is violated.

In our notations this delta-type inhomogeneity is of the form $k^2v(x) = k^2\sum_{m=1}^{M}\overline{v}_m\delta(x-\tilde{z}_m)$.

The scattering theory by the delta-type potentials (see [2]) requires some facts from the theory of selfadjoint extensions of symmetric operators in Hilbert spaces and in this section we will not go into detail (see [28]).

11.3 Justification of the Proposed Method

We start with an exact integral equation equivalent to equation (11.1) with the radiation condition:

$$u(x,y,k) = g(x,y,k) + k^2\sum_{m=1}^{M}\int_{B_m} g(x,z,k)v(z)u(z,y,k)dz. \tag{11.9}$$

For small inhomogeneities the integral on the right-hand side of (11.9) can be approximately written as

$$\begin{aligned} k^2\int_{B_m} g(x,z,k)v(z)u(z,y,k)dz &: \\ &= k^2\int_{B_m} g(x,z,k)v(z)g(z,y,k)dz + \varepsilon^2 \\ &= k^2G(x,y,\overline{z}_m)\int_{B_m} vdz + \varepsilon^2 \\ &= k^2G(\xi,\overline{z}_m)\overline{v}_m + \varepsilon^2, \quad 1\le m\le M \end{aligned} \tag{11.10}$$

where ε^2 is defined by the first equation in formula (11.10), it is the error due to replacing u under the sign of integral in (11.9) by g, and $\overline{z}_m$ is a point close to $\tilde{z}_m$.

One has $|u-g| = O(Mk^2c_0\rho^3/d^2)$ if $x,y \in P$, and $|u-g| = O(Mk^2c_0\rho^2/d)$ if $x \in D$,$y \in P$. Thus, the error term ε^2 in (10.3.10) equals to $O(M^2k^4c_0^2\rho^5/d^2)$ if $x,y \in P$.

Therefore the function $u(z,y,k)$ under the sign of the integral in (11.9) can be replaced by $g(z,y,k)$ with a small relative error $\frac{\varepsilon^2}{|g|}$, where $y \in P$ and $z \in D$, provided that:

$$c_0^2 M^2 k^2 \frac{\rho^5}{d} \ll 1, \quad x,y \in P, \tag{11.11}$$

where $\rho = \max_{1\le m\le M} \rho_m$, $c_0 := \max_{x\in\mathbb{R}^3} |v(x)|$, M is the number of inhomogeneities, d is the minimal distance from B_m, $m = 1,2,\dots,M$ to the surface P.

A sufficient condition for the validity of the Born approximation, that is, the approximation $u(x,y,k) \sim g(x,y,k)$ for $x,y \in D$, is the smallness of the relative error $\frac{|u(x,y,k)-g(x,y,k)|}{|g(x,y,k)|}$ for $x \in D, y \in P$, which holds if:

$$Mk^2 c_0 \rho^2 := \delta \ll 1. \tag{11.12}$$

One has:

$$\varepsilon^2 = O\Big(\frac{M^2k^4c_0^2\rho^5}{d^2}\Big) = O\Big(\frac{\delta^2\rho}{d^2}\Big) \ll 1,$$

if $\rho \ll d$ and if δ is not small, so that the Born approximation may be not applicable. Note that u in (11.9) has dimension L^{-1}, where L is the length, $v(z)$ is dimensionless, and ε^2 has dimension L^{-1}. In many applications it is natural to assume $\rho \ll d$.

If the Born approximation is not valid, for example, if $c_0\rho^3 = V \neq 0$ as $\rho \to 0$, which is the case of scattering by delta-type inhomogeneities, then the error term ε^2 in formula (11.10) can still be negligible: $\varepsilon^2 = O(M^2k^4c_0V\rho^2/d^2)$, so $\varepsilon^2 \ll 1$ if $M^2k^4V\rho^2/d^2 \ll 1$.

If one understands a sufficient condition for the validity of the Born approximation as the condition which guarantees the smallness of ε^2 for all $x,y \in \mathbb{R}^3$ then condition (11.12) is such a condition. However, if one understands a sufficient condition for the validity of the Born approximation as the condition which guarantees the smallness of ε^2 for x,y running only through the region where the scattered field is measured, in our case when $x,y \in P$, then a much weaker condition (11.11) will suffice.

In the limit $\rho \to 0$ and $c_0\rho^3 = V \neq 0$ formula (11.10) takes the form (11.13), (see [28]). It is shown in [28] (see also [2]) that the resolvent kernel of the Schrödinger operator with the delta-type potential supported on a finite set of points (in our case on the set of points $\tilde{z}_1,\dots,\tilde{z}_M$) has the form

$$u(x, y, k) = g(x, y, k) + k^2 \sum_{m=1}^{M} c_{mm'} g(x, \tilde{z}_m) g(y, \tilde{z}_{m'}) \tag{11.13}$$

where $c_{mm'}$ are some constants. These constants are determined by a selfadjoint realization of the corresponding Schrödinger operator with delta-type potential. There is an M^2-parametric family of such realizations (see [28] for more details).

Although in general the matrix $c_{mm'}$ is not diagonal, under a practically reasonable assumption (11.11) one can neglect the off-diagonal terms of the matrix $c_{mm'}$ and then formula (11.13) reduces practically to the form (11.10) with the term ε^2 neglected.

We have assumed in (11.10) that the point $\overline{z}_m$ exists such that

$$\int_{B_m} g(x, z, k) v(z) g(z, y, k) dz = G(x, y, \overline{z}_m) \overline{v}_m.$$

This is an equation of the type of mean-value theorem. However, such a theorem does not hold, in general, for complex-valued functions. Therefore, if one wishes to have a rigorous derivation, one has to add to the error term ε^2 in (11.10) the error which comes from replacing of the integral $\int_{B_m} g(x, z, k) v(z) g(z, y, k) dz$ in (11.10) by the term $G(x, y, \overline{z}_m)\overline{v}_m$. The error of such an approximation can be easily estimated. We do not give such an estimate, because the basic conclusion that the error term is negligible compared with the main term $k^2 G(x, y, \overline{z}_m)\overline{v}_m$ remains valid under our basic assumption $k\rho \ll 1$. From (11.10) and (11.7) it follows that

$$f_j \approx \sum_{m=1}^{M} G_j(\overline{z}_m)\overline{v}_m, \quad G_j(\overline{z}_m) := G(\xi_j, \overline{z}_m, k). \tag{11.14}$$

Therefore, parameters $\tilde{z}_m$ and $\overline{v}_m$ can be estimated by the least-squares method if one finds the global minimum of the function (11.8):

$$\Phi(z_1, \dots, z_M, v_1, \dots, v_M) = \min. \tag{11.15}$$

Indeed, if one neglects the error of the approximation (11.10), then the function (11.8) is a smooth function of several variables, namely, of $z_1, z_2, \dots, z_M, v_1, v_2, \dots, v_M$, and the global minimum of this function is zero and is attained at the actual intensities $\overline{v}_1, \overline{v}_2, \dots, \overline{v}_M$ and at the values $z_i = \overline{z}_i$, $i = 1, 2, \dots, M$.

This follows from the simple argument: if the error of approximation is neglected, then the approximate equality in (11.14) becomes an exact one. Therefore $f_j - \sum_{m=1}^{M} G_j(\overline{z}_m)\overline{v}_m = 0$, so that function (11.8) equals to zero. Since this function is non-negative by definition, it follows that the values $\overline{z}_m$ and $\overline{v}_m$ are global minimizers of the function (11.8). Therefore we take the global minimizers of function (11.8) as approximate values of the positions and intensities of the small inhomogeneities.

In general we do not know that the global minimizer is unique, and in practice it is often not unique. For the case of one small inhomogeneity ($m = 1$) uniqueness of the global minimizer is proved in [46] for all sufficiently small ρ_m for a problem with a different functional. The problem considered in [46] is the (IP) with $M = 1$, and the functional minimized in [46] is specific for one inhomogeneity.

In Chapter 7 analytical formulas for the scattering matrix are derived for acoustic and electromagnetic scattering problems. An important ingredient of our approach from the numerical point of view is the solution of the global minimization problem (11.14). The theory of global minimization is developed extensively and the literature of this subject is quite large (see [129]). In [36] a numerical implementation of the algorithm presented in Chapter 11 is given.

The problem of detection of small inhomogeneities from boundary measurements in impedance tomography is studied in [4] by a quite different approach, see also [46].

Chapter 12

Modified Rayleigh Conjecture and Applications

12.1 Modified Rayleigh Conjecture and Applications

Modified Rayleigh Conjecture (MRC) in scattering theory is proposed and justified. MRC allows one to develop numerical algorithms for solving direct scattering problems related to acoustic wave scattering by soft and hard obstacles of arbitrary shapes. It gives an error estimate for solving the direct scattering problem. It suggests a numerical method for finding the shape of a star-shaped obstacle from the scattering data. Section 12.1 is based on [116]. A numerical implementation of MRC method is given in Section 12.2 and is based on the paper [119]. Section 12.2.1 is based on the paper [117].

12.1.1 *Introduction*

Consider a bounded domain $D \subset \mathbb{R}^n$, $n = 3$ with a boundary S. The exterior domain is $D' = \mathbb{R}^3 \backslash D$. Assume that S is smooth and star-shaped, that is, its equation can be written as

$$r = f(\alpha), \tag{12.1}$$

where $\alpha \in S^2$ is a unit vector and S^2 denotes the unit sphere in R^3. Smoothness of S is used in (12.18) below. For solving the direct scattering problem by the method described in the beginning of Section 12.2, the boundary S can be Lipschitz. The acoustic wave scattering problem by a soft obstacle D consists in finding the (unique) solution to the problem (12.2)-(12.3):

$$(\nabla^2 + k^2)u = 0 \text{ in } D', \quad u = 0 \text{ on } S, \tag{12.2}$$

$$u = u_0 + v := u_0 + A(\alpha', \alpha)\frac{e^{ikr}}{r} + o\left(\frac{1}{r}\right), \quad r := |x| \to \infty, \quad \alpha' := \frac{x}{r}. \tag{12.3}$$

Here $u_0 := e^{ik\alpha\cdot x}$ is the incident field, v is the scattered field, $A(\alpha', \alpha)$ is called the scattering amplitude, its k-dependence is not shown, and $k > 0$ is the wavenumber. Denote

$$A_\ell(\alpha) := \int_{S^2} A(\alpha', \alpha)\overline{Y_\ell(\alpha')}d\alpha', \tag{12.4}$$

where $Y_\ell(\alpha)$ are the orthonormal spherical harmonics, $Y_\ell = Y_{\ell m}$, $-\ell \leq m \leq \ell$. Let $h_\ell(r)$ be the spherical Hankel functions, normalized so that $h_\ell(kr) \sim \frac{e^{ikr}}{r}$ as $r \to +\infty$. Let the ball $B_R := \{x : |x| \leq R\}$ contain D.

In the region $r > R$ the solution to (12.2)-(12.3) is:

$$u(x, \alpha) = e^{ik\alpha\cdot x} + \sum_{\ell=0}^{\infty} A_\ell(\alpha)\psi_\ell, \quad \psi_\ell := Y_\ell(\alpha')h_\ell(kr), \quad r > R, \quad \alpha' = \frac{x}{r}, \tag{12.5}$$

summation includes summation with respect to m, $-\ell \leq m \leq \ell$, and $A_\ell(\alpha)$ are defined in (12.4).

Rayleigh conjecture (RC): the series (12.5) converges up to the boundary S (originally RC dealt with periodic structures, gratings). This conjecture is wrong for many domains, although it holds for some, for example, for a ball (see [5], [133], [120]). If $n = 2$ and D is an ellipse, then the series analogous to (12.5) converges in the region $r > a$, where $2a$ is the distance between the foci of the ellipse [5]. In the engineering literature there are numerical algorithms, based on the Rayleigh conjecture. Our aim is to give a formulation of a modified Rayleigh conjecture (MRC) which is correct and can be used in numerical solution of the direct and inverse scattering problems. We discuss the Dirichlet condition but similar argument is applicable to the Neumann boundary condition, corresponding to acoustically hard obstacles.

Fix $\epsilon > 0$, an arbitrary small number.

Lemma 12.1 *There exist $L = L(\epsilon)$ and $c_\ell = c_\ell(\epsilon)$ such that*

$$\left\| u_0 + \sum_{\ell=0}^{L(\epsilon)} c_\ell(\epsilon)\psi_\ell \right\|_{L^2(S)} \leq \epsilon. \tag{12.6}$$

If (12.6) and the boundary condition (12.2) hold, then

$$\|v_\epsilon - v\|_{L^2(S)} \le \epsilon, \quad v_\epsilon := \sum_{\ell=0}^{L(\epsilon)} c_\ell(\epsilon)\psi_\ell. \tag{12.7}$$

Lemma 12.2 *If* (12.7) *holds then*

$$\|v_\epsilon - v\| = O(\epsilon) \quad \epsilon \to 0, \tag{12.8}$$

where $\|\cdot\| := \|\cdot\|_{H^m_{loc}(D')} + \|\cdot\|_{L^2(D';(1+|x|)^{-\gamma})}$, $\gamma > 1$, $m > 0$ *is an arbitrary integer, and* H^m *is the Sobolev space.*

In particular, (12.8) implies

$$\|v_\epsilon - v\|_{L^2(S_R)} = O(\epsilon) \quad \epsilon \to 0. \tag{12.9}$$

Lemma 12.3 *One has:*

$$c_\ell(\epsilon) \to A_\ell(\alpha)\,\forall\ell, \quad \epsilon \to 0. \tag{12.10}$$

The modified Rayleigh conjecture (MRC) is formulated as a theorem, which follows from the above three lemmas:

Theorem 12.1 *(**MRC**): For an arbitrary small* $\epsilon > 0$ *there exist* $L(\epsilon)$ *and* $c_\ell(\epsilon), 0 \le \ell \le L(\epsilon)$, *such that* (12.6), (12.8) *and* (12.10) *hold.*

The difference between RC and MRC is: (12.7) does not hold if one replaces v_ϵ by $\sum_{\ell=0}^{L} A_\ell(\alpha)\psi_\ell$, and let $L \to \infty$ (instead of letting $\epsilon \to 0$).

For the Neumann boundary condition one minimizes the function $\|\frac{\partial[u_0+\sum_{\ell=0}^{L} c_\ell\psi_\ell]}{\partial N}\|_{L^2(S)}$ with respect to c_ℓ. Analogs of Lemmas 12.1–12.3 are valid and their proofs are essentially the same.

In Section 12.1.2 we discuss the usage of MRC in solving the direct scattering problem, in Section 12.1.3 its usage in solving the inverse scattering problem, and in Section 12.1.4 proofs are given.

12.1.2 *Direct scattering problem and MRC*

The direct problem consists in finding the scattered field v given S and u_0. To solve it using MRC, fix a small $\epsilon > 0$ and find $L(\epsilon)$ and $c_\ell(\epsilon)$ such that (12.6) holds. This is possible by Lemma 12.1 and can be done numerically by minimizing $\|u_0 + \sum_0^L c_\ell\psi_\ell\|_{L^2(S)} := \phi(c_1, \dots, c_L)$. If the minimum of ϕ is larger than ϵ, then increase L and repeat the minimization. Lemma 12.1 guarantees the existence of such L and c_ℓ that the minimum is less than ϵ. Choose the smallest L for which this happens and define

$v_\epsilon := \sum_{\ell=0}^{L} c_\ell \psi_\ell(x)$. Then v_ϵ is the approximate solution to the direct scattering problem with the accuracy $O(\epsilon)$ in the norm $||\cdot||$ by Lemma 12.2

In [106] representations of v and v_ϵ are proposed, which greatly simplified minimization of ϕ. Namely, let Ψ_ℓ solve problem

$$(\nabla^2 + k^2)\Psi_\ell = 0 \text{ in } D', \quad \Psi_\ell = f_\ell \text{ on } S, \tag{12.11}$$

and Ψ_ℓ satisfies the radiation condition. Here $\{f_\ell\}_{\ell \geq 0}$ is an arbitrary orthonormal basis of $L^2(S)$. Denote

$$v(x) := \sum_{\ell=0}^{\infty} c_\ell \Psi_\ell(x), \quad u(x) := u_0 + v(x), \quad c_l := \big(-u_0, f_\ell\big)_{L^2(S)}. \tag{12.12}$$

The series (12.12) on S is a Fourier series which converges in $L^2(S)$. It converges pointwise in D' by the argument given in the proof of Lemma 12.2. A possible choice of f_ℓ for star-shaped S is $f_\ell = Y_\ell/\sqrt{w}$ where $w := dS/d\alpha$. Here dS and $d\alpha$ are respectively the elements of the surface areas of the surface S and of the unit sphere S^2.

12.1.3 *Inverse scattering problem and MRC*

Inverse obstacle scattering problems (IOSPa) and (IOSPb) consist of finding S and the boundary condition on S from the knowledge of:

(IOSPa): the scattering data $A(\alpha', \alpha, k_0)$ for all α', $\alpha \in S^2$, $k = k_0 > 0$ being fixed,

or,

(IOSPb): $A(\alpha', \alpha_0, k)$, known for all $\alpha' \in S^2$ and all $k > 0$, $\alpha = \alpha_0 \in S^2$ being fixed.

Uniqueness of the solution to (IOSPa) is proved by the author (1985) for the Dirichlet, Neumann and Robin boundary conditions, and of (IOSPb) by M.Schiffer (1964), who assumed a priori the Dirichlet boundary condition. The proofs are given in [133], [107]. The author has also proved that not only S but the boundary condition as well is uniquely defined by the above data in both cases, and gave stability estimates for the solution to IOSP [133]. Later he gave a different method of proof of the uniqueness theorems for these problems which covered the rough boundaries (Lipschitz and much rougher boundaries: the ones with finite perimeter [144], see also [104]. In [89] the uniqueness theorem for the solution of inverse scattering problem is proved for a wide class of transmission problems. It is proved that not only the discontinuity surfaces of the refraction coefficient but also the coefficient itself inside the body and the boundary conditions across these surfaces are uniquely determined by the fixed-frequency scattering data. For any strictly convex, smooth, reflecting obstacle D analytical formulas for finding S from the high-frequency asymptotics of the scattering amplitude are proposed by he author, who gave error estimates of his inversion formula also [120]. The uniqueness theorems in the above references hold if the scattering data are given not for all α', $\alpha \in S^2$, but only for α' and α in arbitrary small solid angles, i.e., in arbitrary small open subsets of S^2. The inverse scattering problem with the data $\alpha' \in S^2$, $k = k_0$ and $\alpha = \alpha_0$ being fixed, is open. If a priori one knows that D is sufficiently small, so that $k_0 > 0$ is not a Dirichlet eigenvalue of the laplacian in D, then uniqueness of the solution with the above non-overdetermined data holds (by the usual argument [133]). There are many parameter-fitting schemes for solving IOSP (see [120], [107]).

Let us describe a scheme, based on MRC. Suppose that the scattered field v is observed on a sphere S_R. Calculate $c_\ell := (v, Y_\ell)_{L^2(S^2)}/h_\ell(kR)$. If v is known exactly, then $c_\ell = A_\ell(\alpha)$. If v_δ are noisy data, $\|v - v_\delta\|_{L^2(S_R)} \leq \delta$, then $c_\ell = c_{\ell\delta}$. Choose some L, say $L = 5$, and find $r = r(\alpha')$ as a positive root of the equation $u_0 + v_L := e^{ik\alpha\cdot\alpha' r} + \sum_{\ell=0}^{L} c_{\ell\delta}\psi_\ell(kr, \alpha') := p(r, \alpha', \alpha, k) = 0$. Here α' and $k > 0$ are fixed, and we are looking for the root $r = r(\alpha')$ which is positive and is stable under changes of k and α. In practice equation $p(r, \alpha', \alpha, k) = 0$ may have no such root, the root may have small imaginary part. If for the chosen L such a root (that is, a root which is positive, or has a small imaginary part, and stable with respect to changes of k and α) is not found, then increase L, and/or decrease L, and repeat the search of the root. Stop the search at a smallest L for which such a root is found. The MRC justifies this method: for a suitable L the function $p(r, \alpha', \alpha, k)$ is approximately equals zero on S, that is, for $r = r(\alpha')$, and this $r(\alpha')$ does not depend on k and α. Moreover, by the uniqueness theorem for (IOSPa) and (IOSPb) there is only one such $r = r(\alpha')$. Numerically one expects to find a root of the equation $p(r, \alpha', k) = 0$ which is close to positive semiaxis $r > 0$ and stable with respect to changes of k and α.

If one uses the above scheme for solving the inverse scattering problem for an acoustically hard body (the Neumann boundary condition on S), then one gets not a transcendental equation $p(r, \alpha', \alpha, k) = 0$ for finding the equation of S, $r = r(\alpha')$, but a differential equation for $r = r(\alpha')$, which comes from the equation $\frac{\partial p(r,\alpha',\alpha,k)}{\partial N} = 0$ at $r = r(\alpha')$. One has to write the normal derivative on S in spherical coordinates and then substitute

$r = r(\alpha')$ into the result to get a differential equation for the unknown function $r = r(\alpha')$. For example, if $n = 2$ (the two-dimensional case), then the role of α' plays the polar angle φ' and the equation for $r = r(\varphi')$ takes the form $\frac{dr}{d\varphi'} = (r^2 \frac{dp}{dr} / \frac{dp}{d\varphi'})|_{r=r(\varphi')}$, and $r(\varphi') = r(\varphi' + 2\pi)$.

12.1.4 *Proofs*

Proof of Lemma 12.1. This Lemma follows from the results in [133], (p.162, Lemma 1). □

Proof of Lemma 12.2. By Green's formula one has

$$v_\epsilon(x) = \int_S v_\epsilon(s) G_N(x, s) ds, \quad \|v_\epsilon(s) + u_0\|_{L^2(S)} < \epsilon, \tag{12.13}$$

where N is the unit normal to S, pointing into D', and G is the Dirichlet Green's function of the Laplacian in D':

$$\left(\nabla^2 + k^2\right) G = -\delta(x - y) \text{ in } D', \quad G = 0 \text{ on } S, \tag{12.14}$$

$$\lim_{r\to\infty} \int_{|x|=r} \left| \frac{\partial G}{\partial |x|} - ikG \right|^2 ds = 0 \tag{12.15}$$

From (12.13) one gets (12.8) with $H^m_{loc}(D')$-norm immediately by the Cauchy inequality, and with the weighted norm from the estimate

$$|G_N(x, s)| \le \frac{c}{1 + |x|}, \quad |x| \ge R, \tag{12.16}$$

and from local elliptic estimates for $w_\epsilon := v_\epsilon - v$, which imply that

$$\|w_\epsilon\|_{L^2(B_R \setminus D)} \le c\epsilon. \tag{12.17}$$

Let us recall the elliptic estimate we use. Let $D'_R := B_R \setminus D$, S_R be the boundary of B_R, and choose R such that k^2 is not a Dirichlet eigenvalue of $-\Delta$ in D'_R. The elliptic estimate we have used is ([62], p.189):

$$\|w_\epsilon\|_{H^m(D'_R)} \le c\Big[\|(\Delta + k^2) w_\epsilon\|_{\mathcal{H}^{m-2}(D'_R)} + \|w_\epsilon\|_{H^{m-0.5}(S_R)} + \|w_\epsilon\|_{H^{m-0.5}(S)}\Big]. \tag{12.18}$$

Take $m = 0.5$ in (12.18), use the equation $(\Delta + k^2) w_\epsilon = 0$ in D', the estimate $\|w_\epsilon\|_{H^m(S_R)} = O(\epsilon)$, proved above, the estimate $\|w_\epsilon\|_{H^0(S)} = O(\epsilon)$, and get (12.8). For $m = 0.5$ the space in the first term on the right-hand side of formula (4.6) differs from the usual Sobolev space $H^{m-2}(D'_R)$ (cf [62], p.189), but $(\Delta + k^2)w_\epsilon = 0$, so this term vanishes anyway.

Lemma 12.2 is proved. □

Proof of Lemma 12.3. Lemma 12.2 yields convergence of v_ϵ to v in the norm $||\cdot||$. In particular, $||v_\epsilon - v||_{L^2(S_R)} \to 0$ as $\epsilon \to 0$. On S_R one has $v = \sum_{\ell=0}^{\infty} A_\ell(\alpha)\psi_\ell$ and $v_\epsilon = \sum_{\ell=0}^{L(\epsilon)} c_\ell \psi_\ell$. Multiply $v_\epsilon(R,\alpha') - v(R,\alpha')$ by $\overline{Y_\ell(\alpha')}$, integrate over S^2 and then let $\epsilon \to 0$. The result is (12.10). □

12.2 Modified Rayleigh Conjecture Method for Multidimensional Obstacle Scattering Problems

The Rayleigh conjecture on the representation of the scattered field in the exterior of an obstacle D is widely used in applications. However this conjecture is false for some obstacles. In this section, based on [119], [34], numerical algorithms, based on the MRC, are implemented for various 2D and 3D obstacle scattering problems. The 3D obstacles include a cube and an ellipsoid. The MRC method is easy to implement for both simple and complex geometries. It is shown to be a viable alternative for other obstacle scattering methods.

12.2.1 *Introduction*

The basic theoretical foundation of the MRC method was developed in [116]. The MRC has the appeal of an easy implementation for obstacles of complicated geometry, e.g. having edges and corners. In the numerical experiments ([34], [119], [37], [125]) the method is proved to be a competitive alternative to the BIEM (boundary integral equations method). Unlike the BIEM, the MRC-based algorithm can be applied to different obstacles with very little additional effort. In this Section we describe, following [119], Random Multi-point MRC implementation, which made it possible to successfully solve numerically some 3D obstacle scattering problems. Different implementations of MRC method are given in [125] and [37].

Earlier, in [34] the Multi-point MRC method was used for 2D obstacles of a relatively simple geometry. In this Section an implementaion of MRC method for 3D problems is proposed, and an improvement of our earlier results is obtained.

We formulate the obstacle scattering problem in a 3D setting with the Dirichlet boundary condition, but the MRC method can also be used for the Neumann and Robin boundary conditions.

Consider a bounded domain $D \subset \mathbb{R}^3$, with a boundary S which is assumed to be Lipschitz continuous. Denote the exterior domain by $D' = \mathbb{R}^3 \backslash D$. Let $\alpha, \alpha' \in S^2$ be unit vectors, and S^2 be the unit sphere in $\mathbb{R}^3$.

The acoustic wave scattering problem by a soft obstacle D consists in finding the (unique) solution to the problem (12.2)-(12.3).

Informally, the Random Multi-point MRC algorithm can be described as follows.

Fix a $J > 0$. Let $x_j, j = 1, 2, \dots, J$ be a batch of points randomly chosen inside the obstacle D. For $x \in D'$, let

$$\alpha' = \frac{x - x_j}{|x - x_j|}, \quad \psi_\ell(x, x_j) = Y_\ell(\alpha') h_\ell\Big(k\big|x - x_j\big|\Big). \tag{12.19}$$

Let $g(x) = u_0(x)$, $x \in S$, and minimize the discrepancy

$$\Phi(\mathbf{c}) = \left\| g(x) + \sum_{j=1}^{J} \sum_{\ell=0}^{L} c_{\ell,j} \psi_\ell(x, x_j) \right\|_{L^2(S)}, \tag{12.20}$$

over $\mathbf{c} \in \mathbb{C}^N$, where $\mathbf{c} = \{c_{\ell,j}\}$. That is, the total field $u = g(x) + v$ is desired to be as close to zero as possible at the boundary S, to satisfy the required condition for the soft scattering. If the resulting residual $r^{min} = \min \Phi$ is smaller than the prescribed tolerance ϵ, than the procedure is finished, and the sought scattered field is

$$v_\epsilon(x) = \sum_{j=1}^{J} \sum_{\ell=0}^{L} c_{\ell,j} \psi_\ell(x, x_j), \quad x \in D'.$$

If, on the other hand, the residual $r^{min} > \epsilon$, than we continue by trying to improve on the already obtained fit in (12.23). Adjust the field on the boundary by letting $g(x) := g(x) + v_\epsilon(x)$, $x \in S$. Create another batch of J points randomly chosen in the interior of D, and minimize (12.20) with this new $g(x)$. Continue with the iterations until the required tolerance ϵ on the boundary S is attained, at the same time keeping the track of the changing field v_ϵ.

The minimization in (12.20) is always done over the same number of points J. However, the points x_j are sought to be different in each iteration to assure that the minimal values of Φ are decreasing in consequent iterations. Thus, computationally, the size of the minimization problem remains the same. This is the new feature of the Random multi-point

MRC method, which allows it to solve scattering problems untreatable by previously developed in [34] MRC methods.

Below is the description of the algorithm.

Random Multi-point MRC

For $x_j \in D$, and $\ell \geq 0$ functions $\psi_\ell(x, x_j)$ are defined as in (12.22).

(1) **Initialization.** Fix $\epsilon > 0,\ L \geq 0,\ J > 0,\ N_{max} > 0$. Let $n = 0$, $v_\epsilon = 0$ and $g(x) = u_0(x),\ x \in S$.

(2) **Iteration.**

(a) Let $n := n + 1$. Randomly choose J points $x_j \in D$, $j = 1, 2, \ldots, J$.

(b) Minimize

$$\Phi(\mathbf{c}) = \left\| g(x) + \sum_{j=1}^{J} \sum_{\ell=0}^{L} c_{\ell,j} \psi_\ell(x, x_j) \right\|_{L^2(S)}$$

over $\mathbf{c} \in \mathbb{C}^N$, where $\mathbf{c} = \{c_{\ell,j}\}$.

Let the minimal value of Φ be r^{min}.

(c) Let

$$v_\epsilon(x) := v_\epsilon(x) + \sum_{j=1}^{J} \sum_{\ell=0}^{L} c_{\ell,j} \psi_\ell(x, x_j), \quad x \in D'.$$

(3) **Stopping criterion.**

(a) If $r^{min} \leq \epsilon$, then stop.

(b) If $r^{min} > \epsilon$, and $n \neq N_{max}$, let

$$g(x) := g(x) + \sum_{j=1}^{J} \sum_{\ell=0}^{L} c_{\ell,j} \psi_\ell(x, x_j), \quad x \in S$$

and repeat the iterative step (2).

(c) If $r^{min} > \epsilon$, and $n = N_{max}$, then the procedure failed.

Direct cattering problems and the Rayleigh conjecture

Let a ball $B_R := \{x : |x| \leq R\}$ contain the obstacle D. In the region $r > R$ the scattering solution is:

$$u(x,\alpha) = e^{ik\alpha\cdot x} + \sum_{\ell=0}^{\infty} A_\ell(\alpha)\psi_\ell, \quad \psi_\ell := Y_\ell(\alpha')h_\ell(kr), \quad r > R, \quad \alpha' = \frac{x}{r}, \tag{12.21}$$

where the sum includes the summation with respect to m, $-\ell \leq m \leq \ell$, and $A_\ell(\alpha)$ are defined in (12.4).

*T*he Rayleigh conjecture (RC) is: the series (12.21) converges up to the boundary S (originally RC dealt with periodic structures, gratings). This conjecture is false for many obstacles, but is true for some ([5], [67], [133]). For example, if $n = 2$ and D is an ellipse, then the series analogous to (12.21) converges in the region $r > a$, where $2a$ is the distance between the foci of the ellipse [5]. In the engineering literature there are numerical algorithms, based on the Rayleigh conjecture. Our aim is to give a formulation of a *Modified Rayleigh Conjecture* (MRC) which holds for any Lipschitz obstacle and can be used in numerical solution of the direct and inverse scattering problems. We discuss the Dirichlet condition but similar argument is applicable to the Neumann boundary condition, corresponding to acoustically hard obstacles.

The difference between RC and MRC is: (12.7) does not hold if one replaces v_ϵ by $\sum_{\ell=0}^{L} A_\ell(\alpha)\psi_\ell$, and lets $L \to \infty$ (instead of letting $\epsilon \to 0$). Indeed, the series $\sum_{\ell=0}^{\infty} A_\ell(\alpha)\psi_\ell$ diverges at some points of the boundary for many obstacles. Note also that the coefficients in (12.7) depend on ϵ, so (12.7) is *not* a partial sum of a series.

For the Neumann boundary condition one minimizes

$$\left\| \frac{\partial[u_0 + \sum_{\ell=0}^{L} c_\ell\psi_\ell]}{\partial N} \right\|_{L^2(S)}$$

with respect to c_ℓ. Analogs of Lemmas 12.1–12.3 are valid and their proofs are essentially the same.

See [118] for an extension of these results to scattering by periodic structures.

12.2.2 *Numerical Experiments*

In this section we desribe numerical results obtained by the Random Multi-point MRC method for 2D and 3D obstacles. We also compare the 2D results to the ones obtained by our earlier method introduced in [34]. The method that we used previously can be described as a Multi-point MRC. Its difference from the Random Multi-point MRC method is twofold: It is just the first iteration of the Random method, and the interior points x_j, $j = 1, 2, \ldots, J$ were chosen deterministically, by an *ad hoc* method according to the geometry of the obstacle D. The number of points J was limited by the size of the resulting numerical minimization problem, so the accuracy of the scattering solution (i.e. the residual r^{min}) could not be made small for many obstacles. The method was not capable of treating 3D obstacles. These limitations were removed by using the Random Multi-point MRC method. As we mentioned previously, [34] contains a favorable comparison of the Multi-point MRC method with the BIEM, in spite of the fact that the numerical implementation of the MRC method there is considerably less efficient than the one presented in this paper.

A numerical implementation of the Random Multi-point MRC method follows the same outline as for the Multi-point MRC, which was described in [34]. In 2D case one has:

$$\psi_l(x, x_j) = H_l^{(1)}\Big(k|x - x_j|\Big)e^{il\theta_j},$$

where $(x - x_j)/|x - x_j| = e^{i\theta_j}$.

For a numerical implementation choose M nodes $\{t_m\}$ on the surface S of the obstacle D. After the interior points x_j, $j = 1, 2, \ldots, J$ are chosen, form N vectors

$$\mathbf{a}^{(n)} = \Big\{\psi_l(t_m, x_j)\Big\}_{m=1}^{M},$$

$n = 1, 2, \ldots, N$ of length M. Note that $N = (2L + 1)J$ for a 2D case, and $N = (L+1)^2 J$ for a 3D case. It is convenient to normalize the norm in $\mathbb{R}^M$ by

$$\|\mathbf{b}\|^2 = \frac{1}{M}\sum_{m=1}^{M} \big|b_m\big|^2, \quad \mathbf{b} = \big(b_1, b_2, \ldots, b_M\big).$$

Then $\|u_0\| = 1$.

Now let $\mathbf{b} = \{g(t_m)\}_{m=1}^{M}$, in the Random Multi-point MRC (see section 1), and minimize

$$\Phi(\mathbf{c}) = \|\mathbf{b} + A\mathbf{c}\|, \tag{12.22}$$

for $\mathbf{c} \in \mathbb{C}^N$, where A is the matrix containing vectors $\mathbf{a}^{(n)}$, $n = 1, 2, \ldots, N$ as its columns.

We used the Singular Value Decomposition (SVD) method (see e.g. [80]) to minimize (12.22). Small singular values $s_n < w_{min}$ of the matrix A are used to identify and delete linearly dependent or almost linearly dependent combinations of vectors $\mathbf{a}^{(n)}$. This spectral cut-off makes the minimization process stable, see the details in [34].

Let r^{min} be the residual, i.e. the minimal value of $\Phi(\mathbf{c})$ attained after N_{max} iterations of the Random Multi-point MRC method (or when it is stopped). For a comparison, let r^{min}_{old} be the residual obtained in [34] by an earlier method.

We conducted 2D numerical experiments for four obstacles: two ellipses of different eccentricity, a kite, and a triangle. The $M = 720$ nodes t_m were uniformly distributed on the interval $[0, 2\pi]$, used to parametrize the boundary S. Each case was tested for wave numbers $k = 1.0$ and $k = 5.0$. Each obstacle was subjected to incident waves corresponding to $\alpha = (1.0, 0.0)$ and $\alpha = (0.0, 1.0)$.

The results for the Random Multi-point MRC with $J = 1$ are shown in the first Table, in the last column r^{min}. In every experiment the target residual $\epsilon = 0.0001$ was obtained in under 6000 iterations, in about 2 minutes run time on a 2.8 MHz PC.

In [34], we conducted numerical experiments for the same four 2D obstacles by a Multi-point MRC, as described in the beginning of this section. The interior points x_j were chosen differently in each experiment. Their choice is indicated in the description of each 2D experiment. The column J shows the number of these interior points. Values $L = 5$ and $M = 720$ were used in all the experiments. These results are shown in the first Table, column r^{min}_{old}.

Thus, the Random Multi-point MRC method achieved a significant improvement over the earlier Multi-point MRC.

Experiment 2D-I. The boundary S is an ellipse described by

$$\mathbf{r}(t) = (2.0 \cos t,\ \sin t), \quad 0 \le t < 2\pi\,. \tag{12.23}$$

Normalized residuals attained in the numerical experiments for 2D obstacles, $\|\mathbf{u_0}\| = 1$.

Experiment	J	k	α	r_{old}^{min}	r^{min}
I	4	1.0	$(1.0, 0.0)$	0.000201	0.0001
	4	1.0	$(0.0, 1.0)$	0.000357	0.0001
	4	5.0	$(1.0, 0.0)$	0.001309	0.0001
	4	5.0	$(0.0, 1.0)$	0.007228	0.0001
II	16	1.0	$(1.0, 0.0)$	0.003555	0.0001
	16	1.0	$(0.0, 1.0)$	0.002169	0.0001
	16	5.0	$(1.0, 0.0)$	0.009673	0.0001
	16	5.0	$(0.0, 1.0)$	0.007291	0.0001
III	16	1.0	$(1.0, 0.0)$	0.008281	0.0001
	16	1.0	$(0.0, 1.0)$	0.007523	0.0001
	16	5.0	$(1.0, 0.0)$	0.021571	0.0001
	16	5.0	$(0.0, 1.0)$	0.024360	0.0001
IV	32	1.0	$(1.0, 0.0)$	0.006610	0.0001
	32	1.0	$(0.0, 1.0)$	0.006785	0.0001
	32	5.0	$(1.0, 0.0)$	0.034027	0.0001
	32	5.0	$(0.0, 1.0)$	0.040129	0.0001

The Multi-point MRC used $J = 4$ interior points $x_j = 0.7\mathbf{r}(\frac{\pi(j-1)}{2})$, $j = 1, \ldots, 4$. Run time was 2 seconds.

Experiment 2D-II. The kite-shaped boundary S (see [17], Section 3.5) is described by

$$\mathbf{r}(t) = (-0.65 + \cos t + 0.65\cos 2t,\ 1.5\sin t), \quad 0 \le t < 2\pi\,. \tag{12.24}$$

The Multi-point MRC used $J = 16$ interior points $x_j = 0.9\mathbf{r}(\frac{\pi(j-1)}{8})$, $j = 1, \ldots, 16$. Run time was 33 seconds.

Experiment 2D-III. The boundary S is the triangle with vertices at $(-1.0, 0.0)$ and $(1.0, \pm 1.0)$. The Multi-point MRC used the interior points $x_j = 0.9\mathbf{r}(\frac{\pi(j-1)}{8})$, $j = 1, \ldots, 16$. Run time was about 30 seconds.

Experiment 2D-IV. The boundary S is an ellipse described by

$$\mathbf{r}(t) = (0.1\cos t,\ \sin t), \quad 0 \le t < 2\pi\,. \tag{12.25}$$

The Multi-point MRC used $J = 32$ interior points $x_j = 0.95\mathbf{r}(\frac{\pi(j-1)}{16})$, $j = 1, \ldots, 32$. Run time was about 140 seconds.

The 3D numerical experiments were conducted for 3 obstacles: a sphere, a cube, and an ellipsoid. We used the Random Multi-point MRC with $L = 0$, $w_{min} = 10^{-12}$, and $J = 80$. The number M of the points on the boundary S is indicated in the description of the obstacles. The scattered field for each obstacle was computed for two incoming directions $\alpha_i = (\theta, \phi)$, $i = 1, 2$, where ϕ was the polar angle. The first unit vector α_1 is denoted by (1) in the second Table, $\alpha_1 = (0.0, \pi/2)$. The second one is denoted by (2), $\alpha_2 = (\pi/2, \pi/4)$. A typical number of iterations N_{iter} and the run time on a 2.8 MHz PC are also shown in the second Table. For example, in experiment I with $k = 5.0$ it took about 700 iterations of the Random Multi-point MRC method to achieve the target residual $r^{min} = 0.001$ in 7 minutes.

Experiment 3D-I. The boundary S is the sphere of radius 1, with $M = 450$.

Experiment 3D-II. The boundary S is the surface of the cube $[-1, 1]^3$ with $M = 1350$.

Experiment 3D-III. The boundary S is the surface of the ellipsoid $x^2/16 + y^2 + z^2 = 1$ with $M = 450$.

Normalized residuals attained in the numerical experiments for 3D obstacles, $\|\mathbf{u_0}\| = 1$.

Experiment	k	α_i	r^{min}	N_{iter}	run time
I	1.0		0.0002	1	1 sec
	5.0		0.001	700	7 min
II	1.0	(1)	0.001	800	16 min
	1.0	(2)	0.001	200	4 min
	5.0	(1)	0.0035	2000	40 min
	5.0	(2)	0.002	2000	40 min
III	1.0	(1)	0.001	3600	37 min
	1.0	(2)	0.001	3000	31 min
	5.0	(1)	0.0026	5000	53 min
	5.0	(2)	0.001	5000	53 min

In the last experiment the run time could be reduced by taking a smaller value for J. For example, the choice of $J = 8$ reduced the running time to about 6-10 minutes.

Numerical experiments show that the minimization results depend on the choice of such parameters as J, w_{min}, and L. They also depend on the choice of the interior points x_j. It is possible that further versions of the MRC could be made more efficient by finding a more efficient rule for their placement. Numerical experiments in [34] showed that the efficiency of the minimization greatly depended on the deterministic placement of the interior points, with better results obtained for these points placed sufficiently close to the boundary S of the obstacle D, but not very close to it. The current choice of a random placement of the interior points x_j reduced the variance in the obtained results, and eliminated the need to provide a justified algorithm for their placement. The random choice of these points distributes them in the entire interior of the obstacle, rather than in a subset of it. In [37] an optimal (non-random) choice of these points is proposed and implemented numerically.

12.2.3 *Conclusions*

For a 2D, or 3D obstacle, Rayleigh conjectured that the acoustic field u in the exterior of the obstacle is given by

$$u(x,\alpha) = e^{ik\alpha\cdot x} + \sum_{\ell=0}^{\infty} A_\ell(\alpha)\psi_\ell, \quad \psi_\ell := Y_\ell(\alpha')h_\ell(kr), \quad \alpha' = \frac{x}{r}. \tag{12.26}$$

This conjecture, called the Rayleigh hypothesis or Rayleigh Conjecture (RC), is false for many obstacles, but holds for some. The Modified Rayleigh Conjecture (MRC) is Theorem 12.1, which is a basis for efficient algorithms for solving obstacle scattering problems.

The author thinks that MRC-based algorithms are more efficient than the ones currently used for solving obstacle scattering problems, such as boundary integral equations methods, for example.

Further numerical evidence which testifies that the MRC-based algorithms for solving obstacle scattering problems are efficient, one can find in [37] and [125].

In the next section the MRC-based algorithm is described for static problems.

12.3 Modified Rayleigh Conjecture for Static Fields

Consider a bounded domain $D \subset \mathbb{R}^n$, $n = 3$ with a boundary S. The exterior domain is $D' = \mathbb{R}^3 \backslash D$. Assume that S is Lipschitz. Let S^2 denotes the unit sphere in $\mathbb{R}^3$. Consider the problem:

$$\nabla^2 v = 0 \text{ in } D', \quad v = f \text{ on } S, \tag{12.27}$$

$$v := O\left(\frac{1}{r}\right) \quad r := |x| \to \infty. \tag{12.28}$$

Let $\frac{x}{r} := \alpha \in S^2$. Denote by $Y_\ell(\alpha)$ the orthonormal spherical harmonics, $Y_\ell = Y_{\ell m}, -\ell \leq m \leq \ell$. Let $H_\ell := \frac{Y_\ell(\alpha)}{r^{\ell+1}}$, $\ell \geq 0$, be harmonic functions in D'. Let the ball $B_R := \{x : |x| \leq R\}$ contain D.

In the region $r > R$ the solution to (12.27) - (12.28) is:

$$v(x) = \sum_{\ell=0}^{\infty} c_\ell H_\ell, \qquad r > R, \tag{12.29}$$

the summation in (12.29) and below includes summation with respect to m, $-\ell \leq m \leq \ell$, and c_ℓ are some coefficients determined by f.

In general, the series (12.29) does not converge up to the boundary S. Our aim is to give a formulation of an analog of the Modified Rayleigh Conjecture (MRC) from Section 12.1, which can be used in numerical solution of the static boundary-value problems. The author hopes that the MRC method for static problems can be used as a basis for an efficient numerical algorithm for solving boundary-value problems for Laplace equations in domains with complicated boundaries. In Section 12.2 such an algorithm is developed on the basis of MRC for solving boundary-value problems for the Helmholtz equation. Although the boundary integral equation methods and finite elements methods are widely and successfully used for solving these problems, the method, based on MRC, proved to be competitive and often superior to the currently used methods.

We discuss the Dirichlet condition but a similar argument is applicable to the Neumann and Robin boundary conditions. Boundary-value problems and scattering problems in rough domains were studied in Chapter 9.

Let us present the basic results on which the MRC method is based.

Fix $\epsilon > 0$, an arbitrary small number.

Lemma 12.4 *There exist $L = L(\epsilon)$ and $c_\ell = c_\ell(\epsilon)$ such that*

$$||\sum_{\ell=0}^{L(\epsilon)} c_\ell(\epsilon)H_\ell - f||_{L^2(S)} \leq \epsilon. \tag{12.30}$$

If (12.30) and the boundary condition (12.27) hold, then

$$||v_\epsilon - v||_{L^2(S)} \leq \epsilon, \quad v_\epsilon := \sum_{\ell=0}^{L(\epsilon)} c_\ell(\epsilon)H_\ell. \tag{12.31}$$

Lemma 12.5 *If (12.30) holds then*

$$||v_\epsilon - v|| = O(\epsilon) \quad \epsilon \to 0, \tag{12.32}$$

where $||\cdot|| := ||\cdot||_{H^m_{loc}(D')} + ||\cdot||_{L^2(D';(1+|x|)^{-\gamma})}$, $\gamma > 1$, $m > 0$ is an arbitrary integer, and H^m is the Sobolev space.

In particular, (12.32) implies

$$||v_\epsilon - v||_{L^2(S_R)} = O(\epsilon) \quad \epsilon \to 0. \tag{12.33}$$

One can prove similarly to [133], p.41, that if v satisfies (12.28)-(12.29), then $\max_{r \geq R} ||v||_{L^2(S_r)} = ||v||_{L^2(S_R)}$, where $S_r := \{x : |x| = r\}$ is the sphere of radius $r \geq R$. This is an analog of the "integral" maximum principle, which was first established in [133], p.41, for the solutions to Helmholtz equation, for which the "pointwise" maximum principle is not valid.

Let us formulate an analog of the Modified Rayleigh Conjecture (MRC):

Theorem 12.2 (MRC): *For an arbitrary small $\epsilon > 0$ there exist $L(\epsilon)$ and $c_\ell(\epsilon), 0 \leq \ell \leq L(\epsilon)$, such that (12.30) and (12.33) hold.*

Theorem 12.2 follows from Lemmas 12.4 and 12.5

For the Neumann boundary condition one minimizes $||\frac{\partial[\sum_{\ell=0}^{L} c_\ell H_\ell]}{\partial N} - f||_{L^2(S)}$ with respect to c_ℓ. Analogs of Lemmas 1.1-1.2 are valid and their proofs are essentially the same.

If the boundary data $f \in C(S)$, then one can use $C(S)-$ norm in (12.30)-(12.33), and an analog of Theorem 12.2 then follows immediately from the maximum principle.

Below we discuss the MRC method for solving static boundary-value problems and give proofs of the basic results.

12.3.1 *Solving boundary-value problems by MRC*

To solve problem (12.27)-(12.28) using MRC, fix a small $\epsilon > 0$ and find $L(\epsilon)$ and $c_\ell(\epsilon)$ such that (12.30) holds. This is possible by Lemma 12.4 and can be done numerically by minimizing $||\sum_0^L c_\ell H_\ell - f||_{L^2(S)} := \phi(c_1, ..., c_L)$. If the minimum of ϕ is larger than ϵ, then increase L and repeat the minimization. Lemma 12.4 guarantees the existence of such L and c_ℓ that the minimum is less than ϵ. Choose the smallest L for which this happens and define $v_\epsilon := \sum_{\ell=0}^L c_\ell H_\ell$. Then, by Lemma 12.5, v_ϵ is the approximate solution to problem (1.1)-(1.2) with the accuracy $O(\epsilon)$ in the norm $||\cdot||$.

12.3.2 *Proofs*

Proof of Lemma 12.4. We start with the claim:

Claim: *the restrictions of harmonic functions H_ℓ on S form a total set in $L^2(S)$.*

Lemma 12.4 follows from this claim. Let us prove the claim. Assume the contrary. Then there is a function $g \neq 0$ such that $\int_S g(s)H_\ell(s)ds = 0$ $\forall \ell \geq 0$. This implies $V(x) := \int_S g(s)|x-s|^{-1}ds = 0 \; \forall x \in D'$. Thus $V = 0$ on S, and since $\Delta V = 0$ in D, one concludes that $V = 0$ in D. Thus $g = 0$ by the jump formula for the normal derivatives of the simple layer potential V. This contradiction proves the claim. Lemma 1.1 is proved.

□

Proof of Lemma 12.5 By Green's formula one has

$$w_\epsilon(x) = \int_S w_\epsilon(s)G_N(x,s)ds, \quad \|w_\epsilon\|_{L^2(S)} < \epsilon, \quad w_\epsilon := v_\epsilon - v. \tag{12.34}$$

Here N is the unit normal to S, pointing into D', and G is the Dirichlet Green's function of the Laplacian in D':

$$\nabla^2 G = -\delta(x-y) \text{ in } D', \quad G = 0 \text{ on } S, \tag{12.35}$$

$$G = O\left(\frac{1}{r}\right), \quad r \to \infty. \tag{12.36}$$

From (12.34) one gets (12.33) and (12.32) with $H^m_{loc}(D')$-norm immediately by the Cauchy inequality. Estimate (12.32) in the region $B'_R := \mathbb{R}^3 \setminus B_R$ follows from the estimate

$$|G_N(x,s)| \leq \frac{c}{1+|x|}, \quad |x| \geq R. \tag{12.37}$$

In the region $B_R \backslash D$ estimate (12.32) follows from local elliptic estimates for $w_\epsilon := v_\epsilon - v$, which imply that

$$\|w_\epsilon\|_{L^2(B_R \backslash D)} \leq c\epsilon. \tag{12.38}$$

Let us recall the elliptic estimate we have used. Let $D'_R := B_R \setminus D$ and S_R be the boundary of B_R. Let us use the elliptic estimate for the solution to homogeneous Laplace equation in D'_R:

$$||w_\epsilon||_{H^{0.5}(D'_R)} \leq c[||w_\epsilon||_{L^2(S_R)} + ||w_\epsilon||_{L^2(S)}]. \tag{12.39}$$

The estimates $||w_\epsilon||_{L^2(S_R)} = O(\epsilon)$, $||w_\epsilon||_{L^2(S)} = O(\epsilon)$, and (12.39) yield (12.32). Lemma 12.5 is proved. □

Appendix A

Appendix A1

Many-Body Wave Scattering Problems for Small Scatterers and Creating Materials with a Desired Refraction Coefficient

Alexander G. Ramm

Department of Mathematics, Kansas State University, Manhattan, KS 66506-2602, USA

MSC: *35Q60; 74E99; 78A40; 78A45; 78A48* ***PACS:*** *02.30.Rz; 02.30.Mv; 41.20.Jb*

3.1 Introduction

In this chapter, we discuss a method for creating materials with a desired refraction coefficients. This method is proposed and developed by the author and is based on a series of his papers and on his monograph [1]. The author thinks that these results may be new for materials science people although the results were published in mathematical and mathematical physics Journals. This is the basic reason for including this chapter in this book. This chapter should be useful to materials science researchers, physicists and engineers.

Parts of this chapter are taken verbatim from the paper by the author [2]. The author thanks Springer for permission to use verbatim parts of the author's paper, see also monograph [3].

There is a large literature on wave scattering by small bodies, starting from Rayleigh's work (1871), [4–6]. For the problem of wave scattering by one body, an analytical solution was found only for the bodies of special shapes, for example, for balls and ellipsoids. If the scatterer is small, then the scattered field can be calculated analytically for bodies of arbitrary shapes, see [2, 7], and [1] where this theory is presented.

The many-body wave scattering problem was discussed in the literature, mostly numerically, in the cases when the number of scatterers is small or the influence on a particular particle of the waves scattered by other particles is negligible. This corresponds to the case when the distance d between neighboring particles is much larger than the wavelength λ, and the characteristic size a of a small body (particle) is much smaller than λ, that is, $d \gg \lambda$ and $a \ll \lambda$. By $k = \frac{2\pi}{\lambda}$, the wave number is denoted.

In this chapter, the much more difficult case is considered, when $a \ll d \ll \lambda$. In this case, the influence of the scattered field on a particular particle is essential, that is, *multiple scattering effects are essential.*

The derivations of the results, presented in this chapter, are rigorous. They are taken from the earlier papers of the author, cited in the list of references. Many formulas and arguments are taken from these papers, especially from the paper by the author [2]. Large parts of this chapter are taken verbatim, and monograph [1] is also used essentially. In this chapter, we do not discuss electromagnetic wave scattering by small bodies (particles). A detailed discussion of electromagnetic wave scattering by small perfectly conducting and impedance particles of an *arbitrary shape* is given in [1, 8], and also see [7].

A *physically novel* point in our theory is the following one:

While in the classical theory of wave scattering by small body of characteristic size a (e.g., in Rayleigh's theory) the scattering amplitude is $O(a^3)$ as $a \to 0$, in our theory for a small impedance particle the scattering amplitude is *much larger*: it is of the order $O(a^{2-\kappa})$, where $a \to 0$ and $\kappa \in [0, 1)$ are the parameters (see the text below formula (3.22) in this chapter).

Can this result be used in technology?

The practical applications of the theory, presented in this chapter, are immediate provided that the important practical problem of preparing small particles with the prescribed boundary impedance is solved.

The author thinks that an impedance boundary condition (BC) (condition (3.7)) must be physically (experimentally) realizable if this condition guarantees the uniqueness of the solution to the corresponding boundary problem. The impedance BC (3.7) guarantees the uniqueness of the solution to the scattering boundary problem (3.1)–(3.4) provided that $\text{Im}\zeta_1 \leq 0$.

Therefore, there should exist a practical (experimental) method for producing small particles with any boundary impedance ζ_1 satisfying the inequality $Im\zeta_1 \leq 0$.

The author asks the materials science specialists to contact him if they are aware of a method for practical (experimental) preparing (producing) small particles with the prescribed boundary impedance

The materials science researchers are not familiar with the author's papers on creating materials with a desired refraction coefficient because the author's theory was presented in the journals, which are not popular among materials science researchers.

Although the author's results were presented in many of the author's earlier publications, cited in references, the author hopes that *they will be not only new but practically useful for materials science researchers.*

The basic results of this section consist of:

(i) Derivation of analytic formulas for the scattering amplitude for the wave scattering problem by one small ($ka \ll 1$) impedance body *of an arbitrary shape*;

(ii) Solution to *many-body wave scattering problem* by small particles, embedded in an inhomogeneous medium, under the assumptions $a \ll d \ll \lambda$, where d is the minimal distance between neighboring particles;

(iii) Derivation of the equations for the limiting effective (self-consistent) field in an inhomogeneous medium in which many small particles are embedded, when $a \to 0$ and the number $M = M(a)$ of the small particles tends to infinity at an appropriate rate;
(iv) Derivation of linear algebraic system (LAS) for solving many-body wave scattering problems. These systems are not obtained in the standard way from boundary integral equations; they have physical meaning and give an efficient numerical method for solving many-body wave scattering problems in the case of small scatterers. In [8] for the first time, the many-body wave scattering problems were solved for billions of particles. This was not feasible earlier;
(v) Application of our results to creating materials with a desired refraction coefficient.

The order of the error estimates as $a \to 0$ is obtained. Our presentation follows very closely that in [2], but it is essentially self-contained. Our methods give powerful numerical methods for solving many-body wave scattering problems in the case when the scatterers are small but multiple scattering effects are essential [9–11]. In [9], the scattering problem is solved numerically for 10^{10} particles apparently for the first time.

In Sections 3.1–3.4 wave scattering by small impedance bodies is developed.

Let us formulate the wave scattering problems we deal with. First, let us consider a one-body scattering problem. Let D_1 be a bounded domain in $\mathbb{R}^3$ with a sufficiently smooth boundary S_1. The scattering problem consists of finding the solution to the problem:

$$(\nabla^2 + k^2)u = 0 \text{ in } D_1' := \mathbb{R}^3 \backslash D_1, \tag{3.1}$$

$$\Gamma u = 0 \text{ on } S_1, \tag{3.2}$$

$$u = u_0 + v, \tag{3.3}$$

where

$$u_0 = e^{ik\alpha \cdot x}, \quad \alpha \in S^2, \tag{3.4}$$

S^2 is the unit sphere in $\mathbb{R}^3$, u_0 is the incident field, v is the scattered field satisfying the radiation condition

$$v_r - ikv = o\left(\frac{1}{r}\right), \quad r := |x| \to \infty, v_r := \frac{\partial v}{\partial r}, \tag{3.5}$$

Γu is the BC of one of the following types

$$\Gamma u = \Gamma_1 u = u \quad \text{(Dirichlet BC)}, \tag{3.6}$$

$$\Gamma u = \Gamma_2 u = u_N - \zeta_1 u, \quad \text{Im}\zeta_1 \le 0 \text{ (impedance BC)}, \tag{3.7}$$

where ζ_1 is a constant, N is the unit normal to S_1, pointing out of D_1, and

$$\Gamma u = \Gamma_3 u = u_N \quad \text{(Neumann BC)}. \tag{3.8}$$

It is well known [12, 13] that problem (3.1)–(3.3) has a unique solution. We now assume that

$$a := 0.5 \operatorname{diam} D_1, \quad ka \ll 1, \tag{3.9}$$

which is the "smallness assumption" equivalent to $a \ll \lambda$, where λ is the wave length. We look for the solution to problem (3.1)–(3.3) of the form

$$u(x) = u_0(x) + \int_{S_1} g(x,t)\sigma_1(t)dt, \quad g(x,y) := \frac{e^{ik|x-y|}}{4\pi|x-y|}, \tag{3.10}$$

where dt is the element of the surface area of S_1. One can prove that the unique solution to the scattering problem (3.1)–(3.3) with any of the BCs (3.6)–(3.8) can be found in the form (3.10), and the function σ_1 in (3.10) is uniquely defined from the BC (3.2). The scattering amplitude $A(\beta,\alpha) = A(\beta,\alpha,k)$ is defined by the formula

$$v = \frac{e^{ikr}}{r}A(\beta,\alpha,k) + o\left(\frac{1}{r}\right), \quad r \to \infty, \ \beta := \frac{x}{r}. \tag{3.11}$$

The equations for finding σ_1 are:

$$\int_{S_1} g(s,t)\sigma_1(t)dt = -u_0(s), \tag{3.12}$$

$$u_{0N} - \zeta_1 u_0 + \frac{A\sigma_1 - \sigma_1}{2} - \zeta_1 \int_{S_1} g(s,t)\sigma_1(t)dt = 0, \tag{3.13}$$

$$u_{0N} + \frac{A\sigma_1 - \sigma_1}{2} = 0, \tag{3.14}$$

respectively, for conditions (3.6)–(3.8). The operator A is defined as follows:

$$A\sigma := 2\int_{S_1} \frac{\partial}{\partial N_s} g(s,t)\sigma_1(t)dt. \tag{3.15}$$

Equations (3.12)–(3.14) are uniquely solvable, but there are no analytic formulas for their solutions for bodies of arbitrary shapes. However, if the body D_1 is small, $ka \ll 1$, one can rewrite (3.10) as

$$u(x) = u_0(x) + g(x,0)Q_1 + \int_{S_1} [g(x,t) - g(x,0)]\sigma_1(t)dt, \tag{3.16}$$

where

$$Q_1 := \int_{S_1} \sigma_1(t)dt, \tag{3.17}$$

and $0 \in D_1$ is the origin.

If $ka \ll 1$, then we prove that

$$|g(x,0)Q_1| \gg \left| \int_{S_1} [g(x,t) - g(x,0)]\sigma_1(t)dt \right|, \quad |x| > a. \tag{3.18}$$

Therefore, *the scattered field is determined outside D_1 by a single number Q_1.*

This number can be obtained analytically without solving (3.12) and (3.13). The case (3.14) requires a special approach by the reason discussed in detail later.

Let us give the results for (3.12) and (3.13) first. For (3.12), one has

$$Q_1 = \int_{S_1} \sigma_1(t)\mathrm{d}t = -Cu_0(0)[1 + o(1)], \quad a \to 0, \tag{3.19}$$

where C *is the electric capacitance of a perfect conductor with the shape* D_1. For (3.13), one has

$$Q_1 = -\zeta_1 |S_1| u_0(0)[1 + o(1)], \quad a \to 0, \tag{3.20}$$

where $|S_1|$ is the surface area of S_1. The scattering amplitude for problem (3.1)–(3.3) with $\Gamma = \Gamma_1$ (acoustically soft particle) is

$$A_1(\beta, \alpha) = -\frac{C}{4\pi}[1 + o(1)], \tag{3.21}$$

since

$$u_0(0) = e^{ik\alpha\cdot x}|_{x=0} = 1.$$

Therefore, in this case, the scattering is isotropic and of the order $O(a)$, *because the capacitance* $C = O(a)$.

The scattering amplitude for problem (3.1)–(3.3) with $\Gamma = \Gamma_2$ (*small impedance particles*) is:

$$A_2(\beta, \alpha) = -\frac{\zeta_1 |S_1|}{4\pi}[1 + o(1)], \tag{3.22}$$

since $u_0(0) = 1$.

In this case, the scattering is also isotropic, and of the order $O(\zeta |S_1|)$.

If $\zeta_1 = O(1)$, then $A_2 = O(a^2)$, because $|S_1| = O(a^2)$. If $\zeta_1 = O\left(\frac{1}{a^\kappa}\right)$, $\kappa \in (0, 1)$, then $A_2 = O(a^{2-\kappa})$. The case $\kappa = 1$ was considered in [14].

The scattering amplitude for problem (3.1)–(3.3) with $\Gamma = \Gamma_3$ (acoustically hard particles) is

$$A_3(\beta, \alpha) = -\frac{k^2 |D_1|}{4\pi}(1 + \beta_{pq}\beta_p\alpha_q), \qquad \text{if } u_0 = e^{ik\alpha\cdot x}. \tag{3.23}$$

Here and below summation is understood over the repeated indices, $\alpha_q = \alpha \cdot e_q$, $\alpha \cdot e_q$ denotes the dot product of two vectors in $\mathbb{R}^3$, $p, q = 1, 2, 3$, $\{e_p\}$ is an orthonormal Cartesian basis of $\mathbb{R}^3$, $|D_1|$ is the volume of D_1, β_{pq} is the magnetic polarizability tensor defined as follows [7, p. 62]:

$$\beta_{pq} := \frac{1}{|D_1|} \int_{S_1} t_p \sigma_{1q}(t)\mathrm{d}t, \tag{3.24}$$

σ_{1q} is the solution to the equation

$$\sigma_{1q}(s) = A_0 \sigma_{1q} - 2N_q(s), \tag{3.25}$$

$N_q(s) = N(s) \cdot e_q$, $N = N(s)$ is the unit outer normal to S_1 at the point s, that is, the normal pointing out of D_1, and A_0 is the operator A at $k = 0$. For small bodies, $\|A - A_0\| = o(ka)$.

If $u_0(x)$ is an arbitrary field satisfying (3.1), not necessarily the plane wave $e^{ik\alpha \cdot x}$, then

$$A_3(\beta, \alpha) = \frac{|D_1|}{4\pi}\left(ik\beta_{pq}\frac{\partial u_0}{\partial x_q}\beta_p + \Delta u_0\right). \tag{3.26}$$

The above formulas are derived in Section 3.2. In Section 3.3 we develop a theory for many-body wave scattering problem and derive the equations for effective field in the medium, in which many small particles are embedded, as $a \to 0$.

The results, presented in this chapter, are based on the earlier works of the author [1, 2, 7, 9, 12–34]. These results and methods of their derivation differ much from those published by other authors.

Our approach to homogenization-type theory is also different from the approaches of other authors [35, 36]. The differences are:

(i) no periodic structure in the problems is assumed,
(ii) the operators in our problems are non-selfadjoint and have continuous spectrum,
(iii) the limiting medium is not homogeneous and its parameters are not periodic,
(iv) the technique for passing to the limit is different from the one used in homogenization theory.

Let us summarize the results for one-body wave scattering.

Theorem 3.1 The scattering amplitude for the problem (3.1)–(3.4) for small body of an arbitrary shape is given by formulas (3.25)–(3.27), for the BCs Γ_1–Γ_3, respectively.

3.2 Derivation of the Formulas for One-Body Wave Scattering Problems

Let us recall the known result [12]

$$\frac{\partial}{\partial N_s^-}\int_{S_1} g(x,t)\sigma_1(t)\mathrm{d}t = \frac{A\sigma_1 - \sigma_1}{2} \tag{3.27}$$

concerning the limiting value of the normal derivative of single-layer potential from outside. Let $x_m \in D_m$, $t \in S_m$, S_m is the surface of D_m, $a = 0.5\ \mathrm{diam}D_m$.

In this section $m = 1$, and $x_m = 0$ is the origin.

We assume that $ka \ll 1$, $ad^{-1} \ll 1$, so $|x - x_m| = d \gg a$. Then

$$\frac{e^{ik|x-t|}}{4\pi|x-t|} = \frac{e^{ik|x-x_m|}}{4\pi|x-x_m|}e^{-ik(x-x_m)^o\cdot(t-x_m)}\left(1 + O\left(ka + \frac{a}{d}\right)\right), \tag{3.28}$$

$$k|x-t| = k|x-x_m| - k(x-x_m)^o \cdot (t-x_m) + O\left(\frac{ka^2}{d}\right), \tag{3.29}$$

where

$$d = |x-x_m|, \quad (x-x_m)^o := \frac{x-x_m}{|x-x_m|},$$

and

$$\frac{|x-t|}{|x-x_m|} = 1 + O\left(\frac{a}{d}\right). \tag{3.30}$$

Let us derive estimate (3.19). Since $|t| \le a$ on S_1, one has

$$g(s,t) = g_0(s,t)(1 + O(ka)),$$

where $g_0(s,t) = \frac{1}{4\pi|s-t|}$. Since $u_0(s)$ is a smooth function, one has $|u_0(s) - u_0(0)| = O(a)$. Consequently, (3.12) can be considered as an equation for electrostatic charge distribution $\sigma_1(t)$ on the surface S_1 of a perfect conductor D_1, charged to the constant potential $-u_0(0)$ (up to a small term of the order $O(ka)$). It is known that the total charge $Q_1 = \int_{S_1} \sigma_1(t)dt$ of this conductor is equal to

$$Q_1 = -Cu_0(0)(1 + O(ka)), \tag{3.31}$$

where C is the electric capacitance of the perfect conductor with the shape D_1.

Analytic formulas for electric capacitance C of a perfect conductor of an arbitrary shape, which allow to calculate C with a desired accuracy, are derived in [7]. For example, the zeroth approximation formula is:

$$C^{(0)} = \frac{4\pi|S_1|^2}{\int_{S_1}\int_{S_1}\frac{dsdt}{r_{st}}}, \quad r_{st} = |t-s|, \tag{3.32}$$

and we assume in (3.32) that $\epsilon_0 = 1$, where ϵ_0 is the dielectric constant of the homogeneous medium in which the perfect conductor is placed. Formula (3.31) is formula (3.19). If $u_0(x) = e^{ik\alpha\cdot x}$, then $u_0(0) = 1$, and $Q_1 = -C(1 + O(ka))$. In this case,

$$A_1(\beta, \alpha) = \frac{Q_1}{4\pi} = -\frac{C}{4\pi}[1 + O(ka)],$$

which is formula (3.21).

Consider now wave scattering by an impedance particle.

Let us derive formula (3.20). Integrate (3.13) over S_1, use the divergence formula

$$\int_{S_1} u_{0N}ds = \int_{D_1} \nabla^2 u_0 dx = -k^2 \int_{D_1} u_0 dx = k^2|D_1|u_0(0)[1 + o(1)], \tag{3.33}$$

where $|D_1| = O(a^3)$, and the formula

$$-\zeta_1 \int_{S_1} u_0 ds = -\zeta_1 |S_1| u_0(0)[1 + o(1)], \tag{3.34}$$

which is valid because the body D_1 is small: in this case, $u_0(s) \approx u_0(0)$. Furthermore $|\int_{S_1} g(s,t)ds| = O(a)$, so

$$\zeta_1 \int_{S_1} ds \int_{S_1} g(s,t)\sigma_1(t)dt = O(aQ_1). \tag{3.35}$$

Therefore, the term (3.35) is negligible compared with Q_1 as $a \to 0$. Finally, if $ka \ll 1$, then $g(s,t) = g_0(s,t)(1 + ik|s-t| + \cdots)$, and

$$\frac{\partial}{\partial N_s} g(s,t) = \frac{\partial}{\partial N_s} g_0(s,t)[1 + O(ka)]. \tag{3.36}$$

Denote by A_0 the operator

$$A_0 \sigma = 2 \int_{S_1} \frac{\partial g_0(s,t)}{\partial N_s} \sigma_1(t) dt. \tag{3.37}$$

It is known from the potential theory [1] that

$$\int_{S_1} A_0 \sigma_1 ds = -\int_{S_1} \sigma_1(t)dt, \quad 2\int_{S_1} \frac{\partial g_0(s,t)}{\partial N_s} ds = -1, \quad t \in S_1. \tag{3.38}$$

Therefore,

$$\int_{S_1} ds \frac{A\sigma_1 - \sigma_1}{2} = -Q_1[1 + O(ka)]. \tag{3.39}$$

Consequently, from formulas (3.33)–(3.39), one gets formula (3.22).

One can see that *the wave scattering by an impedance particle is isotropic, and the scattered field is of the order $O(\zeta_1|S_1|)$. Since $|S_1| = O(a^2)$, one has $O(\zeta_1|S_1|) = O(a^{2-\kappa})$ if $\zeta_1 = O\left(\frac{1}{a^\kappa}\right)$, $\kappa \in [0, 1)$.*

Consider now wave scattering by an acoustically hard small particle, that is, the problem with the Neumann BC.

In this case, we will prove that:

(i) The scattering is anisotropic,
(ii) It is defined not by a single number, as in the previous two cases, but by a tensor, and
(iii) The order of the scattered field is $O(a^3)$ as $a \to 0$, for a fixed $k > 0$, that is, the scattered field is much smaller than in the previous two cases.

Integrating over S_1 (3.14), one gets

$$Q_1 = \int_{D_1} \nabla^2 u_0 dx = \nabla^2 u_0(0)|D_1|[1 + o(1)], \quad a \to 0. \tag{3.40}$$

Thus, $Q_1 = O(a^3)$. Therefore, the contribution of the term $e^{-ikx^o \cdot t}$ in formula (3.28) with $x_m = 0$ will be also of the order $O(a^3)$ and should be taken into account, *in contrast to the previous two cases.* Namely,

$$u(x) = u_0(x) + g(x,0) \int_{S_1} e^{-ik\beta \cdot t} \sigma_1(t) dt, \quad \beta := \frac{x}{|x|} = x^o. \tag{3.41}$$

One has

$$\int_{S_1} e^{-ik\beta \cdot t} \sigma_1(t) dt = Q_1 - ik\beta_p \int_{S_1} t_p \sigma_1(t) dt, \tag{3.42}$$

where the terms of higher order of smallness are neglected and summation over index p is understood. The function σ_1 solves (3.14):

$$\sigma_1 = A\sigma_1 + 2u_{0N} = A\sigma_1 + 2ik\alpha_q N_q u_0(s), \quad s \in S_1 \tag{3.43}$$

if $u_0(x) = e^{ik\alpha \cdot x}$.

Comparing (3.43) with (3.25), using (3.24), and taking into account that $ka \ll 1$, one gets

$$\begin{aligned} -ik\beta_p \int_{S_1} t_p \sigma_1(t) dt &= -ik\beta_p |D_1| \beta_{pq} (-ik\alpha_q) u_0(0)[1 + O(ka)] \\ &= -k^2 |D_1| \beta_{pq} \beta_p \alpha_q u_0(0)[1 + O(ka)]. \end{aligned} \tag{3.44}$$

From (3.40), (3.42), and (3.44), one gets formula (3.23), because $\nabla^2 u_0 = -k^2 u_0$.

If $u_0(x)$ is an arbitrary function, satisfying (3.1), then $ik\alpha_q$ in (3.43) is replaced by $\frac{\partial u_0}{\partial x_q}$, and $-k^2 u_0 = \Delta u_0$, which yields formula (3.26).

This completes the derivation of the formulas for the solution of scalar wave scattering problem by one small body on the boundary, of which the Dirichlet, or the impedance, or the Neumann boundary condition is imposed.

3.3 Many-Body Scattering Problem

In this section we assume that there are $M = M(a)$ small bodies (particles) D_m, $1 \leq m \leq M$, $a = 0.5 \max \operatorname{diam} D_m$, $ka \ll 1$. The distance $d = d(a)$ between neighboring bodies is much larger than a, $d \gg a$, but *we do not assume that $d \gg \lambda$, so there may be many small particles on the distances of the order of the wavelength λ.*

This means that our medium with the embedded particles is not necessarily diluted.

We assume that the small bodies are embedded in an arbitrary large but finite domain D, $D \subset \mathbb{R}^3$, so $D_m \subset D$. Denote $D' := \mathbb{R}^3 \setminus D$ and $\Omega := \cup_{m=1}^M D_m$, $S_m := \partial D_m$, $\partial\Omega = \cup_{m=1}^M S_m$. By N, we denote a unit normal to $\partial\Omega$, pointing out of Ω; and by $|D_m|$ the volume of the body D_m is denoted.

The scattering problem consists of finding the solution to the following problem

$$(\nabla^2 + k^2)u = 0 \text{ in } \mathbb{R}^3 \setminus \Omega, \tag{3.45}$$

$$\Gamma u = 0 \text{ on } \partial\Omega, \tag{3.46}$$

$$u = u_0 + v, \tag{3.47}$$

where u_0 is the incident field, satisfying (3.45) in $\mathbb{R}^3$, for example, $u_0 = e^{ik\alpha\cdot x}$, $\alpha \in S^2$, and v is the scattered field, satisfying the radiation condition (3.5). The BC (3.46) can be of the types (3.6)–(3.8).

In the case of impedance BC (3.7), we assume that

$$u_N = \zeta_m u \text{ on } S_m, \quad 1 \le m \le M, \tag{3.48}$$

so the impedance may vary from one particle to another. We assume that

$$\zeta_m = \frac{h(x_m)}{a^\kappa}, \quad \kappa \in (0,1), \tag{3.49}$$

where $x_m \in D_m$ is a point in D_m, and $h(x)$, $x \in D$, is a given function, which we can choose as we wish, subject to the condition $\mathrm{Im}h(x) \le 0$. For simplicity, we assume that $h(x)$ is a continuous function.

Let us make the following assumption about the distribution of small particles:

If $\Delta \subset D$ is an arbitrary open subset of D, then the number $\mathcal{N}(\Delta)$ of small particles in Δ, assuming the impedance BC, is:

$$\mathcal{N}_\zeta(\Delta) = \frac{1}{a^{2-\kappa}} \int_\Delta N(x)\mathrm{d}x[1 + o(1)], \quad a \to 0, \tag{3.50}$$

where $N(x) \ge 0$ is a given function.

If the Dirichlet BC is assumed, then

$$\mathcal{N}_D(\Delta) = \frac{1}{a} \int_\Delta N(x)\mathrm{d}x[1 + o(1)], \quad a \to 0. \tag{3.51}$$

The case of the Neumann BC will not be considered in this chapter, see [2].

We look for the solution to problem (3.45)–(3.47) with the Dirichlet BC of the form

$$u = u_0 + \sum_{m=1}^{M} \int_{S_m} g(x,t)\sigma_m(t)\mathrm{d}t, \tag{3.52}$$

where $\sigma_m(t)$ are some functions to be determined from the boundary condition (3.46). It is proved in [14] that problem (3.45)–(3.47) has a unique solution of the form (3.52). For any $\sigma_m(t)$, function (3.52) solves (3.45) and satisfies condition (3.47). The BC (3.46) determines σ_m uniquely. However, if $M \gg 1$, then numerical solution of the system of integral equations for σ_m, where $1 \le m \le M$, which one gets from the BC (3.52), is practically not feasible.

To avoid this principal difficulty, we prove that the solution to scattering problem (3.45)–(3.47) *is determined by M numbers*

$$Q_m := \int_{S_m} \sigma_m(t)\mathrm{d}t, \tag{3.53}$$

rather than M functions $\sigma_m(t)$. This allows one to drastically reduce the complexity of the numerical solution of the many-body scattering problems in the case of small particles.

This is possible to prove that if the particles D_m are small. We derive analytical formulas for Q_m as $a \to 0$.

Let us define the effective (self-consistent) field $u_e(x) = u_e^{(j)}(x)$, acting on the j-th particle, by the formula

$$u_e(x) := u(x) - \int_{S_j} g(x,t)\sigma_j(t)dt, \quad |x - x_j| \sim a. \tag{3.54}$$

Physically, this field acts on the j–th particle and is a sum of the incident field and the fields acting from all other particles:

$$u_e(x) = u_e^{(j)}(x) := u_0(x) + \sum_{m \neq j} \int_{S_m} g(x,t)\sigma_m(t)dt. \tag{3.55}$$

Let us rewrite (3.55) as follows:

$$u_e(x) = u_0(x) + \sum_{m \neq j}^{M} g(x, x_m)Q_m + \sum_{m \neq j}^{M} \int_{S_m} [g(x,t) - g(x, x_m)]\sigma_m(t)dt. \tag{3.56}$$

We want to prove that the last sum is negligible compared with the first one as $a \to 0$.

To prove this, let us give some estimates. One has $|t - x_m| \leq a$, $d = |x - x_m|$,

$$|g(x,t) - g(x,x_m)| = \max\left\{O\left(\frac{a}{d^2}\right), O\left(\frac{ka}{d}\right)\right\}, \quad |g(x,x_m)| = O(1/d). \tag{3.57}$$

Therefore, if $|x - x_j| = O(a)$, then

$$\frac{\left|\int_{S_m}[g(x,t) - g(x,x_m)]\sigma_m(t)dt\right|}{|g(x,x_m)Q_m|} \leq O(ad^{-1} + ka). \tag{3.58}$$

One can also prove that

$$J_1/J_2 = O(ka + ad^{-1}), \tag{3.59}$$

where J_1 is the first sum in (3.56) and J_2 is the second sum in (3.56). Therefore, at any point $x \in \Omega' = \mathbb{R}^3 \setminus \Omega$, one has

$$u_e(x) = u_0(x) + \sum_{m=1}^{M} g(x,x_m)Q_m, \quad x \in \Omega', \tag{3.60}$$

where the terms of higher order of smallness are omitted.

3.3.1 The Case of Acoustically Soft Particles

If (3.46) is the Dirichlet condition, then, as we have proved in Section 3.2 (see formula (3.31)), one has

$$Q_m = -C_m u_e(x_m). \tag{3.61}$$

Thus,

$$u_e(x) = u_0(x) - \sum_{m=1}^{M} g(x, x_m) C_m u_e(x_m), \quad x \in \Omega'. \tag{3.62}$$

One has

$$u(x) = u_e(x) + o(1), \quad a \to 0, \tag{3.63}$$

so the full field and effective field are practically the same.

Let us write a LAS for finding unknown quantities $u_e(x_m)$:

$$u_e(x_j) = u_0(x_j) - \sum_{m \neq j}^{M} g(x_j, x_m) C_m u_e(x_m). \tag{3.64}$$

If M is not very large, say $M = O(10^3)$, then LAS (3.64) can be solved numerically, and formula (3.62) can be used for calculation of $u_e(x)$.

Consider the limiting case, when $a \to 0$. One can rewrite (3.64) as follows:

$$u_e(\xi_q) = u_0(\xi_q) - \sum_{p \neq q}^{P} g(\xi_q, \xi_p) u_e(\xi_p) \sum_{x_m \in \Delta_p} C_m, \tag{3.65}$$

where $\{\Delta_p\}_{p=1}^{P}$ is a union of cubes which forms a covering of D,

$$\max_p diam\Delta_p := b = b(a) \gg a,$$

$$\lim_{a \to 0} b(a) = 0. \tag{3.66}$$

By $|\Delta_p|$ we denote the volume (measure) of Δ_p, and ξ_p is the center of Δ_p, or a point x_p in an arbitrary small body D_p, located in Δ_p. Let us assume that there exists the limit

$$\lim_{a \to 0} \frac{\sum_{x_m \in \Delta_p} C_m}{|\Delta_p|} = C(\xi_p), \quad \xi_p \in \Delta_p. \tag{3.67}$$

For example, one may have

$$C_m = c(\xi_p) a \tag{3.68}$$

for all m such that $x_m \in \Delta_p$, where $c(x)$ is some function in D. If all D_m are balls of radius a, then $c(x) = 4\pi$. We have

$$\sum_{x_m \in \Delta_p} C_m = C_p a \mathcal{N}(\Delta_p) = C_p N(\xi_p) |\Delta_p| [1 + o(1)], \quad a \to 0, \tag{3.69}$$

so limit (3.67) exists, and

$$C(\xi_p) = c(\xi_p)N(\xi_p). \tag{3.70}$$

From (3.65), (3.68)–(3.70), one gets

$$u_e(\xi_q) = u_0(\xi_q) - \sum_{p\neq q} g(\xi_q, \xi_p)c(\xi_p)N(\xi_p)u_e(\xi_p)|\Delta_p|, \quad 1 \le p \le P. \tag{3.71}$$

LAS (3.71) can be considered as the *collocation method for solving integral equation*

$$u(x) = u_0(x) - \int_D g(x, y)c(y)N(y)u(y)\mathrm{d}y. \tag{3.72}$$

It is proved in [30] that

System (3.71) *is uniquely solvable for all sufficiently small* $b(a)$, *and the function*

$$u_P(x) := \sum_{p=1}^{P} \chi_p(x)u_e(\xi_p) \tag{3.73}$$

converges in $L^\infty(D)$ *to the unique solution of equation* (3.72).

The function $\chi_p(x)$ in (3.73) is the characteristic function of the cube Δ_p: it is equal to 1 in Δ_p and vanishes outside Δ_p. Thus, if $a \to 0$, the solution to the many-body wave scattering problem in the case of the Dirichlet BC is well approximated by the unique solution of the integral equation (3.72).

Applying the operator $L_0 := \nabla^2 + k^2$ to (3.72), and using the formula $L_0 g(x, y) = -\delta(x - y)$, where $\delta(x)$ is the delta-function, one gets

$$\nabla^2 u + k^2 u - q(x)u = 0 \text{ in } \mathbb{R}^3, \quad q(x) := c(x)N(x). \tag{3.74}$$

The physical conclusion is:

If one embeds $M(a) = O(1/a)$ *small acoustically soft particles, which are distributed as in* (3.51), *then one creates, as* $a \to 0$, *a limiting medium, which is inhomogeneous and has a refraction coefficient* $n^2(x) = 1 - k^{-2}q(x)$.

It is interesting from the physical point of view to note that

The limit, as $a \to 0$, *of the total volume of the embedded particles is zero.*

Indeed, the volume of one particle is $O(a^3)$, the total number M of the embedded particles is $O(a^3M) = O(a^2)$, and $\lim_{a\to 0} O(a^2) = 0$.

The second observation is: if (3.51) holds, then on a unit length straight line, there are $O\left(\frac{1}{a^{1/3}}\right)$ particles, so the distance between neighboring particles is $d = O(a^{1/3})$. If $d = O(a^\gamma)$ with $\gamma > \frac{1}{3}$, then the number of the embedded particles in a subdomain Δ_p is $O\left(\frac{1}{d^3}\right) = O(a^{-3\gamma})$. In this case, for $3\gamma > 1$, the limit in (3.69) is $C(\xi_p) = \lim_{a\to 0} c_p a O(a^{-3\gamma}) = \infty$. Therefore, the product of this limit by u remains finite only if $u = 0$ in D. Physically, this means that if the distances between neighboring perfectly soft particles are smaller than $O(a^{1/3})$, namely, they are $O(a^\gamma)$ with any $\gamma > \frac{1}{3}$, then $u = 0$ in D.

On the other hand, if $\gamma < \frac{1}{3}$, then the limit $C(\xi_p) = 0$, and $u = u_0$ in D, so that the embedded particles do not change, in the limit $a \to 0$, the properties of the medium.

This concludes our discussion of the scattering problem for many acoustically soft particles.

3.3.2 Wave Scattering by Many Impedance Particles

We assume now that (3.49) and (3.50) hold, use the exact BC (3.46) with $\Gamma = \Gamma_2$, that is,

$$u_{eN} - \zeta_m u_e + \frac{A_m\sigma_m - \sigma_m}{2} - \zeta_m \int_{S_m} g(s,t)\sigma_m(t)\mathrm{d}t = 0, \tag{3.75}$$

and integrate (3.75) over S_m in order to derive an analytical asymptotic formula for $Q_m = \int_{S_m} \sigma_m(t)\mathrm{d}t$.

We have

$$\int_{S_m} u_{eN}\mathrm{d}s = \int_{D_m} \nabla^2 u_e \mathrm{d}x = O(a^3), \tag{3.76}$$

$$\int_{S_m} \zeta_m u_e(s)\mathrm{d}s = h(x_m)a^{-\kappa}|S_m|u_e(x_m)[1 + o(1)], \quad a \to 0, \tag{3.77}$$

$$\int_{S_m} \frac{A_m\sigma_m - \sigma_m}{2}\mathrm{d}s = -Q_m[1 + o(1)], \quad a \to 0, \tag{3.78}$$

and

$$\zeta_m \int_{S_m}\int_{S_m} g(s,t)\sigma_m(t)\mathrm{d}t = h(x_m)a^{1-\kappa}Q_m = o(Q_m), \quad 0 < \kappa < 1. \tag{3.79}$$

From (3.75) to (3.79), one finds

$$Q_m = -h(x_m)a^{2-\kappa}|S_m|a^{-2}u_e(x_m)[1 + o(1)]. \tag{3.80}$$

This yields the formula for the approximate solution to the wave scattering problem for many impedance particles:

$$u(x) = u_0(x) - a^{2-\kappa}\sum_{m=1}^{M} g(x,x_m)b_m h(x_m)u_e(x_m)[1 + o(1)], \tag{3.81}$$

where

$$b_m := |S_m|a^{-2}$$

are some positive numbers which depend on the geometry of S_m and are independent of a. For example, if all D_m are balls of radius a, then $b_m = 4\pi$.

A LAS for $u_e(x_m)$, analogous to (3.64), is

$$u_e(x_j) = u_0(x_j) - a^{2-\kappa}\sum_{m=1, m\neq j}^{M} g(x_j,x_m)b_m h(x_m)u_e(x_m). \tag{3.82}$$

The integral equation for the limiting effective field in the medium with embedded small particles, as $a \to 0$, is

$$u(x) = u_0(x) - b \int_D g(x, y) N(y) h(y) u(y) \mathrm{d}y, \tag{3.83}$$

where

$$u(x) = \lim_{a \to 0} u_e(x), \tag{3.84}$$

and we have assumed in (3.83) for simplicity that $b_m = b$ for all m, that is, all small particles are of the same size.

Applying operator $L_0 = \nabla^2 + k^2$ to equation (3.83), one finds the differential equation for the limiting effective field $u(x)$:

$$(\nabla^2 + k^2 - bN(x)h(x))u = 0 \text{ in } \mathbb{R}^3, \tag{3.85}$$

and u satisfies condition (3.47).

The conclusion is: the limiting medium is inhomogeneous, and its properties are described by the function

$$q(x) := bN(x)h(x). \tag{3.86}$$

This concludes our discussion of the wave scattering problem with many small impedance particles.

3.4 Creating Materials with a Desired Refraction Coefficient

Since the choice of the functions $N(x) \geq 0$ and $h(x)$, $\mathrm{Im}h(x) \leq 0$, is at our disposal, we can create the medium with a desired refraction coefficient by embedding many small impedance particles, with suitable impedances, according to the distribution law (3.50) with a suitable $N(x)$. The function

$$n_0^2(x) - k^{-2}q(x) = n^2(x) \tag{3.87}$$

is the refraction coefficient of the limiting medium, where $n_0^2(x)$ is the refraction coefficient of the original medium (see also Section 3.5). In (3.85), it is assumed that $n_0^2(x) = 1$. If $n_0^2(x) \neq 1$, then the operator $L_0 = \nabla^2 + k^2 n_0^2(x)$.

A recipe for creating material with a desired refraction coefficient can now be formulated.

Given a desired refraction coefficient $n^2(x)$, $\mathrm{Im}n^2(x) \geq 0$, one can find $N(x)$ and $h(x)$ so that (3.87) holds, where $q(x)$ is defined in (3.86), that is, one can create a material with a desired refraction coefficient by embedding into a given material many small particles with suitable boundary impedances and suitable distribution law.

3.5 Scattering by Small Particles Embedded in an Inhomogeneous Medium

Suppose that the operator $\nabla^2 + k^2$ in (3.1) and in (3.45) is replaced by the operator $L_0 = \nabla^2 + k^2 n_0^2(x)$, where $n_0^2(x)$ is a known function,

$$\text{Im } n_0^2(x) \geq 0. \tag{3.88}$$

The function $n_0^2(x)$ is the refraction coefficient of an inhomogeneous medium in which many small particles are embedded. The results, presented in Sections 3.1–3.3 remain valid if one replaces function $g(x,y)$ by the Green's function $G(x,y)$,

$$[\nabla^2 + k^2 n_0^2(x)]G(x,y) = -\delta(x-y), \tag{3.89}$$

satisfying the radiation condition. We assume that

$$n_0^2(x) = 1 \text{ in } D' := \mathbb{R}^3 \setminus D. \tag{3.90}$$

The function $G(x,y)$ is uniquely defined [14]. The derivations of the results remain essentially the same because

$$G(x,y) = g_0(x,y)[1 + O(|x-y|)], \quad |x-y| \to 0, \tag{3.91}$$

where $g_0(x,y) = \frac{1}{4\pi|x-y|}$. Estimates of $G(x,y)$ as $|x-y| \to 0$ and as $|x-y| \to \infty$ are obtained in [14]. Smallness of particles in an inhomogeneous medium with refraction coefficient $n_0^2(x)$ is described by the relation $kn_0 a \ll 1$, where $n_0 := \max_{x\in D}|n_0(x)|$, and $a = \max_{1\leq m\leq M}\text{diam}D_m$.

3.6 Conclusions

Analytic formulas for the scattering amplitudes for wave scattering by a single small particle are derived for small acoustically soft, or hard, or impedance particles.

The equation for the effective field in the medium, in which many small particles are embedded, is derived in the limit $a \to 0$. *The physical assumptions $a \ll d \ll \lambda$ are such that the multiple scattering effects are essential.* The derivations are rigorous.

On the basis of the developed theory, efficient numerical methods are proposed for solving many-body wave scattering problems in the case of small scatterers. These methods allow one to solve the problems, which earlier were not possible to solve.

A method for creating materials with a desired refraction coefficient is given and rigorously justified. Its practical implementation requires development of a method for preparing small particles with prescribed boundary impedances.

The physically novel point, compared with the known results for wave scattering by small bodies, is the dependence on the size a of the small scatterer, which is much larger than $O(a^3)$, the Rayleigh-type dependence, see, for example, formula (3.22), where the dependence on a is $O(\zeta |S_1|) = O(a^{2-\kappa})$. The formulas for the wave scattering by small particles of an arbitrary shape for various types of the boundary conditions are new. The equations for the effective field in the medium, in which many small particles with various BCs are embedded, are new.

In this chapter, we did not discuss the EM (electromagnetic waves) scattering and the related problems of creating materials with a desired refraction coefficient [1, 8, 37].

References

1 Ramm, A.G. (2013) Scattering of acoustic and electromagnetic waves by small bodies of arbitrary shapes, in *Applications to Creating New Engineered Materials*, Momentum Press, New York.

2 Ramm, A.G. (2013) Many-body wave scattering problems in the case of small scatterers. *J. Appl. Math Comput.*, **41** (1–2), 473–500.

3 Ramm, A.G. (2017) Creating materials with a desired refraction coefficient, IOP Concise Physics, Morgan and Claypool Publishers, San Rafael, CA, USA.

4 Rayleigh, J. (1992) *Scientific Papers*, Cambridge University Press, Cambridge.

5 Landau, L. and Lifschitz, L. (1984) *Electrodynamics of Continuous Media*, Pergamon Press, Oxford.

6 van de Hulst, H.C. (1961) *Light Scattering by Small Particles*, Dover Publications, New York.

7 Ramm, A.G. (2005) *Wave Scattering by Small Bodies of Arbitrary Shapes*, World Science Publishers, Singapore.

8 Ramm, A.G. (2015) Scattering of EM waves by many small perfectly conducting or impedance bodies. *J. Math. Phys.*, **56** (N9), 091901.

9 Ramm, A.G. and Tran, N. (2015) A fast algorithm for solving scalar wave scattering problem by billions of particles. *J. Algorithms Optim.*, **3** (1), 1–13.

10 Andriychuk, M. and Ramm, A.G. (2011) Numerical solution of many-body wave scattering problem for small particles and creating materials with desired refraction coefficient, Chapter in the book *Numerical Simulations of Physical and Engineering Processes*, (edited by J. Awrejcewicz), InTech, Vienna, pp. 1–28. ISBN: 978-953-307-620-1.

11 Andriychuk, M. and Ramm, A.G. (2012) Scattering of electromagnetic waves by many thin cylinders: theory and computational modeling. *Opt. Commun.*, **285** (20), 4019–4026.

12 Ramm, A.G. (1986) *Scattering by Obstacles*, D. Reidel, Dordrecht.

13 Ramm, A.G. (2017) Scattering by obstacles and potentials, World Sci. Publishers, Singapore.
14 Ramm, A.G. (2007) Many-body wave scattering by small bodies and applications. *J. Math. Phys.*, **48**, 103511.
15 Ramm, A.G. (2007) Scattering by many small bodies and applications to condensed matter physics. *Eur. Phys. Lett.*, **80**, 44001.
16 Ramm, A.G. (2007) Wave scattering by small particles in a medium. *Phys. Lett. A*, **367**, 156–161.
17 Ramm, A.G. (2007) Wave scattering by small impedance particles in a medium. *Phys. Lett. A*, **368**, 164–172.
18 Ramm, A.G. (2007) Distribution of particles which produces a "smart" material. *J. Stat. Phys.*, **127**, 915–934.
19 Ramm, A.G. (2007) Distribution of particles which produces a desired radiation pattern. *Physica B*, **394**, 253–255.
20 Ramm, A.G. (2008) Creating wave-focusing materials. *LAJSS (Lat.-Am. J. Solids Struct.)*, **5**, 119–127.
21 Ramm, A.G. (2008) Electromagnetic wave scattering by small bodies. *Phys. Lett. A*, **372**, 4298–4306.
22 Ramm, A.G. (2008) Wave scattering by many small particles embedded in a medium. *Phys. Lett. A*, **372**, 3064–3070.
23 Ramm, A.G. (2009) Preparing materials with a desired refraction coefficient and applications, in *the book "Topics in Chaotic Systems: Selected Papers from Chaos 2008 International Conference"* (eds C. Skiadas and I. Dimotikalis), World Science Publishing, pp. 265–273.
24 Ramm, A.G. (2009) Preparing materials with a desired refraction coefficient. *Nonlinear Anal. Theory Methods Appl.*, **70**, e186–e190.
25 Ramm, A.G. (2009) Creating desired potentials by embedding small inhomogeneities. *J. Math. Phys.*, **50**, 123525.
26 Ramm, A.G. (2010) A method for creating materials with a desired refraction coefficient. *Int. J. Mod. Phys. B*, **24**, 5261–5268.
27 Ramm, A.G. (2010) Materials with a desired refraction coefficient can be created by embedding small particles into the given material. *Int. J. Struct. Changes Solids (IJSCS)*, **2**, 17–23.
28 Ramm, A.G. (2011) Wave scattering by many small bodies and creating materials with a desired refraction coefficient. *Afr. Mat.*, **22**, 33–55.
29 Ramm, A.G. (2011) Scattering by many small inhomogeneities and applications, in *the book "Topics in Chaotic Systems: Selected Papers from Chaos 2010 International Conference"* (eds C. Skiadas and I. Dimotikalis), World Science Publishing, pp. 41–52.
30 Ramm, A.G. (2010) Collocation method for solving some integral equations of estimation theory. *Int. J. Pure Appl. Math.*, **62**, 57–65.
31 Ramm, A.G. (2011) Scattering of scalar waves by many small particles. *AIP Adv.*, **1**, 022135.

32 Ramm, A.G. (2011) Scattering of electromagnetic waves by many thin cylinders. *Results Phys.*, **1** (1), 13–16.

33 Ramm, A.G. (2012) Electromagnetic wave scattering by many small perfectly conducting particles of an arbitrary shape. *Opt. Commun.*, **285** (18), 3679–3683.

34 Ramm, A.G. (2013) Wave scattering by many small bodies: transmission boundary conditions. *Rep. Math. Phys.*, **71** (3), 279–290.

35 Jikov, V., Kozlov, S., and Oleinik, O. (1994) *Homogenization of Differential Operators and Integral Functionals*, Springer-verlag, Berlin.

36 Marchenko, V. and Khruslov, E. (2006) *Homogenization of Partial Differential Equations*, Birkhäuser, Boston, MA.

37 Ramm, A.G. (2013) Scattering of electromagnetic waves by many nano-wires. *Mathematics*, **1**, 89–99.

Appendix A2

WAVE SCATTERING BY MANY SMALL IMPEDANCE PARTICLES AND APPLICATIONS

Alexander G. Ramm

Department of Mathematics, Kansas State University,
Manhattan, KS 66506-2602, USA
(e-mail: ramm@ksu.edu)

(Received December 13, 2021 — Revised April 11, 2022)

Formulas are derived for solutions of many-body wave scattering problem by small impedance particles embedded in a homogeneous medium. The limiting case is considered, when the size a of small particles tends to zero while their number tends to infinity at a suitable rate. The basic physical assumption is $a \ll d \ll \lambda$, where d is the minimal distance between neighboring particles, λ is the wavelength, and the particles can be impedance balls $B(x_m, a)$ with centers x_m located on a grid. Equations for the limiting effective (self-consistent) field in the medium are derived. It is proved that one can create material with a desired refraction coefficient by embedding in a free space many small balls of radius a with prescribed boundary impedances. The small balls can be centered at the points located on a grid. A recipe for creating materials with a desired refraction coefficient is formulated. It is proved that materials with a desired radiation pattern, for example, wave-focusing materials, can be created.

PACS: 02.30.Rz; 02.30.Mv; 41.20.Jb

MSC: 35Q60;78A40; 78A45; 78A48;

Keywords: wave scattering by many small bodies, smart materials, grids.

1. Introduction

There is a large literature on wave scattering by small bodies, starting from Rayleigh's work (1871), [1, 2, 36]. For the problem of wave scattering by one body an analytical solution was found only for the bodies of special shapes, for example, for balls and ellipsoids. If the scatterer is small then the scattered field can be calculated analytically for bodies of arbitrary shapes, see [5], where this theory is presented.

The many-body wave scattering problem was discussed in the literature mostly numerically, if the number of scatterers is small, or under the assumption that the influence of the waves, scattered by other particles on a particular particle is negligible (see [3], where one finds a large bibliography, 1386 entries). This corresponds to the case when *the distance d between neighboring particles is much larger than the wavelength λ, and the characteristic size a of a small body (particle) is much smaller than λ.* Theoretically and practically the assumptions $a \ll \lambda$, $d >> \lambda$

Reprinted with permission. First appeared in *Reports on Mathematical Physics*, **90**(2), 2022, 193–202, ISSN 0034-4877.

are the simplest and they allow to neglect multiple scattering. By $k = \frac{2\pi}{\lambda}$ the wave number is denoted.

In contrast, in our theory the basic assumption is $a \ll d \ll \lambda$, and *the multiple scattering is of basic importance.* We give references to our papers and monographs in which the theory of wave scattering by small bodies of arbitrary shapes was developed under the assumption $a \ll d \ll \lambda$, [4–34]. The novelty of the results in this paper is in the location of the small bodies: *they are placed on a grid.* This may be of practical interest. In [35] for the first time the scattering problem for 10 billions small particles is solved numerically and numerical results are presented.

This paper is a presentation of the new results under simplifying assumptions: the small particles $D_m = B(x_m, a)$, $1 \le m \le M$, are impedance balls with prescribed boundary impedances ζ_m; the centers x_m of the balls are placed on a grid and are embedded in a homogeneous space in a bounded domain D, for example, in a box.

The basic results of this paper consist of:

i) Solution to *many-body wave scattering problem* by small impedance particles, embedded in a homogeneous medium, under the assumptions $a \ll d \ll \lambda$, where d is the minimal distance between neighboring particles and λ is the wavelength in this medium.

ii) Derivation of the equations for the limiting effective (self-consistent) field in this medium, in which many small impedance particles are embedded, when $a \to 0$ and the number $M = M(a)$ of the small particles tends to infinity at an appropriate rate.

iii) Derivation of linear algebraic systems (LAS) for solving many-body wave scattering problems. These systems are not obtained by a discretization of boundary integral equations, and they give an efficient numerical method for solving many-body wave scattering problems in the case of small scatterers under the assumption $a \ll d \ll \lambda$.

iv) Formulation of a recipe for creating materials with a desired refraction coefficient.

v) Formulation of a method for creating materials with a desired radiation pattern.

Our methods give powerful numerical methods for solving many-body wave scattering problems in the case when the scatterers are small (see [31]).

Let us formulate the wave scattering problems we deal with. Let D be a bounded domain in $\mathbb{R}^3$ with a sufficiently smooth boundary. The scattering problem consists of finding the solution to the problem:

$$(\nabla^2 + k^2)u = 0 \text{ in } G' := \mathbb{R}^3 \setminus G, \qquad G := \cup_{m=1}^{M} D_m, \qquad k = \text{const} > 0, \tag{1}$$

where $D_m = B(x_m, a)$ is an impedance ball, centered at x_m and of small radius a,

$$u = u_0 + v, \qquad u_0 = e^{ik\alpha\cdot x}, \qquad \alpha \in S^2, \tag{2}$$

S^2 is the unit sphere in $\mathbb{R}^3$, u_0 is the incident field, v is the scattered field satisfying the radiation condition

$$v_r - ikv = o\left(\frac{1}{r}\right), \qquad r := |x| \to \infty, \qquad v_r := \frac{\partial v}{\partial r}, \tag{3}$$

and u satisfies the impedance boundary condition (bc) on the boundary of G,

$$u_N - \zeta_m u = 0, \qquad \text{on } S_m, \qquad \text{Im}\zeta_m \le 0, \tag{4}$$

where ζ_m is a constant, N is the unit normal to $S := \cup_{m=1}^M S_m$, pointing out of $G := \cup_{m=1}^M D_m$, and S_m is the surface of $D_m = B(x_m, a)$.

By refraction coefficient $n(x)$ the coefficient in the equation

$$(\nabla^2 + k^2 n^2(x))u = (\nabla^2 + k^2 - q(x))u = 0 \tag{5}$$

is understood, where $q(x) := k^2(n^2(x) - 1)$.

Let

$$g(x, y) = \frac{e^{ik|x-y|}}{4\pi|x-y|}.$$

Then $(\nabla^2 + k^2)g(x, y) = -\delta(x - y)$, where $\delta(x)$ is the delta function.

Let us distribute small impedance particles $D_m = B(x_m, a)$ in D so that

$$\mathbb{N}(\Delta) = a^{\kappa-2}|\Delta|[1 + o(1)], \qquad a \to 0, \tag{6}$$

where $\Delta \subset D$ is an arbitrary connected open subset of D, $|\Delta|$ is its volume, $\kappa \in (0, 1)$ is a number the experimenter may choose arbitrarily and $\mathbb{N}(\Delta)$ is the number of particles in Δ. Throughout this paper the important assumptions $a \ll d \ll \lambda$ and (6) are satisfied. As $a \to 0$ the number of small particles $\mathbb{N}(\Delta)$ in (6) tends to infinity since $\kappa - 2 < 0$.

The boundary impedances ζ_m are chosen by the formula

$$\zeta_m = a^{-\kappa} h(x_m), \tag{7}$$

where $h(x)$ is a continuous function in D, $\text{Im}h \le 0$.

It will be clear from Section 3 that the function $h(x)$ can be determined by choosing a suitable boundary impedance $\zeta(x)$. When $a \to 0$, the ζ_m and $h(x_m)$ can be considered as continuous functions $\zeta(x)$ and $h(x)$.

The many-body scattering problem (1)–(4) has a solution and this solution is unique, see [31]. In Section 2 a method for solving this problem is given. In Section 3 a recipe for creating materials with a desired refraction coefficient is given. In Section 4 a recipe for creating materials with a desired radiation pattern is given.

2. Solution of many-body scattering problem

We look for the solution of the form

$$u = u_0 + \sum_{m=1}^{M} \int_{S_m} g(x, s)\sigma_m(s)ds = \sum_{m=1}^{M} g(x, x_m)Q_m + J, \tag{8}$$

where $\sigma_m(s)$ are unknown, $Q_m := \int_{S_m} \sigma_m(s)ds$. One may think about σ_m as of charge densities on S_m and of Q_m as of total charge on the surface S_m. We prove that

$$J := \sum_{m=1}^{M} \int_{S_m} [g(x, s) - g(x, x_m)]\sigma_m(s)ds$$

is negligible compared to

$$I := \sum_{m=1}^{M} g(x, x_m)Q_m, \qquad J \ll I$$

as $a \to 0$.

Let us prove this claim. First, we need the following lemma.

LEMMA 1. *One has:*

$$Q_m = -4\pi a^2 \zeta_m u_m = -4\pi a^{2-\kappa} h_m u_m, \qquad h_m := h(x_m), \qquad u_m := u(x_m). \tag{9}$$

Proof: Let us define *the effective field acting on the m-th body,*

$$u_e := u_e^m := u - \int_{S_m} g(x, s)\sigma_m(s)ds.$$

If a is small, then $u(x) \sim u_e(x)$ for any x such that $|x - x_m| \geq d$. Let us use the exact boudary condition (4) for u_e and the known formula for the normal derivative of the single layer potential to get

$$u_{eN} + (A\sigma_m - \sigma_m)/2 - \zeta_m u_{em} - \zeta_m \int_{S_m} g(x, s)\sigma_m(s)ds = 0. \tag{10}$$

Here $A\sigma := \int_{S_m} g_{N_t}(t, s)\sigma_m(s)ds$, $t \in S_m$. Let us integrate (10) over S_m and keep the main term as $a \to 0$. One knows that $\int_{S_m} (A\sigma - \sigma)/2dt = -Q_m$. Furthermore, $\int_{S_m} g(t, s)ds = a$, as one can check by a simple calculation using the fact that S_m is a sphere of radius a. This allows one to conclude that

$$\zeta_m \int_{S_m} ds\sigma_m(s) \int_{S_m} g(t, s)dt = h_m a^{1-\kappa} Q_m, \qquad \zeta_m \int_{S_m} u_e ds = -4\pi a^{2-\kappa} h_m u_{em}$$

and $\int_{S_m} u_{eN} ds = O(a^2)$ as $a \to 0$. From the above estimates the conclusion of Lemma 1 follows. □

Let us now check our claim $J \ll I$ as $a \to 0$. One has

$$g(x, x_m)Q_m = O(a^{2-\kappa} d^{-1})$$

for $|x - x_m| > d$, $a \to 0$. On the other hand, one derives

$$\left| \int_{S_m} [g(t, s) - g(x, x_m)]\sigma_m(s)ds \right| \leq O(ad^{-2}a^{2-\kappa}) = O\left(\frac{a}{d}\right) O(a^{2-\kappa} d^{-1}).$$

This estimate justifies our claim since $a \ll d$. It follows that asymptotically, as $a \to 0$, one has

$$u \sim u_0 + \sum_{m=1}^{M} g(x, x_m)Q_m \sim u_0 - 4\pi a^{2-\kappa} \sum_{m=1}^{M} g(x, x_m)h_m u_m, \tag{11}$$

for $|x - x_m| \geq a$. Note that $M = O(a^{\kappa-2})$. Formula (11) allows one to calculate $u(x)$ at any point x, if the numbers u_m, $1 \leq m \leq M$, are known. One can use the following linear algebraic system (LAS) for finding u_m,

$$u_j = u_{0j} - 4\pi a^{2-\kappa} \sum_{m \neq j}^{M} g(x_j, x_m) h_m u_m, \qquad 1 \leq j \leq M. \tag{12}$$

The order $M = O(a^{\kappa-2})$ of this system is large if a is small. One can reduce this order: consider a covering of D by nonintersecting small cubes Δ_p, $1 \leq p \leq P$, such that $d \ll \text{diam}(\Delta_p) \ll \lambda$, $u_m \sim u_p$, $h_m \sim h_p$ for all $x_m \in \Delta_p$. Then formula (12) can be written as

$$u_q = u_{0q} - 4\pi a^{2-\kappa} \sum_{p \neq q}^{P} g(x_q, x_p) h_p u_p \sum_{x_m \in \Delta_p} 1 = u_{0q} - 4\pi \sum_{p \neq q}^{P} g(x_q, x_p) h_p u_p |\Delta_p|, \tag{13}$$

where

$$a^{2-\kappa} \sum_{x_m \in \Delta_p} 1 = |\Delta_p|$$

by formula (6). As $a \to 0$, $\text{diam}(\Delta_p) \to 0$ and formula (13) yields in the limit the integral equation for u,

$$u(x) = u_0(x) - 4\pi \int_D g(x, y) h(y) u(y) dy. \tag{14}$$

Lemma 2. *Eq.* (14) *has a solution, this solution is unique and it is a limiting value of the solution to the scattering problem* (1)–(4).

Proof: Apply the operator $\nabla^2 + k^2$ to equation (14) and get

$$(\nabla^2 + k^2) u = 4\pi h(x) u(x). \tag{15}$$

This is a Schrödinger equation with potential $q(x) := 4\pi h(x)$; equations (2)–(3) hold. We assumed $\text{Im} h \leq 0$. Therefore (15) has at most one solution. It is a Fredholm-type equation, so it has a solution. Lemma 2 is proved. □

It follows from Lemma 2 that the LAS (13) for u_p is solvable and its solution is unique. Let us write Eq. (15) as

$$\nabla^2 u + k^2 n^2(x) u = 0, \qquad n^2(x) := 1 - 4\pi k^{-2} h(x). \tag{16}$$

Conclusion. *Embedding small impedance balls $B(x_m, a)$ in D results in creating in D a new material with the refraction coefficient*

$$n(x) = (1 - 4\pi k^{-2} h(x))^{1/2}. \tag{17}$$

If one wants to have a material with the refraction coefficient $n(x)$, then one chooses by (17) the function $h(x)$. If $h(x)$ is chosen, then one knows the boundary impedance $\zeta(x)$ which generates the desired $h(x)$. The practical problem is to prepare small particles with the desired boundary impedance.

3. Recipe for creating materials with a desired refraction coefficient

Let us formulate a recipe for creating materials with a desired refraction coefficient. Formula (17) shows that if $h(x)$ is chosen properly, then any $n(x)$ can be obtained in D.

Recipe for creating materials with a desired refraction coefficient:

a) Calculate by formula (17) *the function* $h(x)$*;*

b) Distribute small impedance balls in the domain D *by the distribution law* (6). *The boundary impedances of these balls are defined by the function* $h(x)$.

THEOREM 1. *The refraction coefficient of the resulting medium tends to the desired coefficient* $n(x)$ *as* $a \to 0$.

Let us show that a practically negative refraction coefficient $n(x)$ can be obtained by the above recipe. Denote $b := 4\pi k^{-2} > 0$ and write (17) as $n(x) = (1-bh(x))^{1/2} = |1-bh(x)|^{1/2}e^{\phi/2}$, where ϕ is the argument of $1-bh(x)$. Since the operator in (14) is Fredholm, it remains Fredholm under small perturbations. Therefore one can take $h - i\epsilon$, where $\epsilon > 0$ is sufficiently small and equation (14) will still have a unique solution.

By choosing h so that $\mathrm{Re}(1 - bh) > 0$ and $\mathrm{Im}(1 - bh) < 0$ and small, one gets the argument $\phi = 2\pi - \delta$, where $\delta > 0$ is arbitrarily small if ϵ is sufficiently small. Then $n(x)$ will be nearly negative: its argument will be $\pi - \delta/2$.

4. Creating materials with a desired radiation pattern

Let us define what we mean by radiation pattern. Consider the scattering problem for Eq. (15),

$$\nabla^2 u + k^2 u - q(x)u = 0, \qquad u = e^{ik\alpha\cdot x} + v, \tag{18}$$

where v satisfies the radiation condition. Assume that $k > 0$ and $\alpha \in S^2$ are fixed. Then the scattering amplitude $A(\beta, \alpha, k) = A(\beta)$, where the dependence on k, α is dropped since k and α are fixed. The formula for the scattering amplitude is known, see, e.g. [34],

$$A(\beta) := A_q(\beta) = -\frac{1}{4\pi}\int e^{-ik\beta\cdot y}q(y)u(y)dy. \tag{19}$$

We call $A(\beta)$ the radiation pattern.

Consider an inverse problem (IP):

Given an arbitrary $f(\beta) \in L^2(S^2)$ *and an arbitrary small* $\epsilon > 0$, *can one find a* $q \in L^2(D)$ *such that*

$$\|f(\beta) - A_q(\beta)\|_{L^2(S^2)} < \epsilon. \tag{20}$$

THEOREM 2. *For any* $f(\beta) \in L^2(S^2)$ *and an arbitrary small* $\epsilon > 0$ *there is a* $q \in L^2(D)$ *such that* (20) *holds.*

Since small perturbations of q result in small perturbations of $A(\beta)$, there are infinitely many potentials q for which inequality (20) holds.

The conclusion of Theorem 2 follows from Lemmas 3 and 4.

LEMMA 3. *The set* $\{\int_D e^{-ik\beta\cdot x}h(x)dx\}_{\forall h\in L^2(D)}$ *is dense in* $L^2(S^2)$.

COROLLARY 1. *Given* $f\in L^2(S^2)$ *and* $\epsilon>0$, *one can find* $h\in L^2(D)$ *such that*

$$\|f(\beta)+\frac{1}{4\pi}\int_D e^{-ik\beta\cdot x}h(x)dx\|<\epsilon.$$

LEMMA 4. *The set* $\{q(x)u(x,\alpha)\}_{\forall q\in L^2(D)}$ *is dense in* $L^2(D)$.

COROLLARY 2. *Given* $h\in L^2(D)$ *and* $\epsilon>0$, *one can find* $q\in L^2(D)$ *such that*

$$\|h(x)-q(x)u(x,\alpha)\|_{L^2(D)}<\epsilon.$$

Since the scattering amplitude

$$A(\beta)=-\frac{1}{4\pi}\int_D e^{-ik\beta\cdot x}h(x)dx$$

depends continuously on h, *the inverse problem* **IP** *is solved by Lemmas* 3 *and* 4.

Proof of Lemma 3*:* Assume the contrary. Then $\exists\ \psi\in L^2(S^2)$ such that

$$0=\int_{S^2}d\beta\psi(\beta)\int_D e^{-ik\beta\cdot x}h(x)dx \qquad \forall h\in L^2(D).$$

Thus,

$$\int_{S^2}d\beta\psi(\beta)e^{-ik\beta\cdot x}=0 \qquad \forall x\in\mathbb{R}^3.$$

Therefore,

$$\int_0^\infty d\lambda\lambda^2\int_{S^2}d\beta e^{-i\lambda\beta\cdot x}\psi(\beta)\frac{\delta(\lambda-k)}{k^2}=0 \qquad \forall x\in\mathbb{R}^3.$$

By the injectivity of the Fourier transform, one gets

$$\psi(\beta)\frac{\delta(\lambda-k)}{k^2}=0.$$

Therefore, $\psi(\beta)=0$. Lemma 3 is proved. □

Proof of Lemma 4*:* Given $h\in L^2(D)$, define

$$u:=u_0-\int_D g(x,y)h(y)dy, \qquad g:=\frac{e^{ik|x-y|}}{4\pi|x-y|}, \tag{21}$$

$$q(x):=\frac{h(x)}{u(x)}. \tag{22}$$

If $q\in L^2(D)$, then this q solves the problem, and u, defined in (21), is the scattering solution,

$$u=u_0-\int_D g(x,y)q(y)u(y)dy, \tag{23}$$

and

$$A(\beta) = -\frac{1}{4\pi}\int_D e^{-ik\beta\cdot y}h(y)dy.$$

If q is not in $L^2(D)$, then the null set $N := \{x : x \in D, u(x) = 0\}$ is non-void. Let

$$N_\delta := \{x : |u(x)| < \delta, x \in D\}, \qquad D_\delta := D \setminus N_\delta.$$

Claim 1. $\exists h_\delta = \begin{cases} h, in\ D_\delta, \\ 0, in\ N_\delta, \end{cases}$ *such that* $\|h_\delta - h\|_{L^2(D)} < c\epsilon$,

$$q_\delta := \begin{cases} \frac{h_\delta}{u_\delta}, in\ D_\delta, \\ 0, \quad in\ N_\delta, \end{cases} \quad q_\delta \in L^\infty(D), \qquad u_\delta := u_0 - \int_D gh_\delta dy.$$

Proof of Claim 1*:* The set N is, generically, a line $l = \{x : u_1(x) = 0, u_2(x) = 0\}$, where $u_1 = \Re u$ and $u_2 = \Im u$. Consider a tubular neighborhood of this line, $\rho(x, l) \le \delta$. Let the origin O be chosen on l, s_3 be the Cartesian coordinate along the tangent to l, and $s_1 = u_1$, $s_2 = u_2$ are coordinates in the plane orthogonal to l, s_j-axis is directed along $\nabla u_j|_l$, $j = 1, 2$.

The Jacobian $\mathcal{J}$ of the transformation $(x_1, x_2, x_3) \mapsto (s_1, s_2, s_3)$ is nonsingular, $|\mathcal{J}| + |\mathcal{J}^{-1}| \le c$, because ∇u_1 and ∇u_2 are linearly independent. Define

$$h_\delta := \begin{cases} h, \text{in } D_\delta, \\ 0, \text{in } N_\delta, \end{cases} \qquad u_\delta := u_0 - \int_D g(x,y)h_\delta(y)dy, \qquad q_\delta := \begin{cases} \frac{h_\delta}{u_\delta}, \text{in } D_\delta, \\ 0, \quad \text{in } N_\delta. \end{cases}$$

One has $u_\delta = u_0 - \int_D ghdy + \int_D g(x,y)(h - h_\delta)dy$,

$$|u_\delta(x)| \ge |u(x)| - c\int_{N_\delta} \frac{dy}{4\pi|x-y|} \ge \delta - I(\delta), \qquad x \in D_\delta, \qquad c = \max_{x\in N_\delta}|h(x)|.$$

If one proves that $I(\delta) = o(\delta)$, $\delta \to 0$, $\forall x \in D_\delta$ then $q_\delta \in L^\infty(D)$, and Claim 1 is proved. □

Claim 2.

$$I(\delta) = O(\delta^2|\ln(\delta)|), \qquad \delta \to 0.$$

Proof of Claim 2:

$$\int_{N_\delta} \frac{dy}{|x-y|} \le \int_{N_\delta} \frac{dy}{|y|} = c_1 \int_0^{c_2\delta} \rho \int_0^1 \frac{ds_3}{\sqrt{\rho^2 + s_3^2}} d\rho$$

$$= c_1 \int_0^{c_2\delta} d\rho\rho \ln(s_3 + \sqrt{\rho^2 + s_3^2})|_0^1 \le c_3 \int_0^{c_2\delta} \rho \ln\left(\frac{1}{\rho}\right) d\rho$$

$$\le O(\delta^2|\ln(\delta)|).$$

The condition $|\nabla u_j|_l \geq c > 0$, $j = 1, 2$, implies that a tubular neighborhood of the line l, $N_\delta = \{x : \sqrt{|u_1|^2 + |u_2|^2} \leq \delta\}$, is included in a region $\{x : |x| \leq c_2\delta\}$ and includes a region $\{x : |x| \leq c_2'\delta\}$. This follows from the estimates

$$c_2'\rho \leq |u(x)| = |\nabla u(\xi) \cdot (x - \xi)| \leq c_2\rho.$$

Here $\xi \in l$, x is a point on a plane passing through ξ and orthogonal to l, $\rho = |x - \xi|$, and $\delta > 0$ is sufficiently small, so that the terms of order ρ^2 are negligible, $c_2 = \max_{\xi \in l} |\nabla u(\xi)|$, $\quad c_2' = \min_{\xi \in l} |\nabla u(\xi)|$.

Claim 2, and, therefore, Lemma 4 are proved. □

Therefore, Theorem 2 is proved. □

Let us describe a numerical method for calculation of h given $f(\beta)$ and $\epsilon > 0$. Let $\{\phi_j\}$ be a basis in $L^2(D)$, $h_n = \sum_{j=1}^n c_j^{(n)} \phi_j$, $\psi_j(\beta) := -\frac{1}{4\pi} \int_D e^{-ik\beta \cdot x} \phi_j(x) dx$. Consider the problem

$$\|f(\beta) - \sum_{j=1}^{n} c_j^{(n)} \psi_j(\beta)\| = \min. \tag{24}$$

A necessary condition for (24) is a linear algebraic system for $c_j^{(n)}$.

REFERENCES

[1] H. C. van de Hulst: *Light Scattering by Small Particles*, Dover, New York 1961.
[2] L. Landau, L. Lifschitz: *Electrodynamics of Continuous Media*, Pergamon Press, Oxford 1984.
[3] P. Martin: *Multiple Scattering*, Cambridge Univ. Press, Cambridge 2006.
[4] A. G. Ramm: *Scattering by Obstacles*, D. Reidel, Dordrecht 1986.
[5] A. G. Ramm: *Wave Scattering by Small Bodies of Arbitrary Shapes*, World Sci. Publishers, Singapore 2005.
[6] A. G. Ramm: Scattering by many small bodies and applications to condensed matter physics, *Europ. Phys. Lett.* **80**, 44001 (2007).
[7] A. G. Ramm: Many-body wave scattering by small bodies and applications, *J. Math. Phys.* **48**, 103511 (2007).
[8] A. G. Ramm: Wave scattering by small particles in a medium, *Phys. Lett. A* **367**, 156–161 (2007).
[9] A. G. Ramm: Wave scattering by small impedance particles in a medium, *Phys. Lett. A* **368**, 164–172 (2007).
[10] A. G. Ramm: Distribution of particles which produces a desired radiation pattern, *Communic. in Nonlinear Sci. and Numer. Simulation*, **12**, N7, 1115–1119 (2007).
[11] A. G. Ramm: Distribution of particles which produces a "smart" material, *J. Stat. Phys.* **127**, 915–934 (2007).
[12] A. G. Ramm: Distribution of particles which produces a desired radiation pattern, *Physica B* **394**, 253–255 (2007).
[13] A. G. Ramm: Creating wave-focusing materials, *LAJSS (Latin-American J. Solids Structures)* **5**, 119–127 (2008).
[14] A. G. Ramm: Electromagnetic wave scattering by small bodies, *Phys. Lett. A* **372**, 4298–4306 (2008).
[15] A. G. Ramm: Wave scattering by many small particles embedded in a medium, *Phys. Lett. A* **372**, 3064–3070 (2008).
[16] A. G. Ramm: Preparing materials with a desired refraction coefficient and applications, *In the book "Topics in Chaotic Systems: Selected Papers from Chaos 2008 International Conference"*, Editors C. Skiadas, I. Dimotikalis, Char. Skiadas, World Scientific Publishing, pp. 265–273 (2009).

[17] A. G. Ramm: Preparing materials with a desired refraction coefficient, *Nonlinear Analysis: Theory, Methods Appl.* **70**, e186-e190 (2009).

[18] A. G. Ramm: Creating desired potentials by embedding small inhomogeneities, *J. Math. Phys.* **50**, 123525 (2009).

[19] A. G. Ramm: A method for creating materials with a desired refraction coefficient, *Int. J. Mod. Phys B* **24**, 5261–5268 (2010).

[20] A. G. Ramm: Materials with a desired refraction coefficient can be created by embedding small particles into the given material, *IJSCS* **2**, 17–23 (2010).

[21] A. G. Ramm: Wave scattering by many small bodies and creating materials with a desired refraction coefficient, *Afrika Matematika* **22**, 33–55 (2011).

[22] A. G. Ramm: Scattering by many small inhomogeneities and applications, in: *Topics in Chaotic Systems: Selected Papers from Chaos 2010 International Conference*, Editors C. Skiadas, I. Dimotikalis, Char. Skiadas, World Sci. Publishing, pp. 41–52 (2011).

[23] A. G. Ramm: Collocation method for solving some integral equations of estimation theory, *Int. J. Pure Appl. Math.* **62**, 57–65 (2010).

[24] A. G. Ramm: A method for creating materials with a desired refraction coefficient, *Int. J. Mod. Phys. B* **24**, 27, 5261–5268 (2010).

[25] A. G. Ramm: Electromagnetic wave scattering by a small impedance particle of arbitrary shape, *Opt. Commun.* **284**, 3872–3877 (2011).

[26] A. G. Ramm: Scattering of scalar waves by many small particles, *AIP Advances* **1**, 022135 (2011).

[27] A. G. Ramm: Scattering of electromagnetic waves by many thin cylinders, *Results in Physics* **1**, N1, 13–16 (2011).

[28] A. G. Ramm: Electromagnetic wave scattering by many small perfectly conducting particles of an arbitrary shape, *Opt. Commun.* **285**, N18, 3679–3683 (2012).

[29] A. G. Ramm: Electromagnetic wave scattering by small impedance particles of an arbitrary shape, *JAMC* **43**, N1, 427–444 (2013).

[30] A. G. Ramm: Many-body wave scattering problems in the case of small scatterers, *JAMC* **41**, N1, 473–500 (2013).

[31] A. G. Ramm: *Scattering of Acoustic and Electromagnetic Waves by Small Bodies of Arbitrary Shapes. Applications to Creating New Engineered Materials*, Momentum Press, New York 2013.

[32] A. G. Ramm: *Creating Materials with a Desired Refraction Coefficient*, IOP Publishers, Bristol, UK, 2020 (Second edition).

[33] A. G. Ramm: How can one create a material with a prescribed refraction coefficient? *Sun Text Review of Material Science* **1:1**, 102 (2020).

[34] A. G. Ramm: *Scattering by Obstacles and Potentials*, World Sci. Publ., Singapore 2017.

[35] A. G. Ramm: N. Tran, A fast algorithm for solving scalar wave scattering problem by billions of particles, *J. Algorithms Optimization* **3**, N1 (2015), 1–13.
Open access: http://www.academicpub.org/jao/Issue.aspx?Abstr=false

[36] J. Rayleigh, *Scientific Papers*, Cambridge Univ. Press, Cambridge 1992.

Appendix A3

ORIGINAL RESEARCH ARTICLE

Is creating materials with a desired refraction coefficient practically possible?

Alexander G. Ramm

Department of Mathematics, Kansas State University, Manhattan, KS 66506-2602, USA. E-mail: ramm@ksu.edu

ABSTRACT

A theory of many-body wave scattering is developed under the assumption $a << d << \lambda$, where a is the characteristic size of the small body, d is the distance between neighboring bodies and λ is the wave-length in the medium in which the bodies are embedded. The multiple scattering is essential under these assumptions. The author's theory is used for creating materials with a desired refraction coefficient. This theory can be used in practice. A recipe for creating materials with a desired refraction coefficient is formulated. Materials with a desired radiation pattern, for example, wave-focusing materials, can be created.

PACS: 02.30.Rz; 02.30.Mv; 41.20.Jb

MSC: 35Q60; 78A40; 78A45; 78A48

Keywords: Wave Scattering by Many Small Bodies; Smart Materials

ARTICLE INFO

Received: 8 February 2023
Accepted: 2 March 2023
Available online: 16 March 2023

1. Introduction

The aim of this paper is to give an affirmative answer to the question in the title of this paper. This brings potentially many possibilities for progress in technology.

There is a large literature on wave scattering by small bodies, starting from Rayleigh's work (1871)[1–3]. If the scatterer is small then the scattered field can be calculated analytically for bodies of arbitrary shapes, see reference [4].

The many-body wave scattering problem was discussed in the literature mostly numerically, if the number of scatterers was small, or under the assumption that the influence of the waves, scattered by other particles on a particular particle is negligible[5]. This corresponds to the case when the distance d between neighbouring particles is much larger than the wavelength λ, and the characteristic size a of a small body (particle) is much smaller than λ. Theoretically and practically the assumptions

$$a << \lambda, \quad d >> \lambda, \tag{1}$$

are the simplest ones which allow one to neglect multiple scattering. By $k = \frac{2\pi}{\lambda}$, the wave number is denoted.

In the author's theory, the basic assumptions are

$$a << d << \lambda, \tag{2}$$

and the multiple scattering is of basic importance under these assumptions[4,6–35]. It is clear that assumption (2) can be practically realized. Its

---

importance comes from the fact that the author gave a rigorous asymptotically exact solution of the many-body scattering problem under assumption (2) when $a \to 0$. This solution can be well approximated numerically by the particles of the size $a > 30$ nm. Practically the size of a can be found by comparison of the solution for some a and for $\frac{a}{2}$. If these solutions are practically close, then one considers this a as suitable. The aim of this paper is to show that our theory can be used practically.

In reference [36], for the first time the author's theory was used for solving the scattering problem for 10 billion small particles. This problem was solved numerically and numerical results were presented.

Let us formulate the wave scattering problems we deal with. Let D be a bounded domain in $\mathbb{R}^3$ with a sufficiently smooth boundary. The scattering problem consists of finding the solution to the problem:

$$(\nabla^2 + k^2)u = 0 \text{ in } G' := \mathbb{R}^3 \setminus G,\ G := \cup_{m=1}^{M} D_m,\quad k = const > 0, \tag{3}$$

where $D_m = B(x_m, a)$ is an impedance ball, centered at x_m and of small radius a,

$$u = u_0 + v,\ u_0 = e^{ik\alpha \cdot x},\ \alpha \in S^2, \tag{4}$$

S^2 is the unit sphere in $\mathbb{R}^3$, u_0 is the incident field, v is the scattered field satisfying the radiation condition

$$v_r - ikv = o(\tfrac{1}{r}),\ r := |x| \to \infty,\ v_r := \frac{\partial v}{\partial r}, \tag{5}$$

and u satisfies the impedance boundary condition (bc) on the boundary of G:

$$u_N - \zeta_m u = 0,\quad \text{on } S_m,\quad \mathrm{Im}\zeta_m \le 0, \tag{6}$$

where ζ_m is a constant, N is the unit normal to $S := \cup_{m=1}^{M} S_m$, pointing out of $G := \cup_{m=1}^{M} D_m$, and S_m is the surface of $D_m = B(x_m, a)$.

By refraction coefficient $n(x)$ the coefficient in the equation

$$(\nabla^2 + k^2 n^2(x))u = (\nabla^2 + k^2 - q(x))u = 0 \tag{7}$$

is understood, where $q(x) := k^2(n^2(x) - 1)$.

Let $g(x, y) = \frac{e^{ik|x-y|}}{4\pi|x-y|}$. Then $(\nabla^2 + k^2)g(x, y) = -\delta(x - y)$, where $\delta(x)$ is the delta function.

Let us distribute small impedance particles $D_m = B(x_m, a)$ in D so that

$$\mathbb{N}(\Delta) = a^{\kappa-2}|\Delta|[1 + o(1)],\ a \to 0, \tag{8}$$

where $\Delta \subset D$ is an arbitrary connected open subset of D, $|\Delta|$ is its volume, $\kappa \in (0, 1)$ is a number the experimenter may choose arbitrarily and $\mathbb{N}(\Delta)$ is the number of particles in Δ. Throughout this paper the important assumptions $a \ll d \ll \lambda$ and (8) are satisfied. As $a \to 0$, the number of small particles $\mathbb{N}(\Delta)$ in (8) tends to infinity since $\kappa - 2 < 0$.

We assume in this paper (for simplicity only) that the small particles are distributed in the domain D and the refraction coefficient in D equals to 1. In the monograph [31], it is assumed that D is filled with the material whose refraction coefficient $n_0(x)$ is known and we wanted to create in D the material with the desired refraction coefficient $n(x)$.

The boundary impedances ζ_m are chosen by the formula

$$\zeta_m = a^{-\kappa} h(x_m), \tag{9}$$

where $h(x)$ is a continuous function in D, $\mathrm{Im}h \le 0$.

It will be clear from Section 3 that the function $h(x)$ can be determined by choosing a suitable boundary impedance $\zeta(x)$. When $a \to 0$, the ζ_m and $h(x_m)$ can be considered as continuous functions $\zeta(x)$ and $h(x)$.

2. Solution of many-body scattering problem

We look for the solution of the form

$$u = u_0 + \sum_{m=1}^{M} \int_{S_m} g(x, s)\sigma_m(s)ds = \sum_{m=1}^{M} g(x, x_m)Q_m + J, \tag{10}$$

where $\sigma_m(s)$ are unknown, $Q_m := \int_{S_m} \sigma_m(s)ds$. One may think about σ_m as of charge densities on S_m and of Q_m as of total charge on the surface S_m. We prove that

$$J := \sum_{m=1}^{M} \int_{S_m} [g(x,s) - g(x,x_m)]\sigma_m(s)ds \tag{11}$$

is negligible compared to

$$I := \sum_{m=1}^{M} g(x, x_m)Q_m, \tag{12}$$

so

$$J << I \text{ as } a \to 0. \tag{13}$$

We prove that the field u satisfies the following integral equation as $a \to 0$:

$$u(x) = u_0(x) - 4\pi \int_D g(x,y)h(y)u(y)dy, \tag{14}$$

where $h(x_m) = \frac{\zeta_m}{a^\kappa}$, and, since there are sufficiently many points $x_m \in D$, the function $h(x)$ is uniquely determined in D if the boundary impedances are known.

Apply the operator to $\nabla^2 + k^2$ to both sides of equation (14) and get

$$(\nabla^2 + k^2 - 4\pi h(x))u(x) := (\nabla^2 + k^2 n^2(x))u(x) = 0 \tag{15}$$

Therefore,

$$n^2(x) = 1 - 4\pi k^{-2} h(x). \tag{16}$$

We omit details since they can be found in the author's publications listed in the References, in particular, in monograph [31].

If originally in D were material with the known refraction coefficient $n_0(x)$, then formula (16) were $n^2(x) = n_0^2(x) - 4\pi h(x)N(x)k^{-2}$, where $N(x)$ is the distribution density for the small particles, see reference [31]. In this paper, we assume (for simplicity only) that $N(x) = 1$, see formula (8).

3. Recipe for creating materials with a desired refraction coefficient

Let us formulate a recipe for creating materials with a desired refraction coefficient. Formula (16) shows that if $h(x)$ is chosen properly, then any $n(x)$ can be obtained in D.

Recipe for creating materials with a desired refraction coefficient:

a) Calculate by formula (16) *the function* $h(x)$*;*

b) Distribute small impedance balls in the domain D *by the distribution law* (8)*. The boundary impedances of these balls are defined by the function* $h(x)$.

Theorem 1. *The refraction coefficient of the resulting medium tends to the desired coefficient* $n(x)$ *as* $a \to 0$.

Let us show that practically negative refraction coefficient $n(x)$ can be obtained by the above recipe. Denote $b := 4\pi k^{-2} > 0$ and write equation (16) as

$$n(x) = (1 - bh(x))^{1/2} = |1 - bh(x)|^{1/2} e^{\phi/2}, \tag{17}$$

where ϕ is the argument of $1 - bh(x)$. Since the operator in (14) is of Fredholm type, it remains Fredholm type under small perturbations. Therefore one can take $h - i\epsilon$, where $\epsilon > 0$ is sufficiently small, and equation (14) will still have a unique solution.

By choosing h so that $\mathrm{Re}(1 - bh) > 0$ and $\mathrm{Im}(1 - bh) < 0$ and small, one gets the argument $\phi = 2\pi - \delta$, where $\delta > 0$ is arbitrarily small if ϵ is sufficiently small. Then $n(x)$ will be nearly negative: its argument will be $\pi - \delta/2$.

4. Creating materials with a desired radiation pattern

Let us define what we mean by the radiation pattern. Consider the scattering problem for the equation:

$$\nabla^2 u + k^2 u - q(x)u = 0, \; u = e^{ik\alpha\cdot x} + v, \tag{18}$$

where v satisfies the radiation condition. Assume that $k > 0$ and $\alpha \in S^2$ are fixed. Then the scattering amplitude $A(\beta, \alpha, k) = A(\beta)$, where the dependence on k, α is dropped since k and α are fixed. The formula for the scattering amplitude is known, see, e.g., reference [35]:

$$A(\beta) := A_q(\beta) = -\frac{1}{4\pi}\int e^{-ik\beta\cdot y} q(y)u(y)dy. \tag{19}$$

We call $A(\beta)$ the radiation pattern.

Consider an inverse problem (IP):

Given an arbitrary $f(\beta) \in L^2(S^2)$ *and an arbitrary small* $\epsilon > 0$, *can one find a* $q \in L^2(D)$ *such that*

$$\|f(\beta) - A_q(\beta)\|_{L^2(S^2)} < \epsilon. \tag{20}$$

This inverse problem was not formulated and was not studied in the works of other authors, to our knowledge.

Our result is stated in Theorem 2.

Theorem 2. *For any $f(\beta) \in L^2(S^2)$ and an arbitrary small $\epsilon > 0$ there is a $q \in L^2(D)$ such that* (20) *holds.*

Since small perturbations of q result in small perturbations of $A(\beta)$, there are infinitely many potentials q for which inequality (20) holds.

The conclusion of Theorem 2 follows from lemmas 3 and 4.

Lemma 3. The set

$\left\{\int_D e^{-ik\beta\cdot x}h(x)dx\right\}_{\forall h\in L^2(D)}$ is dense in $L^2(S^2)$.

Corollary 1. Given $f \in L^2(S^2)$ and $\epsilon > 0$, one can find $h \in L^2(D)$ such that

$$\left\|f(\beta) + \frac{1}{4\pi}\int_D e^{-ik\beta\cdot x}h(x)dx\right\| < \epsilon.$$

Lemma 4. The set $\{q(x)u(x,\alpha)\}_{\forall q\in L^2(D)}$ is dense in $L^2(D)$.

Corollary 2. Given $h \in L^2(D)$ and $\epsilon > 0$, one can find $q \in L^2(D)$ such that

$$\|h(x) - q(x)u(x,\alpha)\|_{L^2(D)} < \epsilon.$$

Since the scattering amplitude

$$A(\beta) = -\frac{1}{4\pi}\int_D e^{-ik\beta\cdot x}h(x)dx$$

depends continuously on h, the inverse problem **IP** is solved by Lemmas 3 and 4.

Proofs are omitted. They can be found in reference [31].

5. Discussion

How is the theory, outlined in the previous sections, can be used practically?

To create a material with a desired refraction coefficient, or a material with a refraction coefficient close to the desired, is practically very important. To my knowledge, there were no general methods for creating material with a desired refraction coefficient. To use the theory, outlined in this paper and in the monographs[31–33], one has to solve a technological problem: how to prepare a small particle, say, a ball of radius a, with the prescribed boundary impedance ζ. This problem should be solvable, see reference [33] for arguments supporting this conclusions. If this technological problem is solved, then the recipe outlined in this paper (and in the author's monographs[31–33] can be immediately used in practice.

The problem of creating materials with a desired radiation pattern, the wave focusing materials, for example, was not investigated earlier. This problem is of great practical interest. The usual bodies scatter waves mostly backwards, somewhat sidewise and a little forwards. If one creates a body which scatters waves, for example, in a given solid angle, this would be of great practical interest. Such a body can be created as follows from the theory outlined in the previous Section.

The author wrote this paper in an attempt to draw attention of the specialists in material sciences to the theory he has developed for creating materials with the desired refraction coefficient.

The author is not aware of the experimental results based on his theory. Such results are very desirable. There are numerical results, based on his theory, see references [37] and [38].

Conflict of interest

Author declares no conflict of interest.

References

1. Rayleigh J. Scientific papers. Cambridge: Cambridge University Press; 1992.
2. Van de Hulst HC. Light scattering by small particles. New York: Dover Publications; 1961.
3. Landau L, Lifschitz L. Electrodynamics of continuous media. Oxford: Pergamon Press; 1984.
4. Ramm AG. Wave scattering by small bodies of arbitrary shapes. Singapore: World Scientific Publishing Co. Pte. Ltd.; 2005.
5. Martin P. Multiple scattering. Cambridge: Cambridge University Press; 2006.
6. Ramm AG. Scattering by obstacles. Dordrecht: D. Reidel; 1986.
7. Ramm AG. Scattering by many small bodies and applications to condensed matter physics. Europhysics Letters 2007; 80(4): 44001. doi: 10.1209/0295-5075/80/44001.
8. Ramm AG. Many-body wave scattering by small bodies and applications. Journal of Mathematical Physics 2007; 48: 103511. doi: 10.1063/1.2799258.
9. Ramm AG. Wave scattering by small particles in a medium. Physics Letters A 2007; 367(1-2): 156–161. doi: 10.1016/j.physleta.2007.02.076.
10. Ramm AG. Wave scattering by small impedance particles in a medium. Physics Letters A 2007; 368(1-2): 164–172. doi: 10.1016/j.physleta.2007.04.061.

11. Ramm AG. Distribution of particles which produces a desired radiation pattern. Communications in Nonlinear Science and Numerical Simulation 2007; 12(7): 1115–1119. doi: 10.1016/j.cnsns.2005.11.001.
12. Ramm AG. Distribution of particles which produces a "smart" material. Journal of Statistical Physics 2007; 127: 915–934. doi: 10.1007/s10955-007-9303-3.
13. Ramm AG. Distribution of particles which produces a desired radiation pattern. Physica B: Condensed Matter 2007; 394(2): 253–255. doi: 10.1016/j.physb.2006.12.019.
14. Ramm AG. Creating wave-focusing materials. Latin American Journal of Solids and Structures 2008; 5(2): 119–127.
15. Ramm AG. Electromagnetic wave scattering by small bodies. Physics Letters A 2008; 372(23): 4298–4306. doi: 10.1016/j.physleta.2008.03.010.
16. Ramm AG. Wave scattering by many small particles embedded in a medium. Physics Letters A 2008; 372(17): 3064–3070. doi: 10.1016/j.physleta.2008.01.006.
17. Ramm AG. Preparing materials with a desired refraction coefficient and applications. In: Skiadas C, Dimotikalis I, Skiadas C (editors). Topics on chaotic systems: Selected papers from Chaos 2008 International Conference. Singapore: World Scientific Publishing Co. Pte. Ltd.; 2009. p. 265–273.
18. Ramm AG. Preparing materials with a desired refraction coefficient. Nonlinear Analysis: Theory, Methods & Applications 2009; 71(12): e186–e190. doi: 10.1016/j.na.2008.10.011.
19. Ramm AG. Creating desired potentials by embedding small inhomogeneities. Journal of Mathematical Physics 2009; 50: 123525. doi: 10.1063/1.3267887.
20. Ramm AG. A method for creating materials with a desired refraction coefficient. International Journal of Modern Physics B 2010; 24(27): 5261–5268. doi: 10.1142/S0217979210056074.
21. Ramm AG. Materials with a desired refraction coefficient can be created by embedding small particles into a given material. International Journal of Structural Changes in Solids 2010; 2(2): 17–23.
22. Ramm AG. Wave scattering by many small bodies and creating materials with a desired refraction coefficient. Afrika Matematika 2011; 22: 33–55. doi: 10.1007/s13370-011-0004-3.
23. Ramm AG. Scattering by many small inhomogeneities and applications. In: Skiadas C, Dimotikalis I, Skiadas C (editors). Topics on chaotic systems: Selected papers from Chaos 2010 International Conference. Singapore: World Scientific Publishing Co. Pte. Ltd.; 2011. p. 41–52.
24. Ramm AG. Collocation method for solving some integral equations of estimation theory. International Journal of Pure and Applied Mathematics 2010; 62(1): 57–65. doi: 10.1504/IJCSM.2009.027874.
25. Ramm AG. Electromagnetic wave scattering by a small impedance particle of arbitrary shape. Optics Communications 2011; 284(16-17): 3872–3877. doi: 10.1016/j.optcom.2011.04.035.
26. Ramm AG. Scattering of scalar waves by many small particles. AIP Advances 2011; 1(2): 022135. doi: 10.1063/1.3600704.
27. Ramm AG. Scattering of electromagnetic waves by many thin cylinders. Results in Physics 2011; 1(1): 13–16. doi: 10.1016/j.rinp.2011.05.002.
28. Ramm AG. Electromagnetic wave scattering by many small perfectly conducting particles of an arbitrary shape. Optics Communications 2012; 285(18): 3679–3683. doi: 10.1016/j.optcom.2012.05.010.
29. Ramm AG. Electromagnetic wave scattering by small impedance particles of an arbitrary shape. Journal of Applied Mathematics and Computing 2013; 43(1): 427–444. doi: 10.1007/s12190-013-0671-3.
30. Ramm AG. Many-body wave scattering problems in the case of small scatterers. Journal of Applied Mathematics and Computing 2013; 41(1): 473–500. doi: 10.1007/s12190-012-0609-1
31. Ramm AG. Scattering of acoustic and electromagnetic waves by small bodies of arbitrary shapes. Applications to creating new engineered materials. New York: Momentum Press; 2013. p. 260.
32. Ramm AG. Creating materials with a desired refraction coefficient. doi: 10.1088/978-1-6817-4708-8. San Rafael, California: IOP Concise Physics, Morgan & Claypool Publishers; 2017.
33. Ramm AG. Creating materials with a desired refraction coefficient. 2nd ed. Bristol, UK: IOP Publishing; 2020. doi: 10.1088/978-0-7503-3391-7.
34. Ramm AG. How can one create a material with a prescribed refraction coefficient? Sun Text Review of Material Science 2020; 1(1): 102. doi: 10.51737/2766-5100.2020.002.
35. Ramm AG. Scattering by obstacles and potentials, Singapore: World Scientific Publishing Co. Pte. Ltd.; 2017. p. 620. doi: 10.1142/10473.
36. Ramm AG, Tran N. A fast algorithm for solving scalar wave scattering problem by billions of particles. Journal of Algorithms and Optimization 2015; 3(1): 1–13.
37. Andriychuk MI, Ramm AG. Numerical solution of many-body wave scattering problem for small particles and creating materials with desired refraction coefficient. In: Awrejcewicz J (editor). Numerical simulations of physical and engineering processes. London: IntechOpen; 2011. p. 1–28. doi: 10.5772/24495.
38. Andriychuk M, Ramm AG. Scattering of electromagnetic waves by many thin cylinders: Theory and computational modeling, Optics Communications 2012; 285(20): 4019–4026. doi: 10.1016/j.optcom.2012.06.017.

Appendix A4

MATERIALS WITH A DESIRED REFRACTION COEFFICIENT

ALEXANDER G. RAMM

ABSTRACT. Producing materials with a desired refraction coefficient is of great theoretical and practical interest. There was no general method for creating such materials, except the method, developed by the author. It was not even known that such a method do exist. The theoretical basis of this method is the asymptotic solution of the many-body wave scattering problem for many small bodies with prescribed boundary impedances. Multiple scattering is essential in our theory. The small bodies are embedded in a bounded region D, filled with a material with a known refraction coefficient $n_0(x)$. Our basic physical assumption is $a \ll d \ll \lambda$, where a is the characteristic size of the small particle, d is the minimal distance between neighboring particles, and λ is the wave length in D. The asymptotic of the solution to the above many-body scattering problem is derived for $a \to 0$.

1. INTRODUCTION

Let D be a bounded domain in $\mathbb{R}^3$ filled with a material with a known refraction coefficient $n_0(x)$. Let us embed into D many small particles D_m of a characteristic size a with boundary impedances ζ_m let d be the minimal distance between neighboring particles, and λ be the wavelength in D. We assume that

$$a \ll d \ll \lambda. \tag{1.1}$$

We assume that the boundary impedance of a small body D_m is given by the equation:

$$\zeta_m = \frac{h(x_m)}{a^\kappa}, \quad h(x) \in C(D), \tag{1.2}$$

where x_m is a point inside D_m and $\kappa \in [0, 1)$ is a constant that can be chosen by a researcher. Since D_m is small, its position can be characterized by a point x_m.

Let us formulate a recipe for producing a material with a desired refraction coefficient $n(x)$. The refraction coefficient $n(x)$ is defined by the wave equation

$$\Delta u + k^2 n^2(x) u = 0, \tag{1.3}$$

2020 *Mathematics Subject Classification.* 35Q60, 78A40, 78A45, 78A48.

Key words and phrases. Wave scattering by many small bodies; Smart materials; Materials with a desired radiation pattern.

Received: January 02, 2025. Accepted: March 08, 2025. Published: March 31, 2025.

Reprinted with permission. First appeared in Annals of Communications in Mathematics Volume 8, Number 1 (2025), 38–42.

where $u(x)$ is the wave field, $k > 0$ is the wave number. For simplicity, we assume in this paper that u is a scalar function. We also assume that D is filled in with a material whose refraction coefficient $n_0(x)$ is known.

Problem: *We want to produce in D the material with a desired refraction coefficient $n(x)$.*

Let us formulate a **recipe** for solving this problem.

Step 1. *Calculate*

$$p(x) = k^2[n_0^2(x) - n^2(x)]. \tag{1.4}$$

This step is trivial.

Step 2. *Given $p(x)$, find functions $h(x)$, $Imh(x) \le 0$, and $N(x) \ge 0$ from the equation:*

$$4\pi h(x)N(x) = p(x), \qquad N(x) \ge 0, \quad Im\,p(x) \le 0. \tag{1.5}$$

This step is also trivial. It has many solutions. For example, one can fix $N(x) > 0$ and define $h(x)$ by the formula:

$$h(x) = \frac{p(x)}{4\pi N(x)}, \tag{1.6}$$

provided that $\text{Im}\, p(x) \le 0$.

Step 3. *Distribute $N = O(\frac{1}{a^{2-\kappa}})$ small particles D_m with boundary impedances $\zeta_m = \frac{h(x_m)}{a^\kappa}$ in the domain D according to the law:*

$$\mathcal{N}(\Delta) = \frac{1}{a^{2-\kappa}} \int_\Delta N(x)dx[1 + o(1)], \quad a \to 0, \tag{1.7}$$

where Δ is any open subset of D, $\mathcal{N}(\Delta)$ is the number of small bodies in the subset Δ, $N(x)$ is the function from Step 2, the boundary impedance of the body D_m is chosen by formula (1.2)*, the function $h(x)$ in this formula is defined in Step 2, and x_m is an arbitrary fixed point inside D_m.*

Our main result is the following theorem.

Theorem 1. *The refraction coefficient of the material, obtained in the domain D after Step 3, tends to the desired refraction coefficient $n(x)$ as $a \to 0$.*

A proof of this result is not short. It is presented in the monographs [1],[2], [3], in the review paper [5], and in the author's papers cited in these references. Many other problems, based on similar ideas and methods, are presented in the above monographs: scattering of electromagnetic waves by many small bodies, scattering of heat waves by many small bodies, scattering of quantum-mechanical waves by many potentials with small supports, some inverse scattering problems, and other results.

Remark 1. *One may use the spherical particles $M(x_m, a)$ centered at the points x_m and of radius a for the creating of the materials with a desired refraction coefficient.*

Remark 2. *The total volume V_a of the embedded particles tends to zero when $a \to 0$.*

A proof is easy:

$$V_a = \frac{4}{3}\pi a^3 \times O(\frac{1}{a^{2-\kappa}}) = O(a^{1+\kappa}) \to 0, \quad a \to 0.$$

2. Additional considerations

It is known how to embed many small particles D_m at the prescribed points x_m. The size of these particles can be as small as 20 nanometers. One of the known methods is stereolitography.

The author is not familiar with the method of producing small particles with a prescribed boundary impedance. To use my recipe for creating materials with a desired refraction coefficient practically, it is necessary to develop a method for creating such small particles. Let us give some arguments showing that such particles can be prepared.

The first argument goes as follows. The wave scattering problem for one small body with a prescribed boundary impedance is:

$$\Delta u + k^2 n_0^2(x)u = 0, \tag{2.1}$$

$$u_\nu = \zeta_m u \quad on \quad S_m, \tag{2.2}$$

$$u = u_0 + v, \tag{2.3}$$

$$v_{|x|} - ikv = o(\frac{1}{|x|}), \qquad |x| \to \infty. \tag{2.4}$$

Problem (2.1)–(2.4) has a solution and this solution is unique, see [4], pp. 30–50. Therefore, the small body with the prescribed boundary impedance should exist. The condition $\operatorname{Im} n(x) \le 0$ is used for the proof of the uniqueness of the solution to the scattering problem (2.1)–(2.4).

The second argument for the existence of small bodies with a prescribed boundary impedan goes as follows.

Problem (2.1)–(2.4) with $\zeta = 0$ does exist. The same is true for $\zeta = \infty$. The small particles with any intermediate value of ζ should also exist.

3. Materials with a desired radiation pattern

Suppose that

$$\Delta u + k^2 n^2(x)u := \Delta u + k^2 u - q(x)u = 0, \tag{3.1}$$

where

$$q(x) = k^2[1 - n^2(x)]. \tag{3.2}$$

Recall that $n(x) = 1$ out of the bounded domain D.

The solution u to the scattering problem (3.1) exists and is unique, subject to the condition

$$u = u_0 + v, \tag{3.3}$$

where v satisfies the radiation condition and u_0 is the incident plane wave.

We assume in this section that $k > 0$ and $\alpha \in S^2$, the direction of the incident wave, are fixed. We denote the scattering amplitude $A(\beta)$,

$$A(\beta) := A_q(\beta) = -\frac{1}{4\pi} \int_{\mathbb{R}^3} e^{-ik\beta\cdot y} q(y)u(y)dy, \tag{3.4}$$

where $u(y) := u(y, \alpha, k)$, and the dependence on α and k is omitted since α and k are fixed.

Choose an arbitrary $f(\beta) \in L^2(S^2)$, where S^2 is the unit sphere in $\mathbb{R}^3$, and an arbitrary small fixed number $\epsilon > 0$, and state the following new inverse problem:

Inverse problem. *Given $f(\beta)$ and ϵ, find $q \in L^2(D)$ such that*

$$\|f(\beta) - A_q(\beta)\|_{L^2(S^2)} < \epsilon. \tag{3.5}$$

It was not known if this problem has a solution. This problem was studied and solved in [3]. We formulate the result and refer the reader to [3] for a detailed proof.

Theorem 2. *For any $f(\beta) \in L^2(S^2)$ and an arbitrary small number $\epsilon > 0$, there exists a $q \in L^2(D)$ such that inequality* (3.5) *holds.*

Remark 3. There are infinitely many potentials satisfying (3.5). Indeed, the scattering amplitude depends continuously on the potential in the following sense:

$$\|A_{q_1} - A_{q_2}\|_{L^2(S^2)} \le c\|q_1 - q_2\|_{L^2(D)}, \tag{3.6}$$

where $c > 0$ is a constant depending only on the bound for the norms of the potentials and on D. Therefore, small changes of the potential in $L^2(D)$ norm lead to small changes in the scattering amplitude in $L^2(S^2)$ norm in the sense (3.6). Thus, if inequality (3.5) holds for some $q \in L^(D)$, it will hold for any potential sufficiently close to q in $L^2(D)$ norm.

Remark 4. Theorem 2 can be of practical interest. For example, let $f(\beta) = 1$ in a narrow cone and $f(\beta) = 0$ outside this cone. Then, the body D with such a radiation pattern will have practical interest. The wave, scattered by this body, will be scattered mostly in the above cone. The scattered wave can be directed not as usual to the back of the body and to the front of the body, but mostly to the above cone.

4. Conclusion

A recipe is given for creating materials with a desired refraction coefficient by embedding many small particles with prescribed boundary impedances into a given material. The refraction coefficient can be so chosen that the resulting material will have a desired radiation pattern for a fixed wave number and a fixed direction of the incident plane wave. Materials with a prescribed radiation pattern can be created. For future developments, it is desirable to do many experiments based on the author's theory. One can change the given refraction coefficient $n_0(x)$ in a desired direction. Theoretically, the major advance is the author's (asymptotical as $a \to 0$) solution to the many-body scattering problem under the assumption $a \ll d \ll \lambda$.

5. Disclosure statement.

There are no competing interests to declare. There is no financial support for this work.

References

[1] A. G. Ramm, *Scattering of Acoustic and Electromagnetic Waves by Small Bodies of Arbitrary Shapes. Applications to Creating New Engineered Materials*, Momentum Press, New York, 2013.

[2] A. G. Ramm, *Creating materials with a desired refraction coefficient*, IOP Publishers, Bristol, UK, 2020 (Second edition).

[3] A. G. Ramm, *Wave scattering by small bodies. Creating materials with a desired refraction coefficient and other applications*, World Sci. Publishers, Singapore, 2023.

[4] A. G. Ramm, *Scattering by obstacles and potentials*, World Sci. Publ., Singapore, 2017.

[5] A. G. Ramm, Wave scattering by many small impedance particles and applications, Reports on Math. Phys., (ROMP), 90, N2, (2022), 193-202.

Department of Mathematics, Kansas State University
Manhattan, KS 66506, USA.
ORCID: 0000-0002-5160-4248
Email address: `ramm@ksu.edu`

Appendix 5

How to Create Materials with a Desired Refraction Coefficient? Wave Scattering by Many Small Particles

Alexander G. Ramm
Mathematics Department,
Kansas State University,
ramm@ksu.edu
www.math.ksu.edu/~ramm

Abstract

The novel points in this work include:

(1) Asymptotic and numerical methods for solving wave scattering problem by many small bodies embedded in a nonhomogeneous medium. Basic assumption: $a \ll d \ll \lambda$.
(2) Derivation of the equation for the field in the limit $a \to 0$, where a is the characteristic size of the bodies (particles), and their number $M = M(a)$ tends to infinity at a suitable rate. Multiple scattering is taken into account.
(3) A recipe for creating materials with a desired refraction coefficient by embedding many small particles in a given material.
(4) Some of the many possible applications:
   (a) creating materials with negative refraction,
   (b) creating wave-focusing materials.

Results I have Published but Do Not Have Time to Discuss in This Talk:

(a) electromagnetic wave scattering,
(b) creating materials with a desired magnetic permeability,

(c) wave scattering by small particles of arbitrary shapes, wave scattering by many nanowires,
(d) heat transfer in a medium in which many small bodies are embedded,
(e) quantum-mechanical scattering by many potentials with small supports,
(f) wave scattering by small bodies with transmission boundary conditions, and other problems.

In the author's monograph (*) these problems are discussed in detail and solved; in (**) the bounded domain is located inside perfectly conducting surface. The new monograph (****) is published in 2023.

My Monographs in Which the Theory is Developed:

(*) A.G. Ramm, **Scattering of Acoustic and Electromagnetic Waves by Small Bodies of Arbitrary Shapes. Applications to Creating New Engineered Materials,** Momentum Press, New York, 2013.

(**) A.G. Ramm, **Creating materials with a desired refraction coefficient,** IOP Publishing, Bristol, UK, 2020, Second edition.

(***) A.G. Ramm, **Wave scattering by small bodies of arbitrary shapes,** World Sci. Publishers, Singapore, 2005.

In (**) the problem of creating material with a desired refraction coefficient is discussed in the case when the material is located inside a bounded closed connected surface on which the Dirichlet boundary condition is imposed.

My New Monograph:

(****) A. G. Ramm, **Wave scattering by small bodies. Creating materials with a desired refraction coefficient and other applications,** World Sci. Publishers, Singapore, 2023.

Recipe for Creating Materials with a Desired Refraction Coefficient:

To my knowledge, there are no other general methods for creating materials with a desired refraction coefficient.

If $\nabla^2 u + k^2 n^2(x) u = 0$, $k = const > 0$, then $n(x)$ is called the refraction coefficient.

Step 1. Given the original coefficient $n_0^2(x)$ and the desired coefficient $n^2(x)$, calculate function $p(x)$ by formula

$$p(x) = k^2[n_0^2(x) - n^2(x)].$$

This step is trivial.

Step 2. Given $p(x)$, solve the quation $p(x) = 4\pi h(x)N(x)$ for $h(x)$ and $N(x)$, which satisfy conditions:

$$Im\ h(x) \leq 0, \quad N(x) \geq 0.$$

This step is also trivial and has many solutions.

For example, one can fix an arbitrary $N(x) > 0$, and then find $h(x) = h_1(x) + ih_2(x)$, where $h_1 = \text{Re}\ h, h_2 = \text{Im}\ h$, by the formulas

$$h_1(x) = \frac{p_1(x)}{4\pi N(x)}, \quad h_2(x) = \frac{p_2(x)}{4\pi N(x)},$$

where $p_1 = \text{Re}\ p, p_2 = \text{Im}\ p$. The condition $\text{Im}\, h \leq 0$ holds if $\text{Im}\ p \leq 0$, i.e., $\text{Im}\,[n_0^2(x) - n^2(x)] \leq 0$.

Step 3. Prepare $M = \frac{1}{a^{2-\kappa}} \int_D N(x)dx[1 + o(1)]$ small balls $B_m(x_m, a)$ with the boundary impedances $\zeta_m = \frac{h(x_m)}{a^\kappa}, 0 \leq \kappa \leq 1$, where the points $x_m, 1 \leq m \leq M$, are distributed in D according to the formula $\mathcal{N}(\Delta) = \frac{1}{a^{2-\kappa}} \int_\Delta N(x)dx[1 + o(1)]$, $\Delta \subset D$ is an arbitrary open subset, $\mathcal{N}(\Delta)$ is the number of particles in Δ, $N(x) \geq 0$ is a continuous function of our choice.

Embed (in D) M balls $B_m(x_m, a)$ with boundary impedance ζ_m, $d = O(a^{(2-\kappa)/3})$, $d = \min_{j \neq m} |x_m - x_j|$.

The material, obtained after the embedding of these M small balls, will have the desired refraction coefficient $n^2(x)$ with an error that tends to zero as $a \to 0$.

Step 3 is the only non-trivial step in this recipe from the practical point of view.

Technological problems

The first technological problem is:

How can one embed many, namely $M = M(a)$, small balls in a given material so that the centers of the balls (points x_m) are distributed as desired?

Physicists know how to solve this problem.

The second technological problem is:

How does one prepare a ball B_m of small radius a with a desired boundary impedance $\zeta_m = \frac{h(x_m)}{a^\kappa}$, $0 \le \kappa < 1$, where $h(x)$, $Imh \le 0$, is a given function?

I argue that small particles with a prescribed boundary impedance can be prepared practically.

Why is it possible to prepare small balls with the prescribed boundary impedance?

(a) The scattering problem for such a ball has a solution and this solution is unique. Therefore, such a ball can be practically prepared.
(b) The acoustically hard balls with $\zeta = \infty$ do exist. The acoustically soft balls with $\zeta = 0$ do exist. The balls with an intermediate boundary impedance should be possible to prepare.

Statement of the Scattering Problem in the Absence of the Embedded Particles

$$L_0 u_0 := [\nabla^2 + k^2 n_0^2(x)]u_0 := [\nabla^2 + k^2 - q_0(x)]u_0 = 0 \text{ in } \mathbb{R}^3,$$

$$u_0 = e^{ik\alpha\cdot x} + v_0, \quad \lim_{r\to\infty} r(v_{0r} - ikv_0) = 0.$$

$$\text{Im } n_0^2(x) \ge 0, \quad \alpha \in S^2, \quad k = \text{const} > 0.$$

$$L_0 G(x,y) = -\delta(x-y) \text{ in } \mathbb{R}^3.$$

$$n_0^2(x) = 1 - k^{-2} q_0(x), \quad q_0(x) = k^2 - k^2 n_0^2(x), \quad \text{Im } q_0(x) \le 0,$$

$$n_0^2(x) = 1 \ in\ D' := \mathbb{R}^3 \backslash D, \quad q_0(x) = 0 \ in\ D'.$$

Many-body Scattering Problem

$$\begin{cases} L_0 u_M = 0 \text{ in } \Omega' := \mathbb{R}^3 \backslash \bigcup_{m=1}^{M} D_m; \quad D_m = B_m(x_m, a) \\ \dfrac{\partial u_M}{\partial N} = \zeta_m u_M \text{ on } S_m := \partial D_m, \qquad \zeta_m = \dfrac{h(x_m)}{a^\kappa}, \quad 0 \le \kappa < 1, \\ u_M = u_0 + v_M, \end{cases}$$

where N is the outer unit normal to S_m, and $h(x) \in C(D)$ is an arbitrary function, such that

$$h = h_1 + ih_2, \quad h_2 \leq 0,$$

and ζ_m is boundary impedance,

$$d := \min_{m \neq j} \operatorname{dist}(x_m, x_j).$$

Basic Assumptions

Our basic assumptions are:

$$\boxed{a \ll d \ll \lambda},$$

$$\mathcal{N}(\Delta) := \sum_{x_m \in \Delta} 1 = \frac{1}{a^{2-\kappa}} \int_\Delta N(x)dx[1 + o(1)], \quad a \to 0. \qquad (*)$$

Here $N(x)a^{-(2-\kappa)} \geq 0$ is the density of the distribution of the particles, d is the minimal distance between neighboring particles,

$$d = O(a^{(2-\kappa)/3}). \qquad (**)$$
$$M = M(a) \sim O(a^{-(2-\kappa)}), \quad 0 \leq \kappa < 1.$$

Since $d^{-3} = O(M)$, relation (**) follows from (*).

Representation of the Solution

$$u_M(x) = u_0(x) + \sum_{m=1}^M \int_{S_m} G(x,t)\sigma_m(t)dt$$
$$= u_0(x) + \sum_{m=1}^M G(x, x_m)Q_m + \sum_{m=1}^M J_m.$$

$$Q_m := \int_{S_m} \sigma_m(t)dt, \quad J_m := \int_{S_m} [G(x,t) - G(x,x_m)]\sigma_m(t)dt,$$
$$I_m := |G(x, x_m)Q_m|.$$

Basic Result:

$$\boxed{|J_m| \ll I_m}, \quad a \to 0; \quad |x - x_m| \gg a.$$

Impedance Boundary Condition

For the *impedance boundary condition (bc)* the limiting field $u = u(x)$ solves the integral equation:

$$\boxed{u(x) = u_0(x) - \int_D G(x,y)p(y)u(y)dy},$$

where

$$p(y) = 4\pi N(y)h(y), \quad Imp \le 0; \quad \zeta_m = \frac{h(x_m)}{a^\kappa}, \quad 0 \le \kappa < 1.$$

If the small bodies D_m are of an arbitrary shape and $|S_m| = ca^2$, then the factor 4π is replaced by the factor c. This factor may depend on m, if the small bodies are not identical.

Effective Field 1

If $|J_m| \ll I_m$, then, as $a \to 0$, one has

$$u_M(x) \sim u_0(x) + \sum_{m=1}^{M} G(x, x_m)Q_m, \quad a \ll 1, \quad |x - x_m| \gg a.$$

Define **effective** (self-consistent) field acting on the m-th particle:

$$u_e := u_e^{(m)} := u_M(x) - G(x, x_m)Q_m,$$

If $|x - x_m| \ge a$, then $u_e \sim u_M$ as $a \to 0$.

We prove the following formula:

$$\boxed{Q_m \sim -4\pi u_e(x_m)h(x_m)a^{2-\kappa},} \quad a \to 0.$$

Effective Field 2

The equation for the effective field $u_e(x)$, as $a \to 0$, is

$$u_e(x) = u_0(x) - 4\pi \sum_{m=1}^{M} G(x, x_m)u_e(x_m)h_m a^{2-\kappa},$$

where $h_m := h(x_m)$ and the term with x_m is dropped if $|x - x_m| \leq a$. Here h_m are known, but $u_m := u_e(x_m)$ are *unknown*.

To calculate u_m I use *a linear algebraic system* (LAS):

$$\boxed{u_j = u_{0j} - 4\pi a^{2-\kappa} \sum_{m=1, m\neq j}^{M} G(x_j, x_m) h_m u_m, \quad 1 \leq j \leq M.}$$

The order of this system will be substantially reduced.

Reduction of the Order of the LAS.

To reduce the order M of the LAS, consider a partition of D into a union of small non-intersecting cubes Δ_p, $1 \leq p \leq P$, $P \ll M$, $y_p \in \Delta_p$, *diam* $\Delta_p \gg d$. Then the LAS for u_p is:

$$\boxed{u_q = u_{0q} - 4\pi \sum_{p\neq q}^{P} G(y_q, y_p) h(y_p) u_p N(y_p) |\Delta_p|,} \qquad (*)$$

where $1 \leq q \leq P$, $P \ll M$, $u_q = u(y_q)$, $u_{0q} = u_0(y_q)$,

$$\boxed{a^{2-\kappa} \sum_{x_m \in \Delta_p} 1 = N(y_p)|\Delta_p|}$$

The LAS (*) is used for efficient numerical solution of the many-body scattering problem when the scatterers are small.

How Efficient is This Reduction?

Let the small particles be distributed in a cube with side $L = 10^{-1}$ m, $a = 10^{-8}$ m, $d = 10^{-6}$ m. Then $M \approx \left(\frac{L}{d}\right)^3 = 10^{15}$.

Let **the side** b **of the partition cubes** Δ_p be $b = a^{1/6} = 10^{-\frac{4}{3}}$ m.

Then $P = \left(\frac{L}{b}\right)^3 = 10$. The reduction of the order M of the LAS in this example is from 10^{15} to 10 .

Of course, there is a question of *the accuracy of the approximation* of the solution of the original LAS by the solution of the reduced order LAS.

If $b = a^{1/4} = 10^{-2}$ m, then $P = \left(\frac{L}{b}\right)^3 = 10^3$. In this case the reduction of the order M of the LAS is from 10^{15} to 10^3.

Numerical experiments allow one to find a and b for which P gives the accurate approximation of u.

Derivation of the Asymptotic Formula for Q_m

We start with the exact boundary equation:

$$\boxed{u_{e_N} - \zeta_m u_e + \frac{A_m\sigma_m - \sigma_m}{2} - \zeta_m T_m \sigma_m = 0 \quad \text{on } S_m.}$$

$$A_m\sigma_m := 2\int_{S_m} \frac{\partial G(s,t)}{\partial N_s}\sigma_m(t)dt, \quad T_m\sigma_m := \int_{S_m} G(s,t)\sigma_m(t)dt.$$

$$G(x,y) = \frac{1}{4\pi|x-y|}[1 + O(|x-y|)], \quad |x-y| \to 0.$$

$$\frac{4}{3}\pi a^3 \Delta u_e(x_m) - \zeta_m 4\pi a^2 u_e(x_m) = Q_m + \zeta_m \int_{S_m} ds \int_{S_m} \frac{\sigma_m(t)dt}{4\pi|s-t|},$$

$$\int_{S_m} A_m\sigma_m dt = -\textstyle\int_{S_m} \sigma_m dt,$$

$$\int_{S_m} \frac{ds}{4\pi|s-t|} = a$$

$$\frac{4}{3}\pi a^3 \Delta u_e(x_m) - 4\pi\zeta_m u_e(x_m)a^2 = Q_m(1+\zeta_m a).$$

$$Q_m = \frac{a^3\left[\frac{4\pi}{3}\Delta u_e(x_m) - 4\pi u_e(x_m)\zeta_m a^{-1}\right]}{1+\zeta_m a}.$$

If $\zeta_m = \frac{h(x_m)}{a^\kappa}$, $\kappa < 1$, then

$$\boxed{Q_m \sim -4\pi u_e(x_m)h(x_m)a^{2-\kappa}.}$$

Asymptotic Formula for σ_m

$$Q_m \sim -4\pi u_e(x_m)h(x_m)a^{2-\kappa}.$$

$$Q_m = \int_S \sigma_m ds \sim 4\pi a^2 \sigma_m.$$

If $a \to 0$, then

$$\boxed{\sigma_m \sim -h(x_m)u_e(x_m)a^{-\kappa}}$$

Why is $I_m \gg |J_m|$?

$$|G(x, x_m)Q_m| = I_m = O\left(\frac{a^{2-\kappa}}{d}\right),$$

$$J_m = O\left(\frac{aa^{2-\kappa}}{d^2}\right) = O\left(\frac{a}{d}\frac{a^{2-\kappa}}{d}\right), \quad \frac{a}{d} \ll 1.$$

Thus, $J_m/I_m = O\left(\frac{a}{d}\right)$. Consequently,

$$\boxed{I_m \gg J_m \quad \textit{if } a \ll d \ll \lambda.}$$

Calculating the Wave Field

Formula for calculating the field $u_M(x)$ is:

$$u_M(x) = u_0(x) - 4\pi \sum_{m=1}^{M} G(x, x_m)h(x_m)u_e(x_m)a^{2-\kappa}.$$

This formula is valid everywhere outside small particles. Since the input from one small particle into u_M is not more than $O(a^{2-\kappa})$, this input tends to zero as $a \to 0$, $\kappa \in [0, 1)$.

Limiting Equation for u as $a \to 0$

$$4\pi \sum_{m=1}^{M} G(x, x_m)h(x_m)u_e(x_m)a^{2-\kappa} \sim 4\pi \sum_{p=1}^{P} G(x, y^{(p)})h(y^{(p)})u_e(y^{(p)})$$

$$\times\, a^{2-\kappa} \sum_{x_m \in \Delta_p} 1 = 4\pi \sum_{p=1}^{P} G(x, y^{(p)})h(y^{(p)})u_e(y^{(p)})N(y^{(p)})|\Delta_p|(1 + \varepsilon_p)$$

$$\to \int_D G(x, y)p(y)u(y)dy, \quad p(y) := 4\pi h(y)N(y).$$

$$\boxed{u(x) = u_0(x) - \int_D G(x, y)p(y)u(y)dy.}$$

An Auxiliary Lemma

Lemma. If $f \in C(D)$ and x_m are distributed in D so that

$$\mathcal{N}(\triangle) = \frac{1}{\varphi(a)} \int_{\triangle} N(x)dx[1 + o(1)], \quad a \to 0,$$

for any subdomain $\triangle \subset D$, where $\varphi(a) \geq 0$ is a continuous, monotone, strictly growing function, $\varphi(0) = 0$, then

$$\lim_{a \to 0} \sum_{x_m \in D} f(x_m)\varphi(a) = \int_D f(x)N(x)dx.$$

Remark: *This lemma holds for bounded f with the set of discontinuities of Lebesgue's measure zero. It can be generalized to a class of unbounded f.*

Proof of the Lemma

Proof. Let $D = \cup_p \triangle_p$ be a partition of D into a union of small cubes $\triangle_p$, having no common interior points. Let $|\triangle_p|$ denote the volume of $\triangle_p$, $\delta := \max_p$ *diam* $\triangle_p$, and $y^{(p)}$ be the center of the cube $\triangle_p$. One has

$$\lim_{a \to 0} \sum_{x_m \in D} f(x_m)\varphi(a) = \lim_{a \to 0} \sum_{y^{(p)} \in \triangle_p} f(y^{(p)})\varphi(a) \sum_{x_m \in \triangle_p} 1$$

$$= \lim_{a \to 0} \sum f(y^{(p)})N(y^{(p)})|\triangle_p|[1 + o(1)] = \int_D f(x)N(x)dx.$$

The last equality holds since the preceding sum is a Riemannian sum for the continuous function $f(x)N(x)$ in the bounded domain D. Thus, the Lemma is proved. □

Equations for the Limiting Field *u*

$$\boxed{u(x) = u_0(x) - \int_D G(x,y)p(y)u(y)dy, \quad p(x) = 4\pi h(x)N(x).}$$

$$\boxed{Lu := [\nabla^2 + k^2 n^2(x)]u = 0,} \quad n^2(x) = n_0^2(x) - \frac{p(x)}{k^2}.$$

We have:

$$\boxed{k^2[n_0^2(x) - n^2(x)] = p(x).}$$

Creating Materials with a Desired Refraction Coefficient

Step 1.

$$\{n^2(x), n_0^2(x)\} \Rightarrow p(x) = k^2[n_0^2(x) - n^2(x)].$$

Step 2. Given $p(x) = p_1 + ip_2$, find $\{h(x), N(x)\}$ from the equation $4\pi h(x)N(x) = p(x)$. One has

$$h_1(x) = \frac{p_1(x)}{4\pi N(x)}, \quad h_2(x) = \frac{p_2(x)}{4\pi N(x)}.$$

There are many solutions, because $N(x) \geq 0$ can be arbitrary. For example, one can take $N(x) = const > 0$.

Step 3.

Embed $\mathcal{N}(\Delta_p) = \frac{1}{a^{2-\kappa}} \int_{\Delta_p} N(x)dx$ small particles in Δ_p, where $\bigcup_p \Delta_p = D$.

Physical properties of these particles are given by their boundary impedances $\zeta_m = \frac{h(y^{(p)})}{a^\kappa}$ for all $x_m \in \Delta_p$.

The distance between neighboring particles is $d = O(a^{\frac{2-\kappa}{3}})$.

Main Theorems

Theorem 1. The resulting new material has the desired function $n^2(x)$ with the error which tends to zero as $a \to 0$.

Denote by V_p the total volume of the embedded particles. Then

$$V_p = N(D)\frac{4\pi a^3}{3} = O(a^{-2+\kappa})O(a^3) = O(a^{1+\kappa}), \quad a \to 0.$$

Theorem 2. The total volume V_p of the embedded particles in the limit $a \to 0$ is equal to zero.

Technological problems

The first technological problem is:

How can one embed many, namely $M = M(a)$, small balls in a given material so that the centers of the balls are points x_m distributed as desired?

The stereolitography process and chemical methods for growing small particles solve this problem.

The second technological problem is:

How does one prepare a ball B_m of small radius a with boundary impedance $\zeta_m = \frac{h(x_m)}{a^\kappa}$, $0 \le \kappa \le 1$, which has a desired frequency dependence?

Remark: It is not necessary to have large boundary impedance: if $\kappa = 0$, or $\kappa = O\left(\frac{1}{|\ln a|}\right)$, then ζ_m is bounded. However, if $\kappa = 0$, then $M = O(a^{-2})$, so more particles have to be embedded.

Playing with Numbers

$$\mathcal{N} \sim 10^6; \quad \mathcal{N} \sim \frac{1}{a^{2-\kappa}}; \quad d \sim a^{(2-\kappa)/3}.$$

$$\mathcal{N} = 10^6; \quad \kappa = 2/3, \quad a = 10^{-4.5}, \quad d = 10^{-2}.$$

$$\mathcal{N} = 10^6; \quad \kappa = 1/2, \quad a = 10^{-4}, \quad d = 10^{-2}.$$

The difference between the solution of the limiting integral equation for the effective field and the solution to the linear algebraic system for $u_e(x_m)$ is $O(1/n)$, where $1/n$ is the side of a partition cube.

Spatial Dispersion. Negative Refraction

$$u = \sum_k a(k) e^{i[k\cdot r - \omega(k)t]}, \quad |k - \bar{k}| + |\omega(k) - \omega(\bar{k})| < \delta$$

$$v_{group} := v_g = \nabla_k \omega(k), \quad v_{phase} := v_p = \frac{\omega}{|k|} k^0.$$

$$\nabla_k |k| = k^0 := \frac{k}{|k|}; \quad \frac{\omega^2 n^2}{c^2} = k^2, \quad \frac{\omega n}{c} = |k|.$$

$$\left(\frac{n}{c} + \frac{\omega}{c}\frac{\partial n}{\partial \omega}\right) \nabla_k \omega = k^0.$$

$$\{v_g = -const \cdot v_p, \quad const > 0\} \Longleftrightarrow \text{negative refraction.}$$

$$\boxed{n + \omega \frac{\partial n}{\partial \omega} < 0}$$

Isotropic Medium

If $\omega > 0$, $\omega = \omega(k)$, $k := |\mathbf{k}|$, then $v_p \cdot v_g = \omega'(k)\frac{\omega}{k} < 0$, provided that

$$\boxed{\omega'(k) < 0.}$$

Indeed,

$$v_g := \nabla_k \omega(k) = \omega'(k)k^0, \quad v_p := \frac{\omega}{k}k^0$$

$$\nabla_k \omega(k) \cdot v_p = \omega'(k)\frac{\omega}{k}, \quad k^0 := \mathbf{k}/k.$$

Terminology:

Negative refraction means v_g is directed opposite to v_p;
Negative index means that $\epsilon < 0$ and $\mu < 0$.

Wave-focusing Materials

This theory gives a method for preparing materials with a desired radiation pattern, wave-focusing materials.

$$[\nabla^2 + k^2 - q(x)]u = 0 \text{ in } \mathbb{R}^3, \quad u = e^{ik\alpha\cdot x} + v := u_0 + v,$$

$$v = A(\beta)\frac{e^{ikr}}{r} + o\left(\frac{1}{r}\right), \quad r = |x| \to \infty, \quad \frac{x}{r} := \beta,$$

$$A(\beta) = -\frac{1}{4\pi}\int_D e^{-ik\beta\cdot x}h(x)dx, \quad h(x) := q(x)u(x,\alpha).$$

Here α is a unit vector in the direction of propagation of the incident wave, $A(\beta)$ is the scattering amplitude (radiation pattern). **We assume that α and $k > 0$ are fixed.**

Under-determined Inverse Scattering Problem

IP (inverse problem): *Given $f(\beta) \in L^2(S^2)$, $\alpha \in S^2$, $k > 0$, and $\epsilon > 0$,* ($D \subset \mathbb{R}^3$ is a bounded domain), *find $q \in L^2(D)$ such that*

$$\|f(\beta) - A(\beta)\|_{L^2(S^2)} < \epsilon. \tag{1}$$

A priori it is not clear that this problem has a solution. We prove that it has a solution.

If this **IP** has a solution, then **it has infinitely many solutions** because small variations of q lead to small variations of $A(\beta)$.

Theorem A. The set $\{\int_D e^{-ik\beta\cdot x}h(x)dx\}_{\forall h\in L^2(D)}$ is dense in $L^2(S^2)$

Corollary 1. Given $f \in L^2(S^2)$ and $\epsilon > 0$, arbitrarily small, one can find $h \in L^2(D)$ such that

$$\left\| f(\beta) + \frac{1}{4\pi} \int_D e^{-ik\beta\cdot x} h(x)dx \right\| < \epsilon.$$

Theorem B. The set $\{q(x)u(x,\alpha)\}_{\forall q \in L^2(D)}$ is dense in $L^2(D)$.

Corollary 2. Given $h \in L^2(D)$ and $\epsilon > 0$, arbitrarily small, one can find $q \in L^2(D)$ such that

$$\|h(x) - q(x)u(x,\alpha)\|_{L^2(D)} < \epsilon.$$

Since the scattering amplitude $A(\beta) = -\frac{1}{4\pi}\int_D e^{-ik\beta\cdot x}h(x)dx$ depends continuously on h, **the inverse problem IP is solved by Theorems A and B.**

Proof of Theorem A.

Assume the contrary. Then $\exists\, \psi \in L^2(S^2)$ such that

$$0 = \int_{S^2} d\beta \psi(\beta) \int_D e^{-ik\beta\cdot x} h(x)dx \quad \forall h \in L^2(D).$$

Changing the order of integration, one gets:

$$\int_{S^2} d\beta \psi(\beta) e^{-ik\beta\cdot x} = 0 \quad \forall x \in D \subset \mathbb{R}^3,$$

and

$$\int_0^\infty d\lambda \lambda^2 \int_{S^2} d\beta e^{-i\lambda\beta\cdot x} \psi(\beta) \frac{\delta(\lambda - k)}{k^2} = 0 \quad \forall x \in \mathbb{R}^3.$$

By the injectivity of the Fourier transform, one gets $\psi(\beta)\frac{\delta(\lambda-k)}{k^2} = 0$. Therefore, $\psi(\beta) = 0$. Theorem A is proved. □

Proof of Theorem B.

Given $h \in L^2(D)$, define

$$u := u_0 - \int_D g(x,y)h(y)dy, \quad g := \frac{e^{ik|x-y|}}{4\pi|x-y|}, \tag{2}$$

$$q(x) := \frac{h(x)}{u(x)}. \tag{3}$$

If $q \in L^2(D)$, then this q solves the problem, and u, defined in (2), is the scattering solution:

$$u = u_0 - \int_D g(x,y)q(y)u(y)dy, \tag{4}$$

and

$$A(\beta) = -\frac{1}{4\pi}\int_D e^{-ik\beta\cdot y}h(y)dy.$$

If q is not in $L^2(D)$, then the null set

$N := \{x : x \in D, u(x) = 0\}$ is non-void. Let

$$N_\delta := \{x : |u(x)| < \delta, x \in D\}, \quad D_\delta := D \backslash N_\delta.$$

Claim 1. Let $h_\delta = \begin{cases} h, & \text{in } D_\delta, \\ 0, & \text{in } N_\delta. \end{cases}$ Then $\|h_\delta - h\|_{L^2(D)} < o(1)$, $q_\delta := \begin{cases} \frac{h_\delta}{u_\delta}, & \text{in } D_\delta, \\ 0, & \text{in } N_\delta, \end{cases}$ $q_\delta \in L^\infty(D)$, $\quad u_\delta := u_0 - \int_D g h_\delta dy$.

Proof of Claim 1. The set N generically is a line $l = \{x : u_1(x) = 0, u_2(x) = 0\}$, where $u_1 = \Re u$ and $u_2 = \Im u$. Consider a tubular neighborhood of this line, $\rho(x,l) \leq \delta$. Let the origin O be chosen on l, s_3 be the Cartesian coordinate along the tangent to l, and $s_1 = u_1$, $s_2 = u_2$ are coordinates in the plane orthogonal to l, s_j-axis is directed along $\nabla u_j|_l$, $j = 1, 2$.

The Jacobian $\mathcal{J}$ of the transformation $(x_1, x_2, x_3) \mapsto (s_1, s_2, s_3)$ is non-singular, $|\mathcal{J}| + |\mathcal{J}^{-1}| \leq c$, because ∇u_1 and ∇u_2 are linearly independent. Define

$$h_\delta := \begin{cases} h, & \text{in } D_\delta, \\ 0, & \text{in } N_\delta, \end{cases} \quad u_\delta := u_0 - \int_D g(x,y)h_\delta(y)dy,$$

$$q_\delta := \begin{cases} \dfrac{h_\delta}{u_\delta}, & \text{in } D_\delta, \\ 0, & \text{in } N_\delta. \end{cases} \quad \|h - h_\delta\|_{L^2(D)} \leq o(1), \quad \max|h - h_\delta| \leq c.$$

One has $u_\delta = u_0 - \int_D ghdy + \int_D g(x,y)(h - h_\delta)dy$,

$$|u_\delta(x)| \geq |u(x)| - c\int_{N_\delta} \frac{dy}{4\pi|x-y|} \geq \delta - I(\delta), \quad x \in D_\delta, \quad c = \max_{x \in N_\delta}|h(x)|.$$

If one proves, that $I(\delta) = o(\delta), \delta \to 0$, $\forall x \in D_\delta$ then $q_\delta \in L^\infty(D)$, and Claim 1 is proved. □

Claim 2:

$$I(\delta) = \mathcal{O}(\delta^2 |\ln(\delta)|), \quad \delta \to 0.$$

Proof of Claim 2.

$$\begin{aligned}\int_{N_\delta} \frac{dy}{|x-y|} &\le \int_{N_\delta} \frac{dy}{|y|} = c_1 \int_0^{c_2\delta} \rho \int_0^1 \frac{ds_3}{\sqrt{\rho^2+s_3^2}} d\rho \\ &= c_1 \int_0^{c_2\delta} d\rho\rho \ln(s_3 + \sqrt{\rho^2+s_3^2})|_0^1 \le c_3 \int_0^{c_2\delta} \rho \ln\frac{1}{\rho} d\rho \\ &\le \mathcal{O}(\delta^2|\ln(\delta)|).\end{aligned}$$

The condition $|\nabla u_j|_l \ge c > 0$, $j = 1, 2$, implies that a tubular neighborhood of the line l, $N_\delta = \{x : \sqrt{|u_1|^2 + |u_2|^2} \le \delta\}$, is included in a region $\{x : |x| \le c_2\delta\}$ and includes a region $\{x : |x| \le c_2'\delta\}$. This follows from the estimates

$$c_2'\rho \le |u(x)| = |\nabla u(\xi) \cdot (x - \xi)| \le c_2\rho.$$

Here $\xi \in l$, x is a point on a plane passing through ξ and orthogonal to l, $\rho = |x - \xi|$, and $\delta > 0$ is sufficiently small, so that the terms of order ρ^2 are negligible,

$$c_2 = \max_{\xi \in l} |\nabla u(\xi)|, \quad c_2' = \min_{\xi \in l} |\nabla u(\xi)|.$$

Claim 2, and, therefore, Theorem B are proved. □

Calculation of h given $f(\beta)$ and $\epsilon > 0$

Let $\{\phi_j\}$ be a basis in $L^2(D)$,

$$h_n = \sum_{j=1}^n c_j^{(n)} \phi_j,$$

$$\psi_j(\beta) := -\frac{1}{4\pi} \int_D e^{-ik\beta\cdot x} \phi_j(x) dx.$$

Consider the problem:

$$\left\| f(\beta) - \sum_{j=1}^n c_j^{(n)} \psi_j(\beta) \right\| = \min.$$

A necessary condition for the minimum is a linear system for $c_j^{(n)}$.

References

[a] A.G. Ramm, **Scattering of Acoustic and Electromagnetic Waves by Small Bodies of Arbitrary Shapes. Applications to Creating New Engineered Materials,** Momentum Press, New York, 2013.

[a1] A.G. Ramm, **Wave scattering by small bodies. Creating materials with a desired refraction coefficient and other applications**, World Sci. Publishers, Singapore, 2023.

[b] A.G. Ramm, **Creating materials with a desired refraction coefficient**, IOP Publishing, Bristol, UK, 2020, Second edition.

[c] A.G. Ramm, **Scattering by obstacles and potentials**, World Sci. Publishers, Singapore, 2017.

[1] A.G. Ramm, **Wave scattering by small bodies of arbitrary shapes**, World Sci. Publishers, Singapore, 2005.

[2] A.G. Ramm, **Inverse problems**, Springer, New York, 2005.

[3] A.G. Ramm, Distribution of particles which produces a "smart" material, J. Stat. Phys., 127, N5, (2007), 915–934.

[4] A.G. Ramm, Many-body wave scattering by small bodies and applications, J. Math. Phys., 48, N10, 103511-1, (2007), pp. 1–29.

[5] A.G. Ramm, Scattering by many small bodies and applications to condensed matter physics, Europ. Phys. Lett., 80, (2007), 44001.

[6] A.G. Ramm, Inverse scattering problem with data at fixed energy and fixed incident direction, Nonlinear Analysis: Theory, Methods and Applications, 69, N4, (2008), 1478–1484.

[7] A.G. Ramm, A recipe for making materials with negative refraction in acoustics, Phys. Lett A, 372/13, (2008), 2319–2321.

[8] A.G. Ramm, Distribution of particles which produces a desired radiation pattern, Physica B, 394, N2, (2007), 253–255.

[9] A.G. Ramm, Wave scattering by many small particles embedded in a medium, Phys. Lett. A, 372/17, (2008), 3064–3070.

[10] A.G. Ramm, Materials with the desired refraction coefficients can be made by embedding small particles, Phys. Lett. A, 370, 5–6, (2007), 522–527.

[11] A.G. Ramm, Wave scattering by small impedance particles in a medium, Phys. Lett. A 368, N1-2, (2007), 164–172.

[12] A.G. Ramm, Electromagnetic wave scattering by many conducting particles, Journ.Physics A, 41, (2008), 212001.

[13] A.G. Ramm, Electromagnetic wave scattering by many small bodies, Phys. Lett. A, 372/23, (2008), 4298–4306.

[14] A.G. Ramm, Creating wave-focusing materials, LAJSS (Latin-Amer. Jour. Solids and Structures), 5, (2008), 119–127.

[15] A.G. Ramm, Creating materials with desired properties, Oberwolfach Workshop "Material properties", report 58/2007, pp. 10–13.

[16] A.G. Ramm, A method for creating materials with a desired refraction coefficient, Internat. Journ. Mod. Phys B, 24, 27, (2010), 5261–5268.

[17] A.G. Ramm, Materials with a desired refraction coefficient can be created by embedding small particles into the given material, International Journal of Structural Changes in Solids (IJSCS), 2, N2, (2010), 17–23.

[18] A.G. Ramm, Electromagnetic wave scattering by many small bodies and creating materials with a desired refraction coefficient, Progress in Electromagnetic Research M (PIER M), 13, (2010), 203–215.

[19] A.G. Ramm, Electromagnetic wave scattering by many small perfectly conducting particles of an arbitrary shape, Optics Communications, 285, N18, (2012), 3679–3683.

[20] A.G. Ramm, Electromagnetic wave scattering by small impedance particles of an arbitrary shape, J. Appl. Math and Comput., (JAMC), 43, N1, (2013), 427–444.

[21] A.G. Ramm, Many-body wave scattering problems in the case of small scatterers, J. of Appl. Math and Comput., (JAMC), 41, N1, (2013), 473–500.

[22] A.G. Ramm, Scattering of electromagnetic waves by many nano-wires, Mathematics, 1, (2013), 89–99.
Open access Journal: http://www.mdpi.com/journal/mathematics

[23] A.G. Ramm, N.Tran, A fast algorithm for solving scalar wave scattering problem by billions of particles, Jour. of Algorithms and Optimization, 3, N1, (2015), 1–13.
Open access Journal

[24] A.G. Ramm, Wave scattering by many small bodies and creating materials with a desired refraction coefficient, Univ. Journ. of Laser, Optics, Photonics and Censors, 2, N1, (2022), 62–73.
Open access Journal

[25] A.G. Ramm, Wave scattering by many small impedance particles and applications, Reports on Math. Physics (ROMP), 90, N2, (2022), 193–202.

[26] A.G. Ramm, M.I. Andriychuk, Application of the asymptotic solution to EM wave scattering problem to creating medium with a prescribed permeability, Journ. Appl. Math. and Computing, (JAMC), 45, (2014), 461–485.
doi: 10.1007/s12190-013-0732-7

[27] A.G. Ramm, M.I. Andriychuk, Calculation of electromagnetic wave scattering by a small impedance particle of an arbitrary shape, Math. Meth. in Natur. Phenomena, (MMNP), 9, N5, (2014), 254–269.

Appendix B

Optimal with Respect to Accuracy Algorithms for Calculation of Multidimensional Weakly Singular Integrals and Applications to Calculation of Capacitances of Conductors of Arbitrary Shapes

In this appendix cubature formulas, asymptotically optimal with respect to accuracy, are derived for calculating multidimensional weakly singular integrals. They are used for developing a universal code for calculating capacitances of conductors of arbitrary shapes. The presentation follows [10].

B.1 Introduction

Asymptotically optimal and optimal with respect to order (to accuracy and to complexity) algorithms for calculating multidimensional singular integrals have been constructed in [12] on Hölder and Sobolev classes of functions.

Although multidimensional weakly singular integrals are used in many applications, optimal methods for calculating these integrals are not well developed.

In [12] asymptotically optimal with respect to accuracy methods for calculating integrals of the form

$$\int_0^{2\pi}\int_0^{2\pi} f(\sigma_1,\sigma_2)\left|ctg\frac{\sigma_1 - s_1}{2}\right|^{\gamma_1}\left|ctg\frac{\sigma_2 - s_2}{2}\right|^{\gamma_2} d\sigma_1 d\sigma_2,$$

$0 < \gamma_1,\ \gamma_2 < 1$, were constructed on Hölder and Sobolev classes.

Development of optimal methods for calculating multidimensional weakly singular integrals is an important problem. Construction of efficient cubature formulas for calculating weakly singular integrals for calculating capacitances of conductors of arbitrary shapes by iterative methods proposed in [100] and [113] is important in many applications, for example, in wave scattering by small bodies of arbitrary shapes and in antenna theory. A bibliography on methods for calculating capacitances and polarizability tensors is contained in [113].

Here the method proposed in [12] is generalized to multidimensional weakly singular integrals and applications of optimal with respect to order cubature formulas for calculating weakly singular integrals on Lyapunov surfaces are given. The results are used for constructing a universal code for calculating capacitances of conductors of arbitrary shapes.

In the first part of the Appendix optimal methods for calculating integrals of the types:

$$Kf \equiv \int_0^{2\pi}\int_0^{2\pi} \frac{f(\sigma_1,\sigma_2)d\sigma_1 d\sigma_2}{\left(\sin^2\left(\frac{\sigma_1-s_1}{2}\right)+\sin^2\left(\frac{\sigma_2-s_2}{2}\right)\right)^{\lambda}}, \quad 0 \le s_1, s_2 \le 2\pi; \tag{B.1}$$

and

$$Tf \equiv \int_{-1}^{1}\int_{-1}^{1} \frac{f(\tau_1,\tau_2)d\tau_1 d\tau_2}{((\tau_1-t_1)^2+(\tau_2-t_2)^2)^{\lambda}}, \quad -1 \le t_1, t_2 \le 1, \quad 0<\lambda<1, \tag{B.2}$$

are constructed for Hölder and Sobolev classes of functions.

Our results for integrals (B.1) can be generalized to the integrals with other periodic kernels and functions. The development of cubature formulas for integrals (B.1) is of considerable interest because the results are applicable to integrals with weakly singular kernels defined on closed Lyapunov surfaces.

It will be clear from our arguments, that the results can be generalized to multidimensional integrals.

In Section B.9 of the Appendix iterative methods for calculating capacitances of conductors of arbitrary shapes are developed. A general numerical method for calculating these capacitances is developed, and some numerical results are given.

B.2 Definitions of Optimality

Various definitions of optimality of numerical methods and a detailed bibliography can be found in [7], [154]. Let us recall the definitions of algorithms, optimal with respect to accuracy, for calculating weakly singular integrals.

Consider the quadrature formula:

$$\begin{aligned} Tf = &\sum_{k_1=1}^{n_1} \sum_{k_2=1}^{n_2} \sum_{l_1=0}^{\rho_1} \sum_{l_2=0}^{\rho_2} p_{k_1k_2l_1l_2}(t_1,t_2) f^{(l_1,l_2)}(x_{k_1}, y_{k_2}) \\ &+ R_{n_1n_2}(f; p_{k_1k_2l_1l_2}; x_{k_1}, y_{k_2}; t_1, t_2), \end{aligned} \tag{B.3}$$

where coefficients $p_{k_1k_2l_1l_2}(t_1,t_2)$ and nodes (x_{k_1}, y_{k_2}) are arbitrary. Here $f^{(l_1,l_2)}(s_1,s_2) = \partial^{l_1+l_2} f(s_1,s_2)/\partial s_1^{l_1} \partial s_2^{l_2}$.

The error of quadrature formula (B.3) is defined as

$$R_{n_1n_2}(f; p_{k_1k_2l_1l_2}; x_{k_1}, y_{k_2}) = \sup_{(t_1,t_2)\in[-1,1]^2} |R_{n_1n_2}(f; p_{k_1k_2l_1l_2}; x_{k_1}, y_{k_2}; t_1, t_2)|.$$

The error of quadrature formula (B.3) on the class Ψ is defined as

$$R_{n_1n_2}(\Psi; p_{k_1k_2l_1l_2}; x_{k_1}, y_{k_2}) = \sup_{f\in\Psi} R_{n_1n_2}(f, p_{k_1k_2l_1l_2}; x_{k_1}, y_{k_2}).$$

Define the functional

$$\zeta_{n_1n_2}(\Psi) = \inf_{p_{k_1k_2l_1l_2}; x_{k_1}, y_{k_2}} R_{n_1n_2}(\Psi; p_{k_1k_2l_1l_2}; x_{k_1}, y_{k_2}).$$

The quadrature rformula with the coefficients $p^*_{k_1k_2l_1l_2}$ and the nodes $(x^*_{k_1}, y^*_{k_2})$ is optimal, asymptotically optimal, optimal with respect to order on the class Ψ among all quadrature rules of type (B.3) provided that:

$$\frac{R_{n_1n_2}(\Psi; p^*_{k_1k_2l_1l_2}; x^*_{k_1}, y^*_{k_2})}{\zeta_{n_1n_2}(\Psi)} = 1, \sim 1, \asymp 1, \quad n_1, n_2 \to \infty.$$

The symbol $\alpha \asymp \beta$ means $A\alpha \le \beta \le B\alpha$, where $0 < A, B < \infty$.

Consider the quadrature rule

$$Tf = \sum_{k=1}^{n} p_k(t_1,t_2) f(M_k) + R_n(f; p_k; M_k; t_1, t_2), \tag{B.4}$$

where coefficients $p_k(t_1,t_2)$ and nodes (M_k) are arbitrary.

The error of quadrature formula (B.4) is defined as

$$R_n(f; p_k; M_k) = \sup_{(t_1,t_2)\in[-1,1]^2} |R_n(f; p_k; M_k; t_1, t_2)|.$$

The error of quadrature rule (B.4) on the class Ψ is defined as

$$R_n(\Psi; p_k; M_k) = \sup_{f \in \Psi} R_n(f, p_k; M_k).$$

Define the functional

$$\zeta_n(\Psi) = \inf_{p_k; M_k} R_n(\Psi; p_k; M_k).$$

The quadrature rule with the coefficients p_k^* and the nodes (M_k^*) is optimal, asymptotically optimal, optimal with respect to order on the class Ψ among all quadrature rules of the type (B.4) provided that:

$$\frac{R_n(\Psi; p_k^*; M_k^*)}{\zeta_n(\Psi)} = 1, \sim 1, \asymp 1, \quad n \to \infty. \tag{B.5}$$

By $R_{n_1 n_2}(\Psi)$ the error of optimal cubature formulas on the class Ψ is defined. One has $R_{n_1 n_2}(\Psi) = \zeta_{n_1 n_2}(\Psi)$.

B.3 Classes of Functions

In this section, we list several classes of functions which are used below (cf [75], [63]).

A function f is defined on $A = [a, b]$ or on $A = K$, where K is a unit circle, satisfies the Hölder condition with constant M and exponent α, or belongs to the class $H_\alpha(M)$, $M > 0$, $0 < \alpha \leq 1$, if $|f(x') - f(x'')| \leq M|x' - x''|^\alpha$ for any $x', x'' \in A$.

Class H_ω, where $\omega(h)$ is a modulus of continuity, consists of all functions $f \in C(A)$ with the property $|f(x_1) - f(x_2)| \leq M\omega(|x_1 - x_2|), x_1, x_2 \in A$.

Class $W^r(M)$ consists of functions $f \in C(A)$ which have continuous derivatives $f', f'', \ldots, f^{(r-1)}$ on A, and a piecewise-continuous derivative $f^{(r)}$ on A satisfying $max_{x \in [a,b]} |f^{(r)}(x)| \leq M$.

Class $W_p^r(M)$, $r = 1, 2 \ldots, 1 \leq p \leq \infty$, consists of functions $f(t)$, defined on a segment $[a, b]$ or on $A = K$, that have continuous derivatives $f', f'', \ldots, f^{(r-1)}$, and an integrable derivative $f^{(r)}$ such that

$$\left[\int_A |f^{(r)}(x)|^p dx \right]^{1/p} \leq M.$$

Class $W_\alpha^r(M)$, $r = 1, 2 \ldots, 0 < \alpha \leq 1$, consists of functions $f(t)$, defined on a segment [a,b] or on $A = K$, which have continuous derivatives

$f', f'', \dots, f^{(r)}$, such that

$$|f^{(r)}(x_1) - f^{(r)}(x_2)| \le M|x_1 - x_2|^{\alpha}.$$

A function $f(x_1, x_2, \dots, x_l), l = 2, 3, \dots$, defined on $A = [a_1, b_2; a_2, b_2; \dots; a_l, b_l]$ or on $A = K_1 \times K_2 \times \dots \times K_l$, where $K_i, i = 1, 2, \dots, l$, are unit circles, satisfying Hölder conditions with constant M and exponent $\alpha_i, i = 1, 2, \dots, l$, or belongs to the class $H_{\alpha_1,\dots,\alpha_l}(M), M > 0, 0 < \alpha \le 1$, $i = 1, 2, \dots, l$, if

$$|f(x_1, x_2, \dots, x_l) - f(y_1, y_2, \dots, y_l)| \le M(|x_1 - y_1|^{\alpha_1} + \dots + |x_l - y_l|^{\alpha_l}).$$

Let ω, ω_i, where $i = 1, 2, \dots, l, l = 1, 2, \dots$, be moduli of continuity.

Class $H_{\omega_1,\dots,\omega_l}(M)$, consists of all functions $f \in C(A), A = [a_1, b_2; a_2, b_2; \dots; a_l, b_l]$ or $A = K_1 \times K_2 \times \dots \times K_l$, with the property

$$|f(x_1, x_2, \dots, x_l) - f(y_1, y_2, \dots, y_l)| \le M(\omega_1(|x_1 - y_1|) + \dots + \omega_l(|x_l - y_l|)).$$

Let $H_j^{\omega}(A), j = 1, 2, 3, A = [a_1, b_2; a_2, b_2; \dots; a_l, b_l]$ or $A = K_1 \times K_2 \times \dots \times K_l, l = 2, 3, \dots$, be the class of functions $f(x_1, x_2, \dots, x_l)$ defined on A and such that

$$|f(x) - f(y)| \le \omega(\rho_j(x, y)), j = 1, 2, 3,$$

where $x = (x_1, \dots, x_l), y = (y_1, \dots, y_l), \rho_1(x, y) = max_{1 \le i \le l}(|x_i - y_i|), \rho_2(x, y) = \sum_{i=1}^{l} |x_i - y_i|, \rho_3(x, y) = [\sum_{i=1}^{l} |x_i - y_i|^2]^{1/2}$.

Let $H_j^{\alpha}(A), j = 1, 2, 3, A = [a_1, b_2; a_2, b_2; \dots; a_l, b_l]$ or $A = K_1 \times K_2 \times \dots \times K_l, l = 2, 3, \dots$, be the class of functions $f(x_1, x_2, \dots, x_l)$ defined on A and such that

$$|f(x) - f(y)| \le (\rho_j(x, y))^{\alpha}, j = 1, 2, 3.$$

More general is the class $H_{\rho j}^{\alpha}(A), j = 1, 2, 3$. It consists of all functions $f(x)$ which can be represented as $f(x) = \rho(x)g(x)$, where $g(x) \in H_j^{\alpha}(A), j = 1, 2, 3$, and $\rho(x)$ is a nonnegative weight function.

Let $Z_j^{\omega}(A), j = 1, 2, 3$, be the class of functions $f(x_1, x_2, \dots, x_l)$ defined on A and satisfying

$$|f(x) + f(y) - 2f((x + y)/2)| \le \omega(\rho_j(x, y)/2), j = 1, 2, 3.$$

Let $Z_j^{\alpha}(A), j = 1, 2, 3$, be the class of functions $f(x_1, x_2, \dots, x_l)$ defined on A and satisfying

$$|f(x) + f(y) - 2f((x + y)/2)| \le (\rho_j(x, y)/2)^{\alpha}, j = 1, 2, 3.$$

Class $Z^{\alpha}_{\rho j}(A), j = 1,2,3$, consists of all functions $f(x)$ which can be represented as $f(x) = \rho(x)g(x)$, where $g(x) \in Z^{\alpha}_{j}(A), j = 1,2,3$, and $\rho(x)$ is a nonnegative weight function.

Let $W^{r_1,\dots,r_l}(M), l = 1,2,\dots$, be the class of functions $f(x_1, x_2, \dots, x_l)$ defined on a domain A, which have continuous partial derivatives $\partial^{|v|} f(x_1,\dots,x_l)/\partial x_1^{v_1} \cdots \partial x_l^{v_l}, 0 \le |v| \le r-1, |v| = v_1 + \cdots + v_l, r_i \ge v_i \ge 0, i = 1,2,\dots,l, r = r_1 + \cdots + r_l$ and all piece-continuous derivatives of order r, satisfying $\|\partial^r f(x_1,\dots,x_l)/\partial x_1^{r_1} \cdots \partial x_l^{r_l}\|_C \le M$ and $\|\partial^{r_i} f(0,\dots,0,x_i,0,\dots,0)/\partial x_i^{r_i}\|_C \le M, \ i = 1,\dots,l$.

Let $W_p^{r_1,\dots,r_l}(M), l = 1,2,\dots, 1 \le p \le \infty$ be the class of functions $f(x_1, x_2, \dots, x_l)$, defined on a domain $A = [a_1, b_1; \dots; a_l, b_l]$, with continuous partial derivatives $\partial^{|v|} f(x_1,\dots,x_l)/\partial x_1^{v_1} \cdots \partial x_l^{v_l}, 0 \le |v| \le r-1, |v| = v_1 + \cdots + v_l, r_i \ge v_i \ge 0, i = 1,2,\dots,l, r = r_1 + \cdots + r_l$, and all derivatives of order r, satisfying

$$\left\|\partial^r f(x_1,\dots,x_l)/\partial x_1^{r_1}\partial x_2^{r_2} \cdots \partial x_l^{r_l}\right\|_{L_p(A)} \le M,$$

$$\left\|\partial^{r_1+v_2+\cdots+v_l} f(x_1,0,\dots,0)/\partial x_1^{r_1}\partial x_2^{v_2} \cdots \partial x_l^{v_l}\right\|_{L_p([a_1,b_1])}$$
$$\le M, |v_2| + |v_3| + \cdots + |v_l| \le r - r_1 - 1;$$

$$\dots\dots\dots$$

$$\left\|\partial^{v_1+\cdots+v_{l-1}+r_l} f(0,\dots,0,x_l)/\partial x_1^{v_1}\partial x_2^{v_2} \dots \partial x_{l-1}^{v_{l-1}}\partial x_l^{r_l}\right\|_{L_p([a_l,b_l])}$$
$$\le M, |v_1| + |v_2| + \cdots + |v_{l-1}| \le r - r_{l-1} - 1.$$

Let $A = [a_1, b_2; a_2, b_2; \dots; a_l, b_l]$ or $A = K_1 \times K_2 \times \cdots \times K_l$. Let $C^r(M)$ be the class of functions $f(x_1, x_2, \dots, x_l)$ which are defined in A and which have continuous partial derivatives of order r. Partial derivatives of order r satisfy the conditions

$$\|\frac{\partial^{|v|} f(x_1,\dots,x_l)}{\partial x_1^{v_1} \cdots \partial x_l^{v_l}}\|_C \le M$$

for any $v = (v_1,\dots,v_l)$, where $v_i \ge 0, i = 1,2,\dots,l$ are integer and $\sum_{i=1}^{l} v_i = r$.

By $\tilde{\Psi}$ we denote the set of periodic functions of the class Ψ.

The Lyapunov spheres ([38]) are defined as regions bounded by a finite number of closed surfaces satisfying the three Lyapunov conditions:

1. At each point of the surface a tangent plane (and, therefore, a normal) exist.

2. If Θ is the angle between the normals at the points m_1 and m_2, and r is the distance between these points, then

$$\Theta < Ar^{\lambda}, \quad 0 < \lambda \leq 1,$$

where A and λ are positive numbers which do not depend on m_1 and m_2.

3. For all points of the surface, a number $d > 0$ exists such that there is exactly one point at which a straight line, parallel to the normal at the surface point m, intersects the surface inside a sphere of radius d centered at m.

Let S be a Lyapunov sphere, and N be the exterior normal to this sphere. We introduce a local system of Cartesian coordinates (χ, η, ζ), whose origin is located at an arbitrary point m_0 of S, the ζ axis is directed along the normal N_0 at the point m_0, and the χ and η axes lie in the tangential plane. In a sufficiently small neighborhood of m_0, the equation of the surface S in the local coordinates (χ, η, ζ) has the form

$$\zeta = F(\chi, \eta).$$

Definition B.1 The surface S belongs to the class $L_k(B, \alpha)$ if $F(\chi, \eta) \in W_{\alpha}^{k}(B)$, and the constants B and α do not depend on the choice of the point m_0.

B.4 Auxiliary Statements

We need the following known facts from the theory of quadrature and cubature formulas. These facts can be found, for example, in [75], [63], [21], [55].

Lemma B.4 *Let Ψ_1 be the class of functions $W_p^r(1)$, $1, 2, \ldots,$ $1 \leq p \leq \infty$, $0 \leq t \leq 1$, $f(t) \in \Psi_1$, and the quadrature rule*

$$\int_0^1 f(t)dt = \sum_{k=1}^{n} p_k f(t_k) + R_n(f)$$

be exact on all the polynomials of order up to $p-1$, and has error $R_n(\Psi_1)$ on the class Ψ_1. Let Ψ_2 be the class of functions $W_p^r(1)$, $r = 1, 2, \ldots,$ $1 \leq$

$p \leq \infty$, $a \leq x \leq b$, and $g(x) \in W_p^r(1)$. Then the quadrature formula

$$\int_a^b g(x)dx = (b-a)\sum_{k=1}^{n} p_k g(a+(b-a)t_k) + R_n(g)$$

has error $R_n(\Psi_2)$ on the class of functions Ψ_2 and

$$R_n(\Psi_2) = (b-a)^{r+1-1/p} R_n(\Psi_1).$$

Theorem B.2 *([75]) Among quadrature formulas*

$$\int_0^1 f(x)dx = \sum_{k=1}^{m}\sum_{l=0}^{\rho} p_{kl} f^{(l)}(x_k) + R(f) \equiv L(f) + R(f)$$

the best formula for the class $W_p^r(1)$ $(1 \leq p \leq \infty)$ with $\rho = r-1$ and $r = 1, 2, \cdots$, or $\rho = r-2$ and $r = 2, 4, 6, \cdots$, is the unique formula defined by the following nodes x_k^* and coefficients p_{kl}^*:

$$x_k^* = h(2(k-1) + [R_{rq}(1)]^{1/r}), \quad k = 1, 2, \ldots, m,$$

$$p_{kl}^* = (-1)^l p_{ml}^* = h^{l+1}\left\{\frac{(-1)^l}{(l+1)!}[R_{rq}(1)]^{(l+1)/r} + \frac{1}{r!}R_{rq}^{(r-1-1)}(1)\right\},$$

$$(l = 0, 1, \ldots, \rho), \quad p_{k,2v}^* = \frac{2h^{2v+1}}{r!} R_{rq}^{(r-2v-1)}(1),$$
$$\left(k = 2, 3, \ldots, m-1; \quad v = 0, 1, \ldots, \left[\frac{r-1}{2}\right]\right),$$

$$p_{k,2v+1}^* = 0\left(k = 2, 3, \ldots, m-1;\ v = 0, 1, \ldots, \left[\frac{r-2}{2}\right]\right),$$
$$h = 2^{-1}(m-1+[R_{rq}(1)]^{1/r})^{-1},$$

and $R_{rq}(t)$ is the Chebyshev polynomial $t^r + \sum_{i=0}^{r-1}\beta_i t^i$, deviating least from zero in the norm $L_q(-1,1)$, where $p^{-1} + q^{-1} = 1$. Here

$$\zeta_n[W_p^r(1)] = R_n[W_p^r(1)] = \frac{R_{rq}(1)}{2^r r! \sqrt[q]{rq+1}(m-1+[R_{rq}(1)]^{1/r})^r}.$$

Let a function $f(x, y)$ be given on a rectangle $D = [a, b; c, d]$. Consider the cubature formula

$$\iint_D f(x,y)dxdy = \sum_{k=1}^{m}\sum_{i=1}^{n} p_{ki} f(x_k, y_i) + R_{mn}(f), \tag{B.6}$$

defined by a vector (X, Y, P) of a nodes $a \le x_1 < x_2 < \cdots < x_m \le b$, $c \le y_1 < y_2 < \cdots < y_n \le d$, and coefficients p_{ki}.

Theorem B.3 *([75]) Among all quadrature formulas of the form of* (B.6) *the formula*

$$\iint_D f(x,y)dxdy = 4hq \sum_{k=1}^{m}\sum_{i=1}^{n} f(a + (2k-1)h, c + (2i-1)q) + R_{mn}(f),$$

where $h = \frac{b-a}{2m}$, $q = \frac{d-c}{2n}$, *is optimal on the classes* $H_{\omega_1,\omega_2}(D)$ *and* $H_3^\omega(D)$. *In addition*

$$R_{mn}[H_{\omega_1,\omega_2}(D)] = 4mn[q \int_0^h \omega_1(t)dt + h \int_0^q \omega_2(t)dt];$$

$$R_{mn}[H_3^\omega(D)] = 4mn \int_0^q \int_0^h \omega(\sqrt{t^2+\tau^2})dtd\tau.$$

Consider the cubature formulas of the form:

$$\iint_D p(x,y)f(x,y)dxdy = \sum_{k=1}^{N} p_k f(M_k) + R(f), \tag{B.7}$$

where $p(x, y)$ is a nonnegative and bounded on D function, p_k, $M_k (M_k \in D)$ are coefficients and nodes.

Theorem B.4 *([75]) Let* $p(x, y)$ *be a nonnegative bounded weight function. If* $R_N[H_{p,j}^\alpha(D)]$ *and* $R_N[Z_{p,j}^\alpha(D)]$, *where* $j = 1, 2, 3$, *and* $0 < \alpha \le 1$, *are the errors of optimal formulas* (B.7) *on the classes* $H_{p,j}^\alpha(D)$ *and* $Z_{p,j}^\alpha(D)$, *respectively, then*

$$\lim_{N\to\infty} N^{\alpha/2} R_N[H_{p,j}^\alpha(D)] = \lim_{N\to\infty} N^{\alpha/2} 2R_N[Z_{p,j}^\alpha(D)]$$

$$= D_j \left[\int\!\!\int_D (p(x,y))^{2/(2+\alpha)} dxdy \right]^{(2+\alpha)/\alpha}, \quad j = 1,2,3,$$

where $D_1 = \frac{12}{2+\alpha}(\frac{1}{2\sqrt{3}})^{(2+\alpha)/2} \int\limits_0^{\pi/6} \frac{d\varphi}{\cos^{2+\alpha}\varphi}$, $D_2 = 2^{1-\alpha}/(2+\alpha)$, *and* $D_3 = 2^{1-0.5\alpha}/(2+\alpha)$.

If $j = 2$, *then the conclusion holds for* n*-dimensional cubature formulas.*

Remark B.1 *Theorem B.3 is generalized to the case of unbounded weights* $p(x,y)$ *in [11].*

We will use the following result (see, e.g., [8]):

Lemma B.5 *Let* H *be a linear metric space,* F *be a bounded, closed, convex, centrally symmetric set with center of symmetry* θ *at the origin, and* $L(f), l_1(f), \ldots, l_N(f)$, *be some linear functionals. Let* $S(l_1(f), \ldots, l_N(f))$ *be some method for calculating the functional* $L(f)$ *using functionals* $(l_1(f), \ldots, l_N(f))$, *and* $\mathcal{S}$ *be the set of all such methods. Then the numbers* $D_1, \ldots, D_N$ *exist such that*

$$\sup_{f\in F} |L(f) - \sum_{k=1}^{N} D_k l_k(f)| = \inf_{\mathcal{S}} \sup_{f\in F} |L(f) - S(l_1(f), \ldots, l_N(f))|. \quad \text{(B.8)}$$

This means that among the best methods for calculating functional $L(f)$:

$$L(f) \approx S(l_1(f), \ldots, l_N(f)), \quad \text{(B.9)}$$

there is a linear method.

Proof. Let us associate with each $f \in F$ a point $(L(f), l_1(f), \ldots, l_N(f))$. Let Y be a set of all such points $(y_0, \ldots, y_N)$ for $f \in F$.

From our assumptions, it follows that Y is a closed centrally symmetric set with the center of symmetry at the origin.

Let $(y_0, 0, \ldots, 0)$ be an extremal point of the set Y, and

$$D_0 = \sup_{(z,0,\ldots,0)\in Y} z = y_0.$$

Because F is bounded, one has $D_0 < \infty$, and because F is convex and centrally symmetric with respect to the origin, one has $D_0 > 0$.

Draw the support plane for the set Y through the point $(D_0, 0, \ldots, 0)$:

$$(y_0 - D_0) + \sum_{j=1}^{N} C_j y_j = 0.$$

Since Y is centrally symmetric with respect to the origin, the plane

$$(y_0 + D_0) + \sum_{j=1}^{N} C_j y_j = 0$$

is also a support plane for Y, and Y lies between these two planes.

Hence, we have for the points of Y the inequality:

$$|y_0 - \sum_{j=1}^{N} D_j y_j| \le D_0, \quad D_j = -C_j.$$

The definition of y_i implies

$$\sup_{f \in F} |L(f) - \sum_{j=1}^{N} D_j l_j(f)| \le D_0. \tag{B.10}$$

Let f_0 be an element F corresponding the point $(D_0, 0, \ldots, 0)$. Then $S(l_1(\pm f_0), \ldots, l_N(\pm f_0)) = S(0, \ldots, 0)$. The right-hand side of (B.8) is not less than

$$\begin{aligned} &\inf_{S} \max\{|L(f_0) - S(0, \ldots, 0)|, |L(-f_0) - S(0, \ldots, 0)|\} \\ &\qquad = \inf_{a} \max\{|D_0 - a|, |D_0 + a|\} = D_0. \end{aligned}$$

This and (B.10) imply that the right-hand side in (B.8) is not less that the left-hand one. But the right-hand side of (B.8) can not be more than the left-hand side of (B.8) because a set of methods $\mathcal{S}$ contains linear methods. Lemma B.5 is proved. □

Corollary B.1 *Among all functions for which the optimal method for calculating $L(t)$ has the greatest error for a given set of functionals, there exists a function satisfying the conditions $l_1(f) = \cdots = l_N(f) = 0$.*

It follows from the proof that such a function is the function f_0.

B.5 Optimal Methods for Calculating Integrals of the Form (B.1)

B.5.1 *Lower bounds for the functionals ζ_{nm} and ζ_N*

In this section we derive lower bounds for the functionals ζ_{nm} and ζ_N, defined in Section B.2 , for calculating integrals (B.1) by the cubature formulas

$$Kf = \sum_{k_1=1}^{n_1} \sum_{k_2=1}^{n_2} \sum_{l_1=0}^{\rho_1} \sum_{l_2=0}^{\rho_2} p_{k_1k_2l_1l_2}(s_1, s_2) f^{(l_1,l_2)}(x_{k_1}, x_{k_2})$$

$$+R_{n_1n_2}(f; p_{k_1k_2l_1l_2}; x_{k_1}, x_{k_2}; s_1, s_2), \tag{B.11}$$

and

$$Kf = \sum_{k=1}^{N} p_k(s_1, s_2) f(M_k) + R_N(f; p_k; M_k; s_1, s_2) \tag{B.12}$$

on Hölder and Sobolev classes.

Theorem B.5 *Let $\Psi = H_{\omega_1,\omega_2}(D)$ or $\Psi = H_3^{\omega}(D)$, and calculate integral (B.1) by formula (B.11) with $\rho_1 = \rho_2 = 0$. Then the inequality*

$$\zeta_{n_1n_2}[\Psi] \geq \frac{\gamma}{\pi^2} n_1 n_2 [q \int_0^h \omega_1(t)dt + h \int_0^q \omega_2(t)dt],$$

where $q = \frac{\pi}{n_2}$, $h = \frac{\pi}{n_1}$, and

$$\gamma := \int_0^{2\pi}\int_0^{2\pi} \frac{ds_1 ds_2}{(\sin^2(s_1/2) + \sin^2(s_2/2))^{\lambda}} \tag{B.13}$$

is valid.

Corollary B.2 *Let $\Psi = H_{\alpha\alpha}(D)$ or $\Psi = H_3^{\alpha}(D)$, and calculate integral (B.1) by formula (B.11) with $n_1 = n_2 = n$ and $\rho_1 = \rho_2 = 0$. Then the inequality*

$$\zeta_{nn}[\Psi] \geq \frac{2\gamma\pi^{\alpha}}{(1+\alpha)n^{\alpha}}$$

is valid.

Proof of Theorem B.5. Denote by $\psi(s_1, s_2)$ a nonnegative function belonging to the class $H_{\omega_1\omega_2}(1)$ and vanishing at the nodes (x_{k_1}, x_{k_2}), $1 \le k_1 \le n_1$, $1 \le k_2 \le n_2$.

One has:

$$R_{n_1 n_2}(\psi; p_{k_1 k_2}; x_{k_1}, x_{k_2})$$
$$\ge \frac{1}{4\pi^2} \int_0^{2\pi}\int_0^{2\pi} \left(\int_0^{2\pi}\int_0^{2\pi} \frac{\psi(\sigma_1, \sigma_2) d\sigma_1 d\sigma_2}{[\sin^2((\sigma_1 - s_1)/2) + \sin^2((\sigma_2 - s_2)/2))]^\lambda} \right) ds_1 ds_2$$
$$= \frac{1}{4\pi^2} \int_0^{2\pi}\int_0^{2\pi} \psi(\sigma_1,\sigma_2) \left(\int_0^{2\pi}\int_0^{2\pi} \frac{ds_1 ds_2}{[\sin^2((\sigma_1 - s_1)/2) + \sin^2((\sigma_2 - s_2)/2))]^\lambda} \right) d\sigma_1 d\sigma_2$$
$$= \frac{1}{4\pi^2} \int_0^{2\pi}\int_0^{2\pi} \frac{ds_1 ds_2}{[\sin^2(s_1/2) + \sin^2(s_2/2)]^\lambda} \int_0^{2\pi}\int_0^{2\pi} \psi(s_1, s_2) ds_1 ds_2. \tag{B.14}$$

From Lemma B.5 and Theorem B.3 one concludes that the following inequality

$$R_{n_1 n_2}(\psi; p_{k_1 k_2}; x_{k_1}, x_{k_2}) \ge \frac{\gamma}{\pi^2} n_1 n_2 \left[q \int_0^h \omega_1(t) dt + h \int_0^q \omega_2(t) dt \right],$$
$$h = \frac{\pi}{n_1}, \quad q = \frac{\pi}{n_2}$$

holds for arbitrary weights $p_{k_1 k_2}$ and nodes (x_{k_1}, x_{k_2}) and

$$\zeta_{nn}(\Psi) \ge \frac{\gamma}{\pi^2} n_1 n_2 \left[q \int_0^h \omega_1(t) dt + h \int_0^q \omega_2(t) dt \right].$$

Theorem B.5 is proved. □

Theorem B.6 *Let* $\Psi = H_i^\alpha$ *or* $\Psi = Z_i^\alpha$, $i = 1, 2, 3$, *and calculate the integral* Kf *by cubature formula* (B.12). *Then*

$$\zeta_N[H_i^\alpha] = 2\zeta_N[Z_i^\alpha] = (1 + o(1))\gamma(4\pi^2)^{2/\alpha} D_i N^{-\alpha/2},$$

where $D_1 = \frac{12}{2+\alpha}\left(\frac{1}{2\sqrt{3}}\right)^{(\alpha+2)/2} \int_0^{\pi/6} \frac{d\varphi}{\cos^{2+\alpha}\varphi}$, $D_2 = \frac{2}{2^\alpha(2+\alpha)}$, *and* $D_3 = \frac{2^{1-\alpha/2}}{2+\alpha}$.

Proof. The proof of Theorem B.6 is similar to the proof of Theorem 5.1, with some difference in the estimation of the integral $\int_0^{2\pi}\int_0^{2\pi} \psi(s_1, s_2) ds_1 ds_2$, where the function $\psi(s_1, s_2)$ belongs to the class H_i^α (or Z_i^α), is nonnegative in the domain $D = [0, 2\pi]^2$, and vanishes at N nodes M_k, $k = 1, 2, \ldots, N$.

Using Lemma B.5 and Theorem B.4, one checks that the inequalities

$$\inf_{M_k} \sup_{\psi \in H_i^\alpha, \psi(M_k)=0} \int_0^{2\pi}\int_0^{2\pi} \psi(s_1, s_2) ds_1 ds_2 = (1 + o(1)) D_i (4\pi^2)^{(2+\alpha)/\alpha} N^{-\alpha/2},$$

$$\inf_{M_k} \sup_{\psi \in Z_i^\alpha, \psi(M_k)=0} \int_0^{2\pi}\int_0^{2\pi} \psi(s_1, s_2) ds_1 ds_2 = (1 + o(1)) \frac{1}{2} D_i (4\pi^2)^{(2+\alpha)/\alpha} N^{-\alpha/2}$$

hold for arbitrary $M_k \in D$, $k = 1, 2, \ldots, N$.

Substituting these values into inequality (B.14), we complete the proof of Theorem B.6. □

Theorem B.7 *Let $\Psi = \tilde{C}_2^r(1)$, and calculate the integral Kf by formula (B.11) with $\rho_1 = \rho_2 = 0$, and $n_1 = n_2 = n$. Then*

$$\zeta_{nn}[\Psi] \geq (1 + o(1)) \frac{2\gamma K_r}{n^r},$$

where $K_r := \frac{4}{\pi} \sum_{j=0}^{\infty} (-1)^{j(r+1)} (2j+1)^{-r-1}$ is the Favard constant.

Proof. Let

$$\psi(s_1, s_2) = \psi_1(s_1) + \psi_2(s_2),$$

where $0 \leq \psi_1(s) \in W^r(1)$ vanishes at the nodes x_k, $k = 1, 2, \ldots, n$, and $0 \leq \psi_2(s) \in W^r(1)$ vanishes at the nodes y_k, $k = 1, 2, \ldots, n$.

According to [75], for arbitrary nodes x_k, $k = 1, 2, \ldots, n$ one has:

$$\int_0^{2\pi} \psi_i(s) ds \geq \frac{2\pi K_r}{n^r}, \quad i = 1, 2.$$

Thus, the inequality

$$\int_0^{2\pi}\int_0^{2\pi} \psi(s_1, s_2) ds_1 ds_2 \geq \frac{8\pi^2 K_r}{n^r}$$

holds for arbitrary nodes $(x_1, \ldots, x_n)$ and $(y_1, \ldots, y_n)$.

The conclusion of Theorem B.7 follows from this inequality and from (B.14). □

Theorem B.8 *Let* $\Psi = W_p^{r,r}(1)$, $r = 1, 2, \ldots$, $1 \le p \le \infty$, *and calculate the integral* Kf *by formula* (B.11) *with* $\rho_1 = \rho_2 = r - 1$ *and* $n_1 = n_2 = n$. *Then*

$$\zeta_{nn}[\Psi] \ge (1 + o(1)) \frac{2^{1/q} \pi^{r-1/p} R_{rq}(1)}{r!(rq+1)^{1/q}(n - 1 + [R_{rq}(1)]^{1/r})^r} \gamma,$$

where $R_{rq}(t)$ *is a polynomial of degree* r, *least deviating from zero in* $L_q([-1, 1])$.

Proof. Let $L = [\frac{n}{\log n}]$. Take an additional set of nodes (ξ_k, ξ_l), $\xi_k = \frac{2\pi k}{L}$, $k, l = 0, 1, \ldots, L - 1$. By (v_i, w_j), $i, j = 0, 1, \ldots, N - 1$, $N = n + L$, denote the union of the sets (x_k, y_l) and (ξ_i, ξ_j). Let $\psi(s_1, s_2) = \psi_1(s_1) + \psi_2(s_2)$, where $\psi_1(s) \in W_p^r(1)$ vanishes with its derivatives up to the order $r - 1$ at the nodes v_i, $i = 0, 1, \ldots, N - 1$, and $\psi_2(s) \in W_p^{(r)}(1)$ vanishes with its derivatives up to order $r - 1$ at the nodes w_j, $j = 0, 1, \ldots, N - 1$. Assume $\int\limits_{v_i}^{v_{i+1}} \psi_1(s) ds > 0$, $i = 0, 1, \ldots, N - 1$, and $\int\limits_{w_j}^{w_{j+1}} \psi_2(s) ds > 0$, $j = 0, 1, \ldots, N - 1$, where $v_N = 2\pi$ and $w_N = 2\pi$.

Let

$$\mathbf{h}(\mathbf{s_1}, \mathbf{s_2}, \sigma_1, \sigma_2) := \begin{cases} 0, & \text{if } (\sigma_1, \sigma_2) = (s_1, s_2), \\ \frac{1}{(\sin^2((\sigma_1 - s_1)/2) + \sin^2((\sigma_2 - s_2)/2))^\lambda}, & \textit{otherwise}, \end{cases}$$

$$\psi^+(\mathbf{s_1}, \mathbf{s_2}) = \begin{cases} \psi(s_1, s_2), & \text{if } \psi(s_1, s_2) \ge 0, \\ 0, & \text{if } \psi(s_1, s_2) < 0, \end{cases}$$

$$\psi^-(\mathbf{s_1}, \mathbf{s_2}) = \begin{cases} 0, & \text{if } \psi(s_1, s_2) \ge 0, \\ -\psi(s_1, s_2), & \text{if } \psi(s_1, s_2) < 0. \end{cases}$$

For each value (ξ_i, ξ_j), $i, j = 0, 1, \ldots, N - 1$, we have (with $N = N_1 = N_2 = L$):

$$\int\limits_0^{2\pi} \int\limits_0^{2\pi} h(\xi_i, \xi_j, \sigma_1, \sigma_2) \psi(\sigma_1, \sigma_2) d\sigma_1 d\sigma_2$$

$$= \sum_{k=0}^{N-1} \sum_{l=0}^{N-1} \int\limits_{\xi_k}^{\xi_{k+1}} \int\limits_{\xi_l}^{\xi_{l+1}} h(\xi_i, \xi_j, \sigma_1, \sigma_2) \psi(\sigma_1, \sigma_2) d\sigma_1 d\sigma_2$$

$$= \sum_{k=0}^{N-1} \sum_{l=0}^{N-1} \int_{\xi_k}^{\xi_{k+1}} \int_{\xi_l}^{\xi_{l+1}} h(\xi_i, \xi_j, \sigma_1, \sigma_2) \psi^+(\sigma_1, \sigma_2) d\sigma_1 d\sigma_2$$

$$- \sum_{k=0}^{N-1} \sum_{l=0}^{N-1} \int_{\xi_k}^{\xi_{k+1}} \int_{\xi_l}^{\xi_{l+1}} h(\xi_i, \xi_j, \sigma_1, \sigma_2) \psi^-(\sigma_1, \sigma_2) d\sigma_1 d\sigma_2$$

$$\geq \sum_{k=i+1}^{i+[(N_1-1)/2]} \sum_{l=j+1}^{j+[(N_2-1)/2]} h(\xi_i, \xi_j, \xi_{k+1}, \xi_{l+1}) \int_{\xi_k}^{\xi_{k+1}} \int_{\xi_l}^{\xi_{l+1}} \psi^+(\sigma_1, \sigma_2) d\sigma_1 d\sigma_2$$

$$+ \sum_{k=i+1}^{i+[(N_1-1)/2]} \sum_{l=j-[(N_2-1)/2]}^{j-1} h(\xi_i, \xi_j, \xi_{k+1}, \xi_{l-1}) \int_{\xi_k}^{\xi_{k+1}} \int_{\xi_{l-1}}^{\xi_l} \psi^+(\sigma_1, \sigma_2) d\sigma_1 d\sigma_2$$

$$+ \sum_{k=i-[(N_1-1)/2]}^{i-1} \sum_{l=j+1}^{j+[(N_2-1)/2]} h(\xi_i, \xi_j, \xi_{k-1}, \xi_{l+1}) \int_{\xi_{k-1}}^{\xi_k} \int_{\xi_l}^{\xi_{l+1}} \psi^+(\sigma_1, \sigma_2) d\sigma_1 d\sigma_2$$

$$+ \sum_{k=i-[(N_1-1)/2]}^{i-1} \sum_{l=j-[(N_2-1)/2]}^{j-1} h(\xi_i, \xi_j, \xi_{k-1}, \xi_{l-1}) \int_{\xi_{k-1}}^{\xi_k} \int_{\xi_{l-1}}^{\xi_l} \psi^+(\sigma_1, \sigma_2) d\sigma_1 d\sigma_2$$

$$- \sum_{k=i+1}^{i+[(N_1-1)/2]} \sum_{l=j+1}^{j+[(N_2-1)/2]} h(\xi_i, \xi_j, \xi_k, \xi_l) \int_{\xi_k}^{\xi_{k+1}} \int_{\xi_l}^{\xi_{l+1}} \psi^-(\sigma_1, \sigma_2) d\sigma_1 d\sigma_2$$

$$- \sum_{k=i+1}^{i+[(N_1-1)/2]} \sum_{l=j-[(N_2-1)/2]}^{j-1} h(\xi_i, \xi_j, \xi_k, \xi_l) \int_{\xi_k}^{\xi_{k+1}} \int_{\xi_{l-1}}^{\xi_l} \psi^-(\sigma_1, \sigma_2) d\sigma_1 d\sigma_2$$

$$- \sum_{k=i-[(N_1-1)/2]}^{i-1} \sum_{l=j+1}^{j+[(N_2-1)/2]} h(\xi_i, \xi_j, \xi_k, \xi_l) \int_{\xi_{k-1}}^{\xi_k} \int_{\xi_l}^{\xi_{l+1}} \psi^-(\sigma_1, \sigma_2) d\sigma_1 d\sigma_2$$

$$- \sum_{k=i-[(N_1-1)/2]}^{i-1} \sum_{l=j-[(N_2-1)/2]}^{j-1} h(\xi_i, \xi_j, \xi_k, \xi_l) \int_{\xi_{k-1}}^{\xi_k} \int_{\xi_{l-1}}^{\xi_l} \psi^-(\sigma_1, \sigma_2) d\sigma_1 d\sigma_2$$

$$= \sum_{k=i+1}^{i+[(N_1-1)/2]} \sum_{l=j+1}^{j+[(N_2-1)/2]} h(\xi_i, \xi_j, \xi_{k+1}, \xi_{l+1}) \int_{\xi_k}^{\xi_{k+1}} \int_{\xi_l}^{\xi_{l+1}} \psi(\sigma_1, \sigma_2) d\sigma_1 d\sigma_2$$

$$+ \sum_{k=i+1}^{i+[(N_1-1)/2]} \sum_{l=j-[(N_2-1)/2]}^{j-1} h(\xi_i, \xi_j, \xi_{k+1}, \xi_{l-1}) \int_{\xi_k}^{\xi_{k+1}} \int_{\xi_{l-1}}^{\xi_l} \psi(\sigma_1, \sigma_2) d\sigma_1 d\sigma_2$$

$$+\sum_{k=i-[(N_1-1)/2]}^{i-1}\sum_{l=j+1}^{j+[(N_2-1)/2]} h(\xi_i,\xi_j,\xi_{k-1},\xi_{l+1})\int\limits_{\xi_{k-1}}^{\xi_k}\int\limits_{\xi_l}^{\xi_{l+1}}\psi(\sigma_1,\sigma_2)d\sigma_1 d\sigma_2$$

$$+\sum_{k=i-[(N_1-1)/2]}^{i-1}\sum_{l=j-[(N_2-1)/2]}^{j-1} h(\xi_i,\xi_j,\xi_{k-1},\xi_{l-1})\int\limits_{\xi_{k-1}}^{\xi_k}\int\limits_{\xi_{l-1}}^{\xi_l}\psi(\sigma_1,\sigma_2)d\sigma_1 d\sigma_2$$

$$-\sum_{k=i+1}^{i+[(N_1-1)/2]}\sum_{l=j+1}^{j+[(N_2-1)/2]}(h(\xi_i,\xi_j,\xi_k,\xi_l)-h(\xi_i,\xi_j,\xi_{k+1},\xi_{l+1}))$$

$$\times\int\limits_{\xi_k}^{\xi_{k+1}}\int\limits_{\xi_l}^{\xi_{l+1}}\psi^-(\sigma_1,\sigma_2)d\sigma_1 d\sigma_2$$

$$-\sum_{k=i+1}^{i+[(N_1-1)/2]}\sum_{l=j-[(N_2-1)/2]}^{j-1}(h(\xi_i,\xi_j,\xi_k,\xi_l)-h(\xi_i,\xi_j,\xi_{k+1},\xi_{l-1}))$$

$$\times\int\limits_{\xi_{k+1}}^{\xi_k}\int\limits_{\xi_{l-1}}^{\xi_l}\psi^-(\sigma_1,\sigma_2)d\sigma_1 d\sigma_2$$

$$-\sum_{k=i-[(N_1-1)/2]}^{i-1}\sum_{l=j+1}^{j+[(N_2-1)/2]}(h(\xi_i,\xi_j,\xi_k,\xi_l)-h(\xi_i,\xi_j,\xi_{k-1},\xi_{l+1}))$$

$$\times\int\limits_{\xi_{k-1}}^{\xi_k}\int\limits_{\xi_l}^{\xi_{l+1}}\psi^-(\sigma_1,\sigma_2)d\sigma_1 d\sigma_2$$

$$-\sum_{k=i-[(N_1-1)/2]}^{i-1}\sum_{l=j-[(N_2-1)/2]}^{l=j-1}(h(\xi_i,\xi_j,\xi_k,\xi_l)-h(\xi_i,\xi_j,\xi_{k-1},\xi_{l-1}))$$

$$\times\int\limits_{\xi_{k-1}}^{\xi_k}\int\limits_{\xi_{l-1}}^{\xi_l}\psi^-(\sigma_1,\sigma_2)d\sigma_1 d\sigma_2$$

$$=J_1+J_2+J_3+J_4+I_1+I_2+I_3+I_4.$$

Let us estimate the integral

$$\left|\int\limits_{\xi_k}^{\xi_{k+1}}\int\limits_{\xi_l}^{\xi_{l+1}}\psi^-(\sigma_1,\sigma_2)d\sigma_1 d\sigma_2\right|\le\int\limits_{\xi_k}^{\xi_{k+1}}\int\limits_{\xi_l}^{\xi_{l+1}}|\psi^-(\sigma_1,\sigma_2)|d\sigma_1 d\sigma_2$$

$$\leq \int_{\xi_k}^{\xi_{k+1}} \int_{\xi_l}^{\xi_{l+1}} |\psi(\sigma_1, \sigma_2)| d\sigma_1 d\sigma_2 \leq (\xi_{l+1} - \xi_l)$$

$$\times \int_{\xi_k}^{\xi_{k+1}} |\psi_1(\sigma)| d\sigma + (\xi_{k+1} - \xi_k) \int_{\xi_l}^{\xi_{l+1}} |\psi_2(\sigma)| d\sigma$$

$$\leq 2 \left(\frac{2\pi}{L}\right)^{r+2} \frac{1}{r!},$$

where we have used the fact that the functions $\psi_1(s)$ and $\psi_2(s)$ on the segments $[\xi_k, \xi_{k+1}]$ and $[\xi_l, \xi_{l+1}]$ vanish with derivatives up to order $r-1$.

Now let us estimate the sum:

$$\sum_{k=i+1}^{i+[(L-1)/2]} \sum_{l=j+1}^{j+[(L-1)/2]} |h(\xi_i, \xi_j, \xi_{k+1}, \xi_{l+1}) - h(\xi_i, \xi_j, \xi_k, \xi_l)|$$

$$= \sum_{k=i+1}^{i+[(L-1)/2]} \sum_{l=j+1}^{j+[(L-1)/2]} \left| \frac{1}{\left(\sin^2 \frac{\pi(k+1-i)}{L} + \sin^2 \frac{\pi(l+1-j)}{L}\right)^\lambda} - \frac{1}{\left(\sin^2 \frac{2\pi(k-i)}{L} + \sin^2 \frac{2\pi(l-j)}{L}\right)^\lambda} \right|$$

$$\leq \frac{c}{L} \sum_{k=i+1}^{i+[(L-1)/2]} \sum_{l=j+1}^{j+[(L-1)/2]} \frac{1}{\left(\sin^2 \frac{\pi(k-i)}{L} + \sin^2 \frac{\pi(l-j)}{L}\right)^{1+\lambda}} \left| \frac{k-i}{L} + \frac{l-j}{L} \right|$$

$$\leq \frac{c}{L} \sum_{k=i+1}^{i+[(L-1)/2]} \sum_{l=j+1}^{j+[(L-1)/2]} \frac{L^{2+2\lambda}}{((k-i)^2 + (l-j)^2)^{1+\lambda}} \frac{(k-i) + (l-j)}{L}$$

$$\leq c\,(L)^{2\lambda} \left(\sum_{l=1}^{[(L-1)/2]} \frac{1}{l^{2\lambda}} + \sum_{k=1}^{[(L-1)/2]} \frac{1}{k^{2\lambda}} \right)$$

$$\leq c\,(L)^{2\lambda} \begin{cases} L^{1-2\lambda} & \text{if} \quad \lambda < \frac{1}{2}, \\ \log L & \text{if} \quad \lambda = \frac{1}{2}, \\ 1 & \text{if} \quad \lambda > \frac{1}{2}. \end{cases}$$

By $c > 0$ various estimation constants are denoted. Thus

$$I_1 = o\left(\frac{1}{n^r}\right).$$

The expressions I_2, I_3, and I_4 are esimated similarly.

From the definition of the function $\psi(s_1, s_2)$ it follows that the error of cubature formula (B.11) for $s_1 = \xi_i,\ s_2 = \xi_j$ can be estimated as follows:

$$\begin{aligned}
&R(\psi, , \xi_i, \xi_j) \\
&= \int_0^{2\pi}\int_0^{2\pi} \psi(\sigma_1, \sigma_2) h(\xi_i, \xi_j, \sigma_1, \sigma_2) d\sigma_1 d\sigma_2 \geq o\left(\frac{1}{n^r}\right) \\
&+ \sum_{k=i+1}^{i+[(L-1)/2]} \sum_{l=j+1}^{j+[(L-1)/2]} h(\xi_i, \xi_j, \xi_{k+1}, \xi_{l+1}) \int_{\xi_k}^{\xi_{k+1}}\int_{\xi_l}^{\xi_{l+1}} \psi(\sigma_1, \sigma_2) d\sigma_1 d\sigma_2 \\
&+ \sum_{k=i+1}^{i+[(L-1)/2]} \sum_{l=j-[(L-1)/2]}^{j-1} h(\xi_i, \xi_j, \xi_{k+1}, \xi_{l-1}) \int_{\xi_k}^{\xi_{k+1}}\int_{\xi_{l-1}}^{\xi_l} \psi(\sigma_1, \sigma_2) d\sigma_1 d\sigma_2 \\
&+ \sum_{k=i-[(L-1)/2]}^{i-1} \sum_{l=j+1}^{j+[(L-1)/2]} h(\xi_i, \xi_j, \xi_{k-1}, \xi_{l+1}) \int_{\xi_{k-1}}^{\xi_k}\int_{\xi_l}^{\xi_{l+1}} \psi(\sigma_1, \sigma_2) d\sigma_1 d\sigma_2 \\
&+ \sum_{k=i-[(L-1)/2]}^{i-1} \sum_{l=j-[(L-1)/2]}^{j-1} h(\xi_i, \xi_j, \xi_{k-1}, \xi_{l-1}) \int_{\xi_{k-1}}^{\xi_k}\int_{\xi_{l-1}}^{\xi_l} \psi(\sigma_1, \sigma_2) d\sigma_1 d\sigma_2.
\end{aligned}$$

Averaging the above inequality over i and j, one gets:

$$R_{nn}[\Psi] \geq \sup_{\psi \in \Psi} \max_{i,j} R_{nn}(\psi, \xi_i, \xi_j) \geq \frac{1}{L^2} \sum_{i=0}^{L-1} \sum_{j=0}^{L-1}$$

$$\times \left[\sum_{k=i+1}^{i+[(L-1)/2]} \sum_{l=j+1}^{j+[(L-1)/2]} h(\xi_i, \xi_j, \xi_{k+1}, \xi_{l+1}) \int_{\xi_k}^{\xi_{k+1}}\int_{\xi_l}^{\xi_{l+1}} \psi(\sigma_1, \sigma_2) d\sigma_1 d\sigma_2 \right.$$

$$+\sum_{k=i+1}^{i+[(L-1)/2]}\sum_{l=j-[(L-1)/2]}^{j-1} h(\xi_i,\xi_j,\xi_{k+1},\xi_{l-1})\int\limits_{\xi_k}^{\xi_{k+1}}\int\limits_{\xi_{l-1}}^{\xi_l}\psi(\sigma_1,\sigma_2)d\sigma_1 d\sigma_2$$

$$+\sum_{k=i-[(L-1)/2]}^{i-1}\sum_{l=j+1}^{j+[(L-1)/2]} h(\xi_i,\xi_j,\xi_{k-1},\xi_{l+1})\int\limits_{\xi_{k-1}}^{\xi_k}\int\limits_{\xi_l}^{\xi_{l+1}}\psi(\sigma_1,\sigma_2)d\sigma_1 d\sigma_2$$

$$\left.+\sum_{k=i-[(L-1)/2]}^{i-1}\sum_{l=j-[(L-1)/2]}^{j-1} h(\xi_i,\xi_j,\xi_{k-1},\xi_{l-1})\int\limits_{\xi_{k-1}}^{\xi_k}\int\limits_{\xi_{l-1}}^{\xi_l}\psi(\sigma_1,\sigma_2)d\sigma_1 d\sigma_2\right]$$

$$+o\left(\frac{1}{n^r}\right)\geq o\left(\frac{1}{n^r}\right)+\frac{1}{L^2}$$

$$\times\left[\sum_{k=i+1}^{i+[(L-1)/2]}\sum_{l=j+1}^{j+[(L-1)/2]}\int\limits_{\xi_k}^{\xi_{k+1}}\int\limits_{\xi_l}^{\xi_{l+1}}\psi(\sigma_1,\sigma_2)d\sigma_1 d\sigma_2\right.$$

$$\times\sum_{i=0}^{L-1}\sum_{j=0}^{L-1} h(\xi_i,\xi_j,\xi_{k+1},\xi_{l+1})$$

$$+\sum_{k=i+1}^{i+[(L-1)/2]}\sum_{l=j-[(L-1)/2]}^{j-1}\int\limits_{\xi_k}^{\xi_{k+1}}\int\limits_{\xi_{l-1}}^{\xi_l}\psi(\sigma_1,\sigma_2)d\sigma_1 d\sigma_2$$

$$\times\sum_{i=0}^{L-1}\sum_{j=0}^{L-1} h(\xi_i,\xi_j,\xi_{k+1},\xi_{l-1})$$

$$+\sum_{k=i-[(L-1)/2]}^{i-1}\sum_{l=j+1}^{j+[(L-1)/2]}\int\limits_{\xi_{k-1}}^{\xi_k}\int\limits_{\xi_l}^{\xi_{l+1}}\psi(\sigma_1,\sigma_2)d\sigma_1 d\sigma_2$$

$$\times\sum_{i=0}^{L-1}\sum_{j=0}^{L-1} h(\xi_i,\xi_j,\xi_{k-1},\xi_{l+1})$$

$$+\sum_{k=i-[(L-1)/2]}^{i-1}\sum_{l=j-[(L-1)/2]}^{j-1}\int\limits_{\xi_{k-1}}^{\xi_k}\int\limits_{\xi_{l-1}}^{\xi_l}\psi(\sigma_1,\sigma_2)d\sigma_1 d\sigma_2$$

$$\left.\times\sum_{i=0}^{L-1}\sum_{j=0}^{L-1} h(\xi_i,\xi_j,\xi_{k-1},\xi_{l-1})\right]$$

$$= o(\frac{1}{n^r}) + \frac{1}{4\pi^2}\int_0^{2\pi}\int_0^{2\pi}\psi(\sigma_1,\sigma_2)d\sigma_1 d\sigma_2$$

$$\times\left(\int_0^{2\pi}\int_0^{2\pi}\frac{d\sigma_1 d\sigma_2}{(\sin^2(\sigma_1/2)+\sin^2(\sigma_2/2))^\lambda} + O\left(\left(\frac{logn}{n}\right)^{2-2\lambda}\right)\right),$$

where the following relation was used:

$$\frac{4\pi^2}{L^2}\sum_{i=0}^{L-1}\sum_{j=0}^{L-1}h(\xi_i,\xi_j,\xi_{k-1},\xi_{l-1})$$

$$= O\left(\frac{\log n}{n}\right) + \int_0^{2\pi}\int_0^{2\pi}\frac{d\sigma_1 d\sigma_2}{\left[\sin^2(\sigma_1/2)+\sin^2(\sigma_2/2)\right]^\lambda}.$$

Without loss of generality one may assume $k = 1, l = 1$ in the previous equation. Let us estimate

$$U_0 = \left|\frac{4\pi^2}{L^2}\sum_{i=0}^{L-1}\sum_{j=0}^{L-1}h(\xi_i,\xi_j,\ 0,0) - \int_0^{2\pi}\int_0^{2\pi}\frac{d\sigma_1 d\sigma_2}{\left(\sin^2(\sigma_1/2)+\sin^2(\sigma_2/2)\right)^\lambda}\right|$$

$$\leq \left|\sum_{i=0}^{L-1}\sum_{j=0}^{L-1}{}'\int_{\xi_i}^{\xi_{i+1}}\int_{\xi_j}^{\xi_{j+1}}\left[\frac{1}{\left(\sin^2((\xi_i)/2)+\sin^2((\xi_j)/2)\right)^\lambda}\right.\right.$$

$$\left.\left.-\frac{1}{\left(\sin^2(\sigma_1/2)+\sin^2(\sigma_2/2)\right)^\lambda}\right]d\sigma_1 d\sigma_2\right|$$

$$+\left|\int_0^{\xi_1}\int_0^{\xi_1}\frac{1}{\left(\sin^2(\sigma_1/2)+\sin^2(\sigma_2/2)\right)^\lambda}d\sigma_1 d\sigma_2\right| = u_1 + u_2,$$

where $\sum\sum'$ means summation over $(i,j) \neq (0,0)$.

Let us estimate u_1 and u_2. One has

$$u_1 \leq \left| \sum_{i=0}^{L-1} \sum_{j=0}^{L-1}{}' \int_{\xi_i}^{\xi_{i+1}} \int_{\xi_j}^{\xi_{j+1}} \left[\frac{1}{\left(\sin^2((\sigma_1)/2) + \sin^2((\sigma_2)/2)\right)^{\lambda}} - \frac{1}{\left(\sin^2(\xi_i/2) + \sin^2(\sigma_2/2)\right)^{\lambda}} \right] d\sigma_1 d\sigma_2 \right|$$

$$+ \left| \sum_{i=0}^{L-1} \sum_{j=0}^{L-1}{}' \int_{\xi_i}^{\xi_{i+1}} \int_{\xi_j}^{\xi_{j+1}} \left[\frac{1}{\left(\sin^2((\xi_i)/2) + \sin^2((\sigma_2)/2)\right)^{\lambda}} - \frac{1}{\left(\sin^2(\xi_i/2) + \sin^2(\xi_j/2)\right)^{\lambda}} \right] d\sigma_1 d\sigma_2 \right|$$

$$= u_{11} + u_{12}.$$

The expressions u_{11} and u_{12} can be estimated similarly. Let us estimate u_{11}

$$u_{11} \leq \frac{c}{L^4} \sum_{i=0}^{L} \sum_{j=0}^{L}{}' \frac{1}{\left(\sin^2((\xi_i)/2) + \sin^2((\xi_j)/2)\right)^{1+\lambda}}$$

$$\leq \frac{c}{L^{2-2\lambda}} \sum_{i=0}^{L} \sum_{j=0}^{L}{}' \frac{1}{(i^2 + j^2)^{1+\lambda}} \leq c \frac{1}{L^{2-2\lambda}},$$

where $c > 0$ stands for various estimation constants. Hence

$$u_1 \leq \frac{c}{L^{2-2\lambda}}.$$

Let us estimate u_2 :

$$u_2 = \left| \int_0^{\xi_1} \int_0^{\xi_1} \frac{1}{\left(\sin^2(\sigma_1/2) + \sin^2(\sigma_2/2)\right)^{\lambda}} d\sigma_d \sigma_2 \right|$$

$$\leq c \int_0^{\xi_1} \int_0^{\xi_1} \frac{1}{(\sigma_1^2 + \sigma_2^2)^{\lambda}} d\sigma_1 d\sigma_2.$$

Using polar coordinates, one gets:

$$u_2 \le c \int_0^{1/L} \int_0^{2\pi} \frac{1}{\rho^{2\lambda-1}} d\rho d\phi \le \frac{c}{L^{2-2\lambda}}.$$

Thus:

$$U_0 \le \frac{c}{L^{2-2\lambda}}.$$

From Lemmas B.5, B.4, and Theorem B.2 it follows that

$$\int_0^{2\pi} \psi_1(\sigma_1) d\sigma_1 \ge \frac{(1+o(1))(2\pi)^{r+1/q} R_{rq}(1)}{2^r r!(rq+1)^{1/q}(n-1+[R_{rq}(1)]^{1/r})^r}, \tag{B.15}$$

where $R_{rq}(t)$ is a polynomial of degree r, least deviating from zero in $L_q([-1,1])$.

Theorem B.8 follows from inequalities (B.5.1) and (B.15). □

B.5.2 *Optimal cubature formulas for calculating integrals* (B.1)

Hölder class of functions.

Let $x_k := 2k\pi/n$, $k = 0,1,\ldots,n$, $\Delta_{kl} = [x_k, x_{k+1}, x_l, x_{l+1}]$, $k,l = 0,1,\ldots,n-1$, $x'_k = (x_{k+1}+x_k)/2$, $k = 0,1,\ldots,n-1$, and $(s_1,s_2) \in \Delta_{ij}$, $i,j = 0,1,\ldots,n-1$.

Calculate the integral Kf by the formula:

$$Kf = \sum_{k=0}^{n-1}\sum_{l=0}^{n-1} f(x'_k, x'_l) \int\int_{\Delta_{kl}} \frac{d\sigma_1 d\sigma_2}{\left(\sin^2\left(\frac{\sigma - x'_i}{2}\right) + \sin^2\left(\frac{\sigma - x'_j}{2}\right)\right)^\lambda} + R_{nn}. \tag{B.16}$$

Theorem B.9 *Let* $\Psi = H_{\alpha\alpha}(D), 0 < \alpha < 1$. *Among all cubature formulas* (B.11) *with* $\rho_1 = \rho_2 = 0$, *formula* (B.16), *which has the error*

$$R_{nn}[\Psi] = \frac{(2+o(1))\gamma}{1+\alpha}\left(\frac{\pi}{n}\right)^\alpha,$$

is asymptotically optimal. Here γ *is defined in* (B.13).

Proof. Using the periodicity of the integrand, we estimate the error of cubature formula (B.16) as follows:

$$|R_{nn}| \leq \left|\sum_{k=0}^{n-1}\sum_{l=0}^{n-1}\int\int_{\Delta_{kl}}\left[\frac{f(\sigma_1,\sigma_2)-f(x_i',x_j')}{\left(\sin^2\frac{\sigma_1-s_1}{2}+\sin^2\frac{\sigma_2-s_2}{2}\right)^\lambda} - \frac{f(x_k',x_l')-f(x_i',x_j')}{\left(\left(\sin^2\frac{\sigma_1-x_i'}{2}+\sin^2\frac{\sigma_2-x_j'}{2}\right)^\lambda\right.}\right]d\sigma_1 d\sigma_2\right|$$

$$\leq \left|\sum_{k=0}^{n-1}\sum_{l=0}^{n-1}\int\int_{\Delta_{kl}}\frac{f(\sigma_1,\sigma_2)-f(x_k',x_l')}{\left(\sin^2\frac{\sigma_1-s_1}{2}+\sin^2\frac{\sigma_2-s_2}{2}\right)^\lambda}d\sigma_1 d\sigma_2\right|$$

$$+\left|\sum_{k=0}^{n-1}\sum_{l=0}^{n-1}\int\int_{\Delta_{kl}}(f(x_k',x_l')-f(x_i',x_j')) \times\left[\frac{1}{\left(\sin^2\frac{\sigma_1-s_1}{2}+\sin^2\frac{\sigma_2-s_2}{2}\right)^\lambda}-\frac{1}{\left(\sin^2\frac{\sigma_1-x_i'}{2}+\sin^2\frac{\sigma_2-x_j'}{2}\right)^\lambda}\right] d\sigma_1 d\sigma_2\right|$$

$$= r_1 + r_2.$$

Let us estimate each of the sums r_1 and r_2 separately. One has:

$$r_1 \leq \left|\sum_{k=i-M}^{i+M}\sum_{l=j-M}^{j+M}\int\int_{\Delta_{kl}}\left[\frac{f(\sigma_1,\sigma_2)-f(x_k',x_l')}{\left(\sin^2\frac{\sigma_1-s_1}{2}+\sin^2\frac{\sigma_2-s_2}{2}\right)^\lambda}d\sigma_1 d\sigma_2\right|$$

$$+\left|\sum_{k=0}^{n-1}\sum_{l=0}^{n-1}{}'\int\int_{\Delta_{kl}}\left[\frac{f(\sigma_1,\sigma_2)-f(x_k',x_l')}{\left(\sin^2\frac{\sigma_1-s_1}{2}+\sin^2\frac{\sigma_2-s_2}{2}\right)^\lambda}d\sigma_1 d\sigma_2\right|$$

$$= r_{11} + r_{12},$$

where $\sum\sum'$ means summation over (k,l) such that $\Delta_{kl}\notin\Delta^*$, $\Delta^*=[x_{i-M},x_{i+M+1};x_{j-M},x_{j+M+1}]$, $M=[lnn]$.

Furthermore

$$r_{11} \le \frac{c}{n^\alpha} \iint\limits_{\Delta^*} \frac{d\sigma_1 d\sigma_2}{\left(\sin^2 \frac{\sigma_1 - s_1}{2} + \sin^2 \frac{\sigma_2 - s_2}{2}\right)^\lambda}$$

$$\le \frac{c}{n^\alpha} \int\limits_0^{2\pi M/n} \int\limits_0^{2\pi} \frac{d\rho d\phi}{\rho^{2\lambda-1}} \le \frac{c \log n}{n^{\alpha+2-2\lambda}} = o\left(\frac{1}{n^\alpha}\right).$$

Estimating r_{12}, one can assume without loss of generality $(i,j) = (0,0)$, and get:

$$r_{12} \le 4 \int\limits_0^{\pi/n} \int\limits_0^{\pi/n} (\omega_1(\sigma_1) + \omega_2(\sigma_2)) d\sigma_1 d\sigma_2 \sum_{k=0}^{n-1} \sum_{l=0}^{n-1} h_{kl}(s_1, s_2, \sigma_1, \sigma_2)$$

$$\le 4 \int\limits_0^{\pi/n} \int\limits_0^{\pi/n} (\sigma_1^\alpha + \sigma_2^\alpha) d\sigma_1 d\sigma_2 \sum_{k=0}^{n-1} \sum_{l=0}^{n-1} h_{kl}(s_1, s_2, \sigma_1, \sigma_2)$$

$$\le \frac{8}{1+\alpha} \left(\frac{\pi}{n}\right)^{2+\alpha} \sum_{k=0}^{n-1} \sum_{l=0}^{n-1} h_{kl}(s_1, s_2, \sigma_1, \sigma_2)$$

$$\le \frac{1+o(1)}{1+\alpha} 2 \left(\frac{\pi}{n}\right)^\alpha \int\limits_0^{2\pi} \int\limits_0^{2\pi} \frac{d\sigma_1 d\sigma_2}{\left(\sin^2 \frac{\sigma_1}{2} + \sin^2 \frac{\sigma_2}{2}\right)^\lambda}.$$

Here

$$h_{kl}(s_1, s_2; \sigma_1, \sigma_2) = \sup_{(\sigma_1, \sigma_2) \in \Delta_{kl}} h(s_1, s_2; \sigma_1, \sigma_2).$$

Combining the estimates of r_{11} and r_{12}, one gets:

$$r_1 \le \frac{1+o(1)}{1+\alpha} 2 \left(\frac{\pi}{n}\right)^\alpha \gamma$$

Let us estimate r_2. To this end we estimate the difference

$$r_2(k,l) = \iint\limits_{\Delta_{kl}} |f(x_k', x_l') - f(x_i', x_j')|$$

$$\times \left| \left[\frac{1}{\left(\sin^2 \frac{\sigma_1 - s_1}{2} + \sin^2 \frac{\sigma_2 - s_2}{2}\right)^\lambda} - \frac{1}{\left(\sin^2 \frac{\sigma_1 - x_i'}{2} + \sin^2 \frac{\sigma_2 - x_j'}{2}\right)^\lambda} \right] d\sigma_1 d\sigma_2 \right|.$$

First, we estimate

$$r_2(i,j) \leq \frac{c}{n^\alpha} \int\int_{\Delta_{ij}} d\sigma_1 d\sigma_2$$

$$\times \left| \frac{1}{\left(\sin^2 \frac{\sigma_1 - s_1}{2} + \sin^2 \frac{\sigma_2 - s_2}{2}\right)^\lambda} - \frac{1}{\left(\sin^2 \frac{\sigma_1 - x_i'}{2} + \sin^2 \frac{\sigma_2 - x_j'}{2}\right)^\lambda} \right|$$

$$\leq \frac{c}{n^{2+\alpha-2\lambda}}.$$

The value $r_2(k,l)$ is estimated similarly for $|k-i| \leq 3$ and $|l-j| \leq 3$. Let us estimate $r_2(k,l)$ for other values of k and l.

One has:

$$r_2(k,l) = \int\int_{\Delta_{kl}} |f(x_k', x_l') - f(x_i', x_j')|$$

$$\times \left| \frac{1}{\left(\sin^2 \frac{\sigma_1 - s_1}{2} + \sin^2 \frac{\sigma_2 - s_2}{2}\right)^\lambda} - \frac{1}{\left(\sin^2 \frac{\sigma_1 - x_i'}{2} + \sin^2 \frac{\sigma_2 - x_j'}{2}\right)^\lambda} \right|$$

$$d\sigma_1 d\sigma_2 \leq \frac{c}{n} \int\int_{\Delta_{kl}} \left[|x_k' - x_i'|^\alpha + |x_l' - x_j'|^\alpha\right] \left[\left(\frac{|k-i|}{n}\right) + \left(\frac{|l-j|}{n}\right)\right]$$

$$\times \left| \frac{1}{\left(\sin^2 \frac{\sigma_1 - x_i' + \theta_1(s_1 - x_i')}{2} + \sin^2 \frac{\sigma_2 - s_2}{2}\right)^{1+\lambda}} \right.$$

$$\left. + \frac{1}{\left(\sin^2 \frac{\sigma_1 - x_i' + \theta_1(s_1 - x_i')}{2} + \sin^2 \frac{\sigma_2 - x_j' + \theta_2(s_2 - x_j')}{2}\right)^{1+\lambda}} \right|$$

$$\leq \frac{c}{n^3} \left(\left|\frac{|k-i|}{n}\right|^\alpha + \left|\frac{|l-j|}{n}\right|^\alpha \right) \left(\left|\frac{|k-i|}{n}\right| + \left|\frac{|l-j|}{n}\right| \right)$$

$$\times \left(\frac{n^2}{|k-i|^2 + |l-j|^2} \right)^{1+\lambda} \leq \frac{c}{n^{\alpha+2-2\lambda}} \frac{(|k-i| + |l-j|)^{1+\alpha}}{(|k-i|^2 + |l-j|^2)^{1+\lambda}}$$

$$\leq \frac{c}{n^{\alpha+2-2\lambda}} \frac{(|k-i|^2 + |l-j|^2)^{(1+\alpha)/2}}{(|k-i|^2 + |l-j|^2)^{1+\lambda}}$$

$$\leq \frac{c}{n^{\alpha+2-2\lambda}} \frac{1}{(|k-i|^2 + |l-j|^2)^{1/2-\alpha/2+\lambda}}.$$

To estimate r_2, one sums up the last expression over k and l. Without loss of generality assume $(i, j) = (0, 0)$. Then

$$r_2 \leq \frac{c}{n^{\alpha+2-2\lambda}} \left(16 + 4 \sum_{k=0}^{[n/2]+1} \sum_{l=0}^{[n/2]+1}{}' \frac{1}{(k^2 + l^2)^{\lambda+1/2-\alpha/2}} \right),$$

where $\sum\sum'$ means summation over k and l such that $k > 3$ or $l > 3$.

One has:

$$\sum_{k=0}^{[n/2]+1} \sum_{l=0}^{[n/2]+1}{}' \frac{1}{(k^2 + l^2)^{\lambda+1/2-\alpha/2}}$$

$$\leq A \left[\sum_{k=3}^{[n/l]+1} \frac{1}{k^{2\lambda+1-\alpha}} + \sum_{k=3}^{[n/2]+1} \sum_{l=3}^{[n/2]+1} \frac{1}{(k^2 + l^2)^{\lambda+1/2-\alpha/2}} \right]$$

$$\leq A \begin{cases} 1, & \text{if } \ 2\lambda - \alpha > 1; \\ \log n, & \text{if } \ 2\lambda - \alpha = 1; \\ n^{1-2\lambda+\alpha}, & \text{if } \ 2\lambda - \alpha < 1. \end{cases}$$

Hence

$$\mathbf{r_2} \leq A \begin{cases} n^{-(\alpha+2-2\lambda)}, & \text{if } \ 2\lambda - \alpha > 1; \\ n^{-1} \log n, & \text{if } \ 2\lambda - \alpha = 1; \\ n^{-1}, & \text{if } \ 2\lambda - \alpha < 1. \end{cases}$$

Thus, if $\alpha < 1$, then

$$r_2 \leq o(n^{-\alpha}).$$

Combining the estimates of r_1 and r_2, one gets:

$$R_{nn}[\Psi] \leq \gamma \frac{(2 + o(1))}{1 + \alpha} \left(\frac{\pi}{n} \right)^{\alpha}.$$

Theorem B.9 follows from the comparison of this inequality with the lower bound of the value $\zeta_{nn}[H_{\alpha,\alpha}(D)]$, mentioned in the Corollary to Theorem B.5. □

Remark B.2 *If* $\alpha = 1$, *the cubature formula* (B.16) *is optimal with respect to order.*

The proof of Theorem B.9 yields also the following result

Theorem B.10 *Let $\Psi = H_{\alpha\alpha}(D), 0 < \alpha \leq 1$. Among all possible cubature formulas* (B.11) *with $\rho_1 = \rho_2 = 0$, formula*

$$Kf = \sum_{k=0}^{n-1} \sum_{l=0}^{n-1} f(x'_k, x'_l) \int\int_{\Delta_{kl}} \frac{d\sigma_1 d\sigma_2}{\left(\sin^2\left(\frac{\sigma - s_1}{2}\right) + \sin^2\left(\frac{\sigma - s_2}{2}\right)\right)^\lambda} + R_{nn},$$

which has the error

$$R_{nn}[\Psi] = \frac{(2 + o(1))\gamma}{1 + \alpha} \left(\frac{\pi}{n}\right)^\alpha,$$

is asymptotically optimal.

To apply formula (B.16), one has to calculate the integrals

$$I_{kl} = \int\int_{\Delta_{kl}} \frac{d\sigma_1 d\sigma_2}{\left(\sin^2 \frac{\sigma_1 - x'_i}{2} + \sin^2 \frac{\sigma_2 - x'_j}{2}\right)^\lambda} \tag{B.17}$$

for $k, l = 0, 1, \ldots, n-1$. Exact values of these integrals for arbitrary values λ are apparently unknown. Therefore the procedure of numerical calculation of integrals (B.17) should be given for practical application of formula (B.16).

Let $k = i$ and $l = j$. Then the integral I_{ij} is replaced by the integral

$${p_{ij}}^* = \int_{-\frac{\pi}{n}}^{\frac{\pi}{n}} \int_{-\frac{\pi}{n}}^{\frac{\pi}{n}} \frac{d\sigma_1 d\sigma_2}{\left(\sin^2 \frac{\sigma_1}{2} + \sin^2 \frac{\sigma_2}{2}\right)^\lambda + h}, \quad h > 0,$$

which can be calculated by cubature formulas (in particular, Gauss quadrature rule) with arbitrary degree of accuracy because the function $\frac{1}{(\sin^2 \frac{\sigma_1}{2} + \sin^2 \frac{\sigma_2}{2})^\lambda + h}$, has derivatives up to arbitrary order. The choice of parameter h is discussed in Section 8.

Let $k = i, l \neq j$, and

$$I_{il} = \frac{4\pi^2}{n^2} \left(\sin^2 \frac{x'_l - x'_j}{2}\right)^{-\lambda} = p^*_{il}.$$

Let $k \neq i$, $l = j$, and

$$I_{kj} = \frac{4\pi^2}{n^2} \left(\sin^2 \frac{x'_k - x'_i}{2}\right)^{-\lambda} = p^*_{kj}.$$

Let $k \neq i$, $l \neq j$, and

$$I_{kl} = \frac{4\pi^2}{n^2}\left(\sin^2\frac{x'_k - x'_i}{2} + \sin^2\frac{x'_l - x'_j}{2}\right)^{-\lambda} = p^*_{kl}.$$

The integral Kf is calculated by the formula

$$Kf = \sum_{k=0}^{n-1}\sum_{l=0}^{n-1} p^*_{kl} f(x'_k, x'_l) + R_{nn}(f, p^*_{kl}, x_k, y'_l). \qquad \text{(B.18)}$$

Formula (B.18) is not optimal since it is not exact on constant functions $f(x, y) = const$. But one can estimate the error of this formula:

$$|R_{nn}(f, p^*_{kl}, x'_k, y'_l))| \le M \sum_{k=0}^{n-1}\sum_{l=0}^{n-1} |I_{kl} - p^*_{kl}| + R_{nn}(\Psi),$$

where $M = \max|f(x, y)|$.

The values $|I_{kl} - p^*_{kl}|$ are easily estimated, and one gets the conclusion of Theorem B.10.

Classes of smooth functions

Theorem B.11 *Assume $\varphi \in \tilde{W}^{r,r}(1)$. Let $\Psi = \tilde{W}^{r,r}(1)$, and calculate the integral $K\varphi$ by formula* (B.11) *with $\rho_1 = r - 1$, $\rho_2 = r - 1$, and $n_1 = n_2 = n$. Then the cubature formula*

$$K\varphi = \int_0^{2\pi}\int_0^{2\pi} \frac{\varphi_{mn}(\sigma_1, \sigma_2) d\sigma_1 d\sigma_2}{(\sin^2(\sigma_1 - s_1)/2 + \sin^2(\sigma_2 - s_2)/2)^\lambda} + R_{mn}(\varphi) \qquad \text{(B.19)}$$

is asymptotically optimal.

Before proving Theorem B.11, let us describe the construction of the spline φ_{mn}. Let $x_k = 2k\pi/n$, $k = 0, 1, \dots, n$. Divide the sides of the squares $\Omega = [0, 2\pi; 0, 2\pi]$ into n equal parts. Denote by Δ_{kl} the rectangle $\Delta_{kl} = [2k\pi/n, 2(k+1)\pi/n; 2l\pi/n, 2(l+1)\pi/n], k, l = 0, 1, \dots, n-1$. Let $(s_1, s_2) \in \Delta_{ij}$. First we approximate $\varphi(\sigma_1, \sigma_2)$ as a function of σ_2, and construct a spline $\varphi_n(\sigma_1, \sigma_2)$ by the following rule. Let σ_1 be an arbitrary fixed number, $0 \le \sigma_1 \le 2\pi$. On the segments $[2k\pi/n, 2(k+1)\pi/n]$ for $k \neq j-2, \dots, j+1$, one has:

$$\varphi_n(\sigma_1, \sigma_2) = \sum_{l=0}^{r-1}\left[\frac{\varphi^{(0,l)}(\sigma_1, 2k\pi/n)}{l!}(\sigma_2 - 2k\pi/n)^l + B_l \delta^{(l)}(\sigma_1, (k+1)/n)\right],$$

where

$$\delta(\sigma_1, \sigma_2) := \varphi(\sigma_1, \sigma_2) - \sum_{l=0}^{r-1} \frac{\varphi^{(0,l)}(\sigma_1, 2k\pi/n)}{l!}(\sigma_2 - 2k\pi/n)^l.$$

The coefficients B_l are defined by the equation

$$(2(k+1)\pi/n - \sigma_2)^r - \sum_{l=0}^{r-1} \frac{B_l r!}{(r-l-1)!}\frac{2\pi}{n}\left(2\pi(k+1)/n - \sigma_2\right)^{r-l-1}$$

$$= (-1)^r R_{r1}\left(2\pi(2k+1)/2n; \pi/n; \sigma_2\right),$$

where $R_{r1}(a, h, x)$ is a polynomial of degree r, least deviating from zero in the norm of the space L on the segment $[a-h, a+h]$. On the segment $[2\pi(j-2)/n, 2\pi(j+2)/n]$ the function $\varphi_n(\sigma_1, \sigma_2)$ is defined by the partial sum of the Taylor series:

$$\varphi_n(\sigma_1, \sigma_2) = \varphi(\sigma_1, 2\pi j/n) + \frac{\varphi^{(0,1)}(\sigma_1, 2\pi j/n)}{1!}(\sigma_2 - j/n) + \cdots$$

$$+ \frac{\varphi^{(0,r-1)}(\sigma_1, 2\pi j/n)}{(r-1)!}(\sigma_2 - 2\pi j/n)^{r-1}.$$

We define the function $\varphi_{nn}(\sigma_1, \sigma_2)$ by analogy with the function $\varphi_n(\sigma_1, \sigma_2)$.

Proof of Theorem B.11. Let $(s_1, s_2) \in \Delta_{ij}$. The error of formula (B.19) we estimate by the inequality

$$|R_{nn}| \leqslant \sum_{k=0}^{n-1} {\sum_{l=0}^{n-1}}' \left| \iint\limits_{\Delta_{kl}} \frac{\varphi(\sigma_1, \sigma_2) - \varphi_{nn}(\sigma_1, \sigma_2)}{\left(\sin^2 \frac{\sigma_1 - s_1}{2} + \sin^2 \frac{\sigma_2 - s_2}{2}\right)^\lambda} d\sigma_1 d\sigma_2 \right|$$

$$+ \sum_{k=0}^{n-1} {\sum_{l=0}^{n-1}}'' \left| \int\limits_{\Delta_{kl}} \frac{\varphi(\sigma_1, \sigma_2) - \varphi_{nn}(\sigma_1, \sigma_2)}{\left(\sin^2 \frac{\sigma_1 - s_1}{2} + \sin^2 \frac{\sigma_2 - s_2}{2}\right)^\lambda} d\sigma_1 d\sigma_2 \right| = r_1 + r_2, \qquad \text{(B.20)}$$

where ${\sum_{k,l}}'$ means summation over (k, l) such that $i - 1 \leqslant k \leqslant i + 1$, $0 \leqslant l \leqslant n - 1$ or $0 \leqslant k \leqslant n - 1$, $j - 1 \leqslant l \leqslant j + 1$, and ${\sum_{k,l}}''$ means summation over the other values of (k, l).

Let us estimate each of the sums r_1 and r_2 separately. In addition without loss of generality assume that $\iint\limits_{\Delta_{kl}} (\varphi(\sigma_1, \sigma_2) - \varphi_{nn}(\sigma_1, \sigma_2)) d\sigma_1 d\sigma_2 \geqslant 0$. Then

$$r_1 \leqslant \sum_{k=0}^{n-1} {\sum_{l=0}^{n-1}}' |\varphi(\sigma_1, \sigma_2) - \varphi_{nn}(\sigma_1, \sigma_2)| \iint\limits_{\Delta_{kl}} \frac{d\sigma_1 d\sigma_2}{\left(\sin^2 \frac{\sigma_1 - s_1}{2} + \sin^2 \frac{\sigma_2 - s_2}{2}\right)^\lambda}$$

$$\leqslant A \begin{cases} n^{-(r+1)}, & \lambda \leq 1/2 \\ n^{-(r+2-2\lambda)}, & \lambda > 1/2; \end{cases} \tag{B.21}$$

$$r_2 \leqslant 4 \sum_{k=i+2}^{i+1+[(n-1)/2]} \sum_{l=j+2}^{j+1+[(n-1)/2]} \frac{1}{\left(\sin^2 \frac{x_k - s_1}{2} + \sin^2 \frac{x_l - s_2}{2}\right)^\lambda}$$

$$\times \iint\limits_{\Delta_{kl}} \psi(\sigma_1, \sigma_2) d\sigma_1 d\sigma_2 - 4 \sum_{k=i+2}^{i+1+[(n-1)/2]} \sum_{l=j+2}^{j+1+[(n-1)/2]}$$

$$\times \left[\frac{1}{\left(\sin^2 \frac{x_k - s_1}{2} + \sin^2 \frac{x_l - s_2}{2}\right)^\lambda} - \frac{1}{\left(\sin^2 \frac{x_{k+1} - s_1}{2} + \sin^2 \frac{x_{l+1} - s_2}{2}\right)^\lambda} \right]$$

$$\cdot \iint\limits_{\Delta_{kl}} \psi^-(\sigma_1, \sigma_2) d\sigma_1 d\sigma_2 = r_{21} + r_{22}, \tag{B.22}$$

where $\psi(\sigma_1, \sigma_2) = \varphi(\sigma_1, \sigma_2) - \varphi_{nn}(\sigma_1, \sigma_2)$,

$$\psi^+(\sigma_1, \sigma_2) = \begin{cases} \psi(\sigma_1, \sigma_2), & \text{if } \psi(\sigma_1, \sigma_2) \geqslant 0 \\ 0, & \text{if } \psi(\sigma_1, \sigma_2) < 0; \end{cases}$$

$$\psi^-(\sigma_1, \sigma_2) = \begin{cases} 0, & \text{if } \psi(\sigma_1, \sigma_2) \geqslant 0; \\ -\psi(\sigma_1, \sigma_2), & \text{if } \psi(\sigma_1, \sigma_2) < 0. \end{cases}$$

One has:

$$4\sum_{k=i+2}^{i+1+[(n-1)/2]}\sum_{l=j+2}^{j+1+[(n-1)/2]}\frac{1}{\left(\sin^2\frac{x_k-s_1}{2}+\sin^2\frac{x_l-s_2}{2}\right)^{\lambda}}$$
$$\leqslant\frac{1+o(1)}{4\pi^2}\int_0^{2\pi}\int_0^{2\pi}\frac{d\sigma_1 d\sigma_2}{\left(\sin^2\frac{\sigma_1}{2}+\sin^2\frac{\sigma_2}{2}\right)^{\lambda}} \tag{B.23}$$

Let us estimate the integral

$$i:=\iint_{\Delta_{kl}}\psi(\sigma_1,\sigma_2)d\sigma_1 d\sigma_2\leqslant\left|\iint_{\Delta_{kl}}\Big(\varphi(\sigma_1,\sigma_2)-\varphi_n(\sigma_1,\sigma_2)\Big)d\sigma_1 d\sigma_2\right|$$

$$+\left|\iint_{\Delta_{kl}}\Big(\varphi_n(\sigma_1,\sigma_2)-\varphi_{nn}(\sigma_1,\sigma_2)\Big)d\sigma_1 d\sigma_2\right|=i_1+i_2. \tag{B.24}$$

Since the expressions i_1 and i_2 are estimated similarly, we estimate only i_1. One has

$$i_1\leqslant\frac{2\pi}{n}\max_{s_1}\left|\int_{x_l}^{x_{l+1}}\Big(\varphi(s_1,\sigma_2)-\varphi_n(s_1,\sigma_2)\Big)d\sigma_2\right|.$$

This integral is a continuous function of s_1, which attains its maximum at a point s^*, and

$$i_1\leqslant\frac{2\pi}{n}\left|\int_{x_l}^{x_{l+1}}\Big(\varphi(s^*,\sigma_2)-\varphi_n(s^*,\sigma_2)\Big)d\sigma_2\right|$$

$$\leqslant \frac{2\pi}{r!n} \int_{x_l}^{x_{l+1}} \left|\varphi^{(0,r)}(s^*, \sigma_2)\right| \Bigg| (x_{l+1} - \sigma_2)^r$$

$$- \sum_{j=0}^{r-1} \frac{B_{lj}(x_{l+1} - x_l)r!}{(r-1-j)!} (x_{l+1} - \sigma_2)^{r-j-1} \Bigg| d\sigma_2$$

$$\leqslant \frac{2\pi}{r!n} \int_{x_l}^{x_{l+1}} \left| (x_{l+1} - \sigma_2)^r - \sum_{j=0}^{r-1} \frac{B_{lj}(x_{l+1} - x_l)r!}{(r-1-j)!} (x_{l+1} - \sigma_2)^{r-j-1} \right| d\sigma_2$$

$$= \frac{2\pi}{r!n} \int_{x_l}^{x_{l+1}} \left|R_{r1}(\sigma_2)\right| d\sigma_2 \leqslant \frac{4}{(r+1)!} \left(\frac{\pi}{n}\right)^{r+2} R_{r1}(1). \tag{B.25}$$

From inequalities (B.24) and (B.25) one gets

$$i \leqslant \frac{8}{(r+1)!} \left(\frac{\pi}{n}\right)^{r+2} R_{r1}(1)$$

and

$$r_{21} \leqslant \frac{2 + o(1)}{(r+1)!} \left(\frac{\pi}{n}\right)^r R_{r1}(1) \int_0^{2\pi} \int_0^{2\pi} \frac{d\sigma_1 d\sigma_2}{\left(\sin^2 \frac{\sigma_1}{2} + \sin^2 \frac{\sigma_2}{2}\right)^\lambda}. \tag{B.26}$$

One has:

$$r_{22} = o(n^{-r}). \tag{B.27}$$

Estimate (B.27) follows from the inequalities:

$$\left| \iint_{\Delta_{kl}} \psi^-(\sigma_1, \sigma_2) d\sigma_1 d\sigma_2 \right| \leqslant \iint_{\Delta_{kl}} \left|\psi(\sigma_1, \sigma_2)\right| d\sigma_1 d\sigma_2 = O(n^{-r-2})$$

and

$$\sum_{k=i+2}^{i+1+[(n-1)/2]} \sum_{l=j+2}^{j+1+[(n-1)/2]}$$

$$\times \left| \frac{1}{\left(\sin^2 \frac{x_k - s_1}{2} + \sin^2 \frac{x_l - s_2}{2}\right)^\lambda} - \frac{1}{\left(\sin^2 \frac{x_{k+1} - s_1}{2} + \sin^2 \frac{x_{l+1} - s_2}{2}\right)^\lambda} \right|$$

$$\leqslant An^{2\lambda}\sum_k\sum_l \frac{(k-i)+(l-j)}{\left((k-i)^2+(l-j)^2\right)^{\lambda+1}} \leqslant c\begin{cases} n, & \lambda<1/2\\ n\log n, & \lambda=1/2\\ n^{2\lambda}, & \lambda>1/2.\end{cases}$$

The estimate

$$R_{nn}(\Psi)\leq (1+o(1))\frac{2\pi^r R_{r1}(1)}{(r+1)!(n-1+[R_{r1}(1)]^{1/r})^r}\gamma$$

follows from inequalities (B.20), (B.21), (B.26), and (B.27).

Theorem B.9 follows from the comparison of the values $\zeta_{nn}[\Psi]$ and $R_{nn}[\Psi]$. □

Let us construct cubature formulas for calculating integrals Kf on classes of functions $W^{rr}(1)$. These formulas will be less accurate than the ones in Theorem B.7, but they will be optimal with respect to order, and easier to apply.

First, we investigate the smooth function

$$\psi(t_1,t_2)=\int_0^{2\pi}\int_0^{2\pi}\frac{f(\tau_1,\tau_2)d\tau_1 d\tau_2}{\left(\sin^2\frac{\tau_1-t_1}{2}+\sin^2\frac{\tau_2-t_2}{2}\right)^\lambda}$$

assuming $f(t_1,t_2)\in\tilde{W}^{r,r}$. Changing the variables $\tau_1=\tau_1-t$, $\tau_2=\tau_2-t$, in the last integral, one gets:

$$\psi(t_1,t_2)=\int_0^{2\pi}\int_0^{2\pi}\frac{f(\tau_1+t_1,\tau_2+t_2)d\tau_1 d\tau_2}{\left(\sin^2\frac{\tau_1}{2}+\sin^2\frac{\tau_2}{2}\right)^\lambda}$$

Thus, $\psi(t_1,t_2)\in W^{r,r}$.

Remark B.3 *It is known (see, e.g., [6]) that Kolmogorov and Babenko widths of the class of functions $W^{r,r}(1)$ are $\delta_n(W^{r,r}(1))\asymp d_n(W^{r,r}(1),C)\asymp \frac{1}{n^{r/2}}$. Hence the recovery of the function $\psi(t_1,t_2)$ using n functionals is not possible with accuracy greater than $O(\frac{1}{n^{r/2}})$. More precise conclusions are obtained in Theorems 5.3 and 5.4.*

Thus, for recovery of a function $\psi(t_1,t_2)$, $(t_1,t_2)\in[0,2\pi]^2$ with the accuracy $O(n^{-r/2})$, it is sufficient to calculate the value of the function $\psi(t_1,t_2)$ at the nodes (v_k,v_l), where $v_k=2k\pi/N$, $k,l=0,1,\dots,N$, and $N^2=n$, and to use the local spline $\psi_N(t_1,t_2)$ of degree r with respect to each variable.

Let us describe the construction of such spline.

Assume for simplicity that $M := N/r$ is an integer, and cover the domain $[0, 2\pi]^2$ with the squares $\Delta_{kl} = [w_k, w_l]$, $k, l = 0, 1, \ldots, M-1$, here $w_k = 2k\pi/M$, $k = 0, \ldots, M$. Approximate the function $\psi(t_1, t_2)$ in each domain Δ_{kl} by the interpolation polynomial $\psi_N(t_1, t_2, \Delta_{kl})$ constructed on the nodes (x_i^k, x_j^l), $i, j = 0, 1, \ldots, r$, $x_i^k = w_k + \frac{2\pi}{Mr} i$, $i = 0, 1, \ldots, r$.

Denote the local spline, which is defined by the polynomials $\psi_N(t_1, t_2, \Delta_{kl})$, by $\psi_N(t_1, t_2)$.

If the values $\psi(v_k, v_l)$ are calculated by formula (B.19) with the accuracy $O(n^{-r/2})$, then

$$\|\psi(t_1, t_2) - \psi_N(t_1, t_2)\|_C \leq O(n^{-r/2}).$$

Therefore the spline $\psi_N(t_1, t_2)$ is optimal with respect to order, and a method for recovery of the function $\psi(t_1, t_2)$, which has the error $O(n^{-r/2})$ (in the sup −norm) is constructed.

B.6 Optimal Methods for Calculating Integrals of the Form Tf

B.6.1 *Lower bounds for the functionals ζ_{mn} and ζ_N*

First we get a lower bound for the error of formula (B.3) with $\rho_1 = \rho_2 = 0$ and $n_1 = n_2 = n$, on Hölder classes.

Theorem B.12 *Let $\Psi = H_{\alpha\alpha}(D)$, and calculate the integral Tf by formula* (B.3) *with $n_1 = n_2 = n$ and $\rho_1 = \rho_2 = 0$. Then the estimate*

$$\zeta_{nn}[\Psi] \geq \frac{(1 + o(1))}{2^{2\lambda}(1+\alpha)n^\alpha} \int_{-1}^{1}\int_{-1}^{1} \frac{dt_1 dt_2}{(\tau_1^2 + t_1^2)^\lambda} \tag{B.28}$$

holds.

Proof. Let $n > 0$ be an integer, $L = [n/\log n]$. Let $v_k := -1 + 2k/L$, $k = 0, 1, \ldots, L$. By (ξ_k, η_l) we denote a set which is the union of nodes (x_i, y_j), $i, j = 1, 2, \ldots, n$ of formula (B.3) and the nodes (v_i, v_j), $i, j = 1, 2, \ldots, L$. Let $\Delta_{kl} = [v_k, v_{k+1}; v_l, v_{l+1}]$, $k, l = 0, 1, \ldots, L-1$. Let $0 \leq \psi(t_1, t_2) \in H_{\alpha\alpha}(D)$, where $D = [-1, 1]^2$, vanishing at the nodes (ξ_k, η_l),

$k, l = 0, 1, \dots, N$, where $N = n + L$. Consider the integral

$$
\begin{aligned}
&(T\psi)(v_i, v_j) \\
&= \int_{-1}^{1}\int_{-1}^{1} \frac{\psi(\tau_1, \tau_2)d\tau_1 d\tau_2}{((\tau_1 - v_i)^2 + (\tau_2 - v_j)^2)^\lambda} \\
&= \left(\sum_{k=i}^{L-1}\sum_{l=j}^{L-1} + \sum_{k=i}^{L-1}\sum_{l=0}^{j-1} + \sum_{k=0}^{i-1}\sum_{l=j}^{L-1} + \sum_{k=0}^{i-1}\sum_{l=0}^{j-1}\right) \\
&\times \iint_{\Delta_{kl}} \frac{\psi(\tau_1, \tau_2)d\tau_1 d\tau_2}{((\tau_1 - v_i)^2 + (\tau_2 - v_j)^2)^\lambda} \\
&\geq \sum_{k=0}^{L-i-1}\sum_{l=0}^{L-j-1} \left(\frac{L}{2}\right)^{2\lambda} \frac{1}{((k+1)^2 + (l+1)^2)^\lambda} \iint_{\Delta_{k+i,l+j}} \psi(\tau_1, \tau_2)d\tau_1 d\tau_2 \\
&\quad + \sum_{k=0}^{L-i-1}\sum_{l=0}^{j-1} \left(\frac{L}{2}\right)^{2\lambda} \frac{1}{((k+1)^2 + (l+1)^2)^\lambda} \iint_{\Delta_{k+i,j-l-1}} \psi(\tau_1, \tau_2)d\tau_1 d\tau_2 \\
&\quad + \sum_{k=0}^{i-1}\sum_{l=0}^{L-j-1} \left(\frac{L}{2}\right)^{2\lambda} \frac{1}{((k+1)^2 + (l+1)^2)^\lambda} \iint_{\Delta_{i-k-1,j+l}} \psi(\tau_1, \tau_2)d\tau_1 d\tau_2 \\
&\quad + \sum_{k=0}^{i-1}\sum_{l=0}^{j-1} \left(\frac{L}{2}\right)^{2\lambda} \frac{1}{((k+1)^2 + (l+1)^2)^\lambda} \iint_{\Delta_{i-k-1,j-l-1}} \psi(\tau_1, \tau_2)d\tau_1 d\tau_2 \\
&= \sum_{k=0}^{L-1}\sum_{l=0}^{L-1} \left(\frac{L}{2}\right)^{2\lambda} \frac{U(L-i-1-k)U(L-j-1-l)}{((k+1)^2 + (l+1)^2)^\lambda} \iint_{\Delta_{k+i,l+j}} \psi(\tau_1, \tau_2)d\tau_1 d\tau_2 \\
&\quad + \sum_{k=0}^{L-1}\sum_{l=0}^{L-1} \left(\frac{L}{2}\right)^{2\lambda} \frac{U(L-i-1-k)U(j-1-l)}{((k+1)^2 + (l+1)^2)^\lambda} \iint_{\Delta_{k+i,j-l-1}} \psi(\tau_1, \tau_2)d\tau_1 d\tau_2 \\
&\quad + \sum_{k=0}^{L-1}\sum_{l=0}^{L-1} \left(\frac{L}{2}\right)^{2\lambda} \frac{U(i-1-k)U(L-j-1-l)}{((k+1)^2 + (l+1)^2)^\lambda} \iint_{\Delta_{i-k-1,j+l}} \psi(\tau_1, \tau_2)d\tau_1 d\tau_2 \\
&\quad + \sum_{k=0}^{L-1}\sum_{l=0}^{L-1} \left(\frac{L}{2}\right)^{2\lambda} \frac{U(i-1-k)U(j-1-l)}{((k+1)^2 + (l+1)^2)^\lambda} \iint_{\Delta_{i-k-1,j-l-1}} \psi(\tau_1, \tau_2)d\tau_1 d\tau_2.
\end{aligned}
$$

Here $U(k) = 1$ for $k \geq 0$, and $U(k) = 0$ for $k < 0$.

Averaging the above inequality over all i and j, $i, j = 0, 1, \ldots, L-1$, one gets:

$$
\begin{aligned}
&R_{nn}(\Psi, p_{kl}; x_k, y_l) \\
&\geq \frac{1}{L^2}\sum_{i=0}^{L-1}\sum_{j=0}^{L-1} T(\psi)(\xi_i, \eta_j) \geq \frac{1}{L^{2-2\lambda}2^{2\lambda}}\sum_{k=0}^{L-1}\sum_{l=0}^{L-1}\frac{1}{((k+1)^2+(l+1)^2)^\lambda} \\
&\quad \times \Bigg[\sum_{i=0}^{L-1}\sum_{j=0}^{L-1} U(L-i-1-k)\times U(L-j-1-l)\iint_{\Delta_{k+i,l+j}} \psi(\tau_1,\tau_2)d\tau_1 d\tau_2 \\
&\qquad + \sum_{i=0}^{L-1}\sum_{j=0}^{L-1} U(L-i-1-k)U(j-1-l)\int\int_{\Delta_{k+i,j-l-1}} \psi(\tau_1,\tau_2)d\tau_1 d\tau_2 \\
&\qquad + \sum_{i=0}^{L-1}\sum_{j=0}^{L-1} U(i-1-k)U(L-j-1-l)\int\int_{\Delta_{i-k-1,j+l}} \psi(\tau_1,\tau_2)d\tau_1 d\tau_2 \\
&\qquad + \sum_{i=0}^{L-1}\sum_{j=0}^{L-1} U(i-1-k)U(j-1-l)\int\int_{\Delta_{i-k-1,j-l-1}} \psi(\tau_1,\tau_2)d\tau_1 d\tau_2\Bigg] \\
&\geq \frac{1}{L^{2-2\lambda}2^{2\lambda}}\sum_{k=0}^{L-1}\sum_{l=0}^{L-1}\frac{1}{((k+1)^2+(l+1)^2)^\lambda} \\
&\quad \times \Bigg[\int_{v_k}^{1}\int_{v_l}^{1}\psi(\tau_1,\tau_2)d\tau_1 d\tau_2 + \int_{v_k}^{1}\int_{-1}^{v_{L-l-2}}\psi(\tau_1,\tau_2)d\tau_1 d\tau_2 \\
&\qquad + \int_{-1}^{v_{L-k-2}}\int_{v_l}^{1}\psi(\tau_1,\tau_2)d\tau_1 d\tau_2 + \int_{-1}^{v_{L-k-2}}\int_{-1}^{v_{L-l-2}}\psi(\tau_1,\tau_2)d\tau_1 d\tau_2\Bigg] \\
&\geq \frac{1}{L^{2-2\lambda}2^{2\lambda}}\sum_{k=0}^{L-1}\sum_{l=0}^{L-1}\frac{1}{((k+1)^2+(l+1)^2)^\lambda}\int_{-1}^{1}\int_{-1}^{1}\psi(\tau_1,\tau_2)d\tau_1 d\tau_2.
\end{aligned}
\tag{B.29}
$$

From inequality (B.29) it follows that

$$
\zeta_{nn}[H_{\alpha\alpha}(D)] \geq (1+o(1))\frac{1}{L^{2-2\lambda}2^{2\lambda}}\sum_{k=1}^{L-1}\sum_{l=1}^{L-1}\frac{1}{(k^2+l^2)^\lambda}\int_{-1}^{1}\int_{-1}^{1}\psi(\tau_1,\tau_2)d\tau_1 d\tau_2
$$

$$= \frac{1+o(1)}{2^{2\lambda}4} \int_{-1}^{1}\int_{-1}^{1} \frac{dt_1dt_2}{(t_1^2+t_2^2)^\lambda} \int_{-1}^{1}\int_{-1}^{1} \psi(\tau_1,\tau_2)d\tau_1d\tau_2. \tag{B.30}$$

From Theorem B.3 and Lemma B.5 it follows that the inequality

$$\int_{-1}^{1}\int_{-1}^{1} \psi(\tau_1,\tau_2)d\tau_1d\tau_2 \geq \frac{4}{1+\alpha}\frac{1}{n^\alpha} \tag{B.31}$$

is valid for an arbitrary vector of the weights and the nodes (X, Y, P) on the class $H_{\alpha\alpha}(D)$.

Theorem B.12 follows from inequalities (B.30) and (B.31). □

Theorem B.13 *Let* $\Psi = C_2^r(1)$*, and calculate the integral* Tf *by formula* (B.3) *with* $\rho_1 = \rho_2 = 0$*. If* $n_1 = n_2 = n$*, then*

$$\zeta_{nn}[\Psi] \geq (1+o(1))\frac{2K_r}{2^{2\lambda}(\pi n)^r} \int_{-1}^{1}\int_{-1}^{1} \frac{ds_1ds_2}{(s_1^2)+s_2^2))^\lambda},$$

where K_r *is the Faward constant.*

Proof. Let

$$\psi(s_1,s_2) = \psi_1(s_1) + \psi_2(s_2),$$

where $0 \leq \psi_1(s) \in W^r(1)$, vanishes at the nodes x_k, $k = 1, 2, \ldots, n$, and $0 \leq \psi_2(s) \in W^r(1)$ vanishes at the nodes y_k, $k = 1, 2, \ldots, n$.

For arbitrary nodes x_k, $k = 1, 2, \ldots, n$, one has (see [75]):

$$\int_{-1}^{1} \psi_i(s)ds \geq \frac{2K_r}{(\pi n)^r}, \; i = 1, 2.$$

Thus the inequality

$$\int_{-1}^{1}\int_{-1}^{1} \psi(s_1,s_2)ds_1ds_2 \geq \frac{8K_r}{(\pi n)^r}$$

holds for arbitrary nodes $(x_1, \ldots, x_n)$ and $(y_1, \ldots, y_n)$.

Theorem B.13 follows from this estimate and inequality (B.30). □

Theorem B.14 *Let $\Psi = W_p^{r,r}(1)$, $r = 1, 2, \ldots$, $1 \le p \le \infty$, and calculate the integral Tf by formula* (B.3) *with $\rho_1 = \rho_2 = r - 1$ and $n_1 = n_2 = n$. Then the estimate*

$$\zeta_{nn}[\Psi] \ge (1 + o(1)) \frac{2^{1/q} R_{rq}(1)}{2^{2\lambda} r! (rq+1)^{1/q} (n - 1 + [R_{rq}(1)]^{1/r})^r} \int_{-1}^{1} \int_{-1}^{1} \frac{ds_1 ds_2}{(s_1^2 + s_2^2)^\lambda} \tag{B.32}$$

holds, where $R_{rq}(t)$ is a polynomial of degree r, least deviating from zero in $L_q([-1,1])$.

Proof. Let $L = [n/\log n]$. Consider the nodes (v_k, v_l), $v_k = \frac{2k}{L}$, $k, l = 0, 1, \ldots, L-1$. By (ξ_i, η_j), $i, j = 0, 1, \ldots, N-1$, $N = n + L$ denote the union of the nodes (x_k, y_l) and (ξ_i, ξ_j). Let $\psi(s_1, s_2) = \psi_1(s_1) + \psi_2(s_2)$, where $0 \le \psi_1(s) \in W_p^r(1)$ vanishes with its derivatives up to order $r - 1$ at the nodes ξ_i, $i = 0, 1, \ldots, N-1$, and $0 \le \psi_2(s) \in W_p^r(1)$ vanishes with its derivatives up to order $r - 1$ at the nodes η_j, $j = 0, 1, \ldots, N-1$. Assume that $\int_{v_i}^{v_{i+1}} \psi_1(s) ds > 0$, $i = 0, 1, \ldots, N-1$, and $\int_{v_j}^{v_{j+1}} \psi_2(s) ds > 0$, $j = 0, 1, \ldots, N-1$.

Using the argument similar to the one in the proof of Theorem B.12, one gets

$$\zeta_{nn}(\Psi, p_{kl}; v_k, v_l) \ge \frac{1}{L^2} \sum_{i=0}^{L-1} \sum_{j=0}^{L-1} T(\psi)(v_i, v_j)$$

$$\begin{aligned} &\ge \frac{1}{L^{2-2\lambda} 2^{2\lambda}} \sum_{k=0}^{L-1} \sum_{l=0}^{L-1} \frac{1}{((k+1)^2 + (l+1)^2)^\lambda} \int_{-1}^{1} \int_{-1}^{1} \psi(\tau_1, \tau_2) d\tau_1 d\tau_2 \\ &= \frac{1 + o(1)}{2^{2\lambda} 4} \int_{-1}^{1} \int_{-1}^{1} \frac{dt_1 dt_2}{(t_1^2 + t_2^2)^\lambda} \int_{-1}^{1} \int_{-1}^{1} \psi(\tau_1, \tau_2) d\tau_1 d\tau_2. \end{aligned} \tag{B.33}$$

From Theorem B.2 and Lemma B.5 it follows that the inequality

$$\int_{-1}^{1} \int_{-1}^{1} \psi(\tau_1, \tau_2) d\tau_1 d\tau_2 \ge (1 + o(1)) \frac{2^{2+1/q} R_{rq}(1)}{r! (rq+1)^{1/q} (n - 1 + [R_{rq}(1)]^{1/q})^r} \tag{B.34}$$

is valid for arbitrary weights and the nodes (X, Y, P) on the class $H_{\alpha\alpha}(D)$.

Theorem B.14 follows from inequalities (B.33)-(B.34). □

B.6.2 *Cubature formulas*

Let us construct a cubature formula for calculating the integral Tf on the Hölder class $H_{\alpha\alpha}(D)$. Let $x_k := -1 + 2k/n$, $k = 0, 1, \ldots, n$, $x'_k = (x_{k+1} + x_k)/2$, $k = 0, 1, \ldots, n-1$, and $\Delta_{kl} = [x_k, x_{k+1}; x_l, x_{l+1}]$, $k, l = 0, 1, \ldots, n-1$.

Calculate the integral Tf by the formula

$$Tf = \sum_{k=0}^{n-1} \sum_{l=0}^{n-1} f(x'_k, x'_l) \iint\limits_{\Delta_{kl}} \frac{d\tau_1 d\tau_2}{((\tau_1 - t_1)^2 + (\tau_2 - t_2)^2)^\lambda} + R_{nn}(f). \quad \text{(B.35)}$$

Consider another cubature formula for calculating the integral Tf.

Let $(t_1, t_2) \in \Delta_{ij}$. By Δ_* denote the union of the square Δ_{ij} and of those squares Δ_{kl} which have common points with the Δ_{ij}. Consider the formula

$$Tf = f(x'_i, x'_j) \iint\limits_{\Delta_*} \frac{d\tau_1 d\tau_2}{((\tau_1 - t_1)^2 + (\tau_2 - t_2)^2)^\lambda}$$

$$+ \sum_{k=0}^{n-1} {\sum_{l=0}^{n-1}}' f(x'_k, x'_l) \iint\limits_{\Delta_{kl}} \frac{d\tau_1 d\tau_2}{((\tau_1 - t_1)^2 + (\tau_2 - t_2)^2)^\lambda} + R_{nn}(f), \quad \text{(B.36)}$$

where $\sum\sum'$ means summation over the squares which do not belong to Δ_*.

Theorem B.15 *Among all cubature formulas* (B.3) *with* $\rho_1 = \rho_2 = 0$ *and* $n_1 = n_2 = n$, *formula* (B.35), *with the error estimate* (B.42), *is optimal with respect to order.*

Remark B.4 *Similar statement holds for formula* (B.36).

Proof of Theorem B.15. Let us estimate errors of formulas (B.35) and (B.36).

The error of formula (B.35) can be estimated as follows:

$$|R_{nn}(f)| \le \sum_{k=0}^{n-1} {\sum_{l=0}^{n-1}}' \iint\limits_{\Delta_{kl}} \frac{|f(\tau_1, \tau_2) - f(x'_i, x'_j)|}{((\tau_1 - t_1)^2 + (\tau_2 - t_2)^2)^\lambda} d\tau_1 d\tau_2$$

$$+ \sum_{k=0}^{n-1} {\sum_{l=0}^{n-1}}'' \iint\limits_{\Delta_{kl}} \frac{|f(\tau_1, \tau_2) - f(x'_k, x'_l)|}{((\tau_1 - t_1)^2 + (\tau_2 - t_2)^2)^\lambda} d\tau_1 d\tau_2 = r_1 + r_2, \quad \text{(B.37)}$$

where $\sum\sum'$ means summation over k and l such that the squares Δ_{kl} belong to Δ_*, and $\sum\sum''$ means summation over the other squares.

Let us estimate r_1 and r_2:

$$r_1 \leq \frac{2}{n^\alpha} \iint\limits_{\Delta_*} \frac{d\tau_1 d\tau_2}{((\tau_1 - t_1)^2 + (\tau_2 - t_2)^2)^\lambda} \leq \frac{c}{n^{2-2\lambda+\alpha}} = o(n^{-\alpha}), \quad \text{(B.38)}$$

$$r_2 \leq \frac{4}{1+\alpha} \frac{1}{n^{2+\alpha}} \sum_{k=0}^{n-1} \sum_{l=0}^{n-1}{}'' h(\Delta_{kl}). \quad \text{(B.39)}$$

Here $h(\Delta_{kl})$ denotes the maximum value of the function $((\tau_1 - t_1)^2 + (\tau_2 - t_2)^2)^{-\lambda}$ in the square Δ_{kl}.

One has:

$$\left| \iint\limits_{\Delta_{kl}} \left[\frac{1}{((\tau_1 - t_1)^2 + (\tau_2 - t_2)^2)^\lambda} - h(\Delta_{kl}) \right] d\tau_1 d\tau_2 \right|$$

$$= \left| \iint\limits_{\Delta_{kl}} \left[\frac{1}{((\tau_1 - t_1)^2 + (\tau_2 - t_2)^2)^\lambda} - \frac{1}{((x_k - t_1)^2 + (x_l - t_2)^2)^\lambda} \right] d\tau_1 d\tau_2 \right|$$

$$\leq \iint\limits_{\Delta_{kl}} \left| \frac{2\lambda(x_k - t_1 + q_1(\tau_1 - x_k))(\tau_1 - x_k))}{((x_k - t_1 + q_1(\tau_1 - x_k))^2 + (\tau_2 - t_2)^2)^{\lambda+1}} d\tau_1 d\tau_2 \right|$$

$$+ \iint\limits_{\Delta_{kl}} \left| \frac{2\lambda(x_l - t_2 + q_2(\tau_2 - x_l))(\tau_2 - x_l))}{((x_k - t_1) + (x_l - t_2 + q_2(\tau_2 - x_l))^2)^\lambda} d\tau_1 d\tau_2 \right|$$

$$\leq \iint\limits_{\Delta_{kl}} \frac{2\lambda(\tau_1 - x_k)}{((x_k - t_1 + q(\tau_1 - x_k))^2 + (\tau_2 - t_2)^2)^{\lambda+1/2}} d\tau_1 d\tau_2 \Bigg|$$

$$+ \iint\limits_{\Delta_{kl}} \frac{2\lambda(\tau_2 - x_l))}{((\tau_1 - t_1)^2 + (x_l - t_2 + q_2(\tau_2 - x_l))^2)^{\lambda+1/2}} d\tau_1 d\tau_2 \Bigg|$$

$$\leq \frac{2^4\lambda}{n^3} \frac{n^{2\lambda+1}}{(k^2 + l^2)^{\lambda+1/2}} = \frac{2^4\lambda}{n^{2-2\lambda}} \frac{1}{(k^2 + l^2)^{\lambda+1/2}},$$

where it was assumed that $k \geq i+1$, and $l \geq j+1$. Estimates for the other combinations of k and l are similar. Thus:

$$r_2 \leq \frac{1}{(1+\alpha)}\frac{1}{n^\alpha}\sum_{k=0}^{n-1}\sum_{l=0}^{n-1}{}'' \iint_{\Delta_{kl}} \frac{d\tau_1 d\tau_2}{((\tau_1-t_1)^2+(\tau_2-t_2)^2)^\lambda} + \frac{1}{(1+\alpha)}\frac{2^3\lambda}{n^{2-2\lambda+\alpha}}\sum_{k=0}^{n-1}\sum_{l=0}^{n-1}{}'' \frac{1}{(k^2+l^2)^{\lambda+1/2}}. \tag{B.40}$$

Let us estimate the last term in the above inequality.

One has:

$$\sum_{k=0}^{n-1}\sum_{l=0}^{n-1}{}'' \frac{1}{(k^2+l^2)^{\lambda+1/2}} \leq \sum_{k=-[n/2]}^{n/2}\sum_{l=-[n/2]}^{n/2}{}^{*} \frac{1}{(k^2+l^2)^{\lambda+1/2}} \leq \mathbf{c}\begin{cases} 1, & \lambda > 1/2 \\ \log n, & \lambda = 1/2 \\ n^{1-2\lambda}, & \lambda < 1/2 \end{cases} \tag{B.41}$$

where $\sum\sum^*$ means summation over k and l, $(k,l) \neq (0,0)$.

In deriving (B.41) we have used the known result ([14], Theorem 56) which says that a number of points with integer-value coordinates, situated in the circle $x^2+y^2=r^2$, is equal to $\pi r^2 + O(r)$.

From inequalities (B.40) and (B.41) it follows that

$$r_2 \leq \frac{(1+o(1))}{(1+\alpha)}\frac{1}{n^\alpha}\sum_{k=0}^{n-1}\sum_{l=0}^{n-1}{}'' \iint_{\Delta_{kl}} \frac{d\tau_1 d\tau_2}{((\tau_1-t_1)^2+(\tau_2-t_2)^2)^\lambda}.$$

This and (B.38) yield

$$R_{nn}[H_{\alpha\alpha}(D)] \leq \frac{1+o(1)}{(1+\alpha)n^\alpha}\sup_{(t_1,t_2)\in D}\int_{-1}^{1}\int_{-1}^{1}\frac{d\tau_1 d\tau_2}{((\tau_1-t_1)^2+(\tau_2-t_2)^2)^\lambda} \leq \frac{1+o(1)}{(1+\alpha)n^\alpha}\int_{-1}^{1}\int_{-1}^{1}\frac{d\tau_1 d\tau_2}{(\tau_1^2+\tau_2^2)^\lambda}. \tag{B.42}$$

Theorem B.15 follows from a comparison the estimates of $\zeta[H_{\alpha,\alpha}(D)]$ and $R_{nn}[H_{\alpha,\alpha}(D)]$. □

Let us construct optimal with respect to order cubature formula for calculating integrals Tf on the classes W^{rr}. In the derivation of formula

(B.19) the local spline $\varphi_n(t_1, t_2)$, approximating the function $\varphi(t_1, t_2)$ in the domain $[0, 2\pi; 0, 2\pi]$, was constructed. A spline $f_{nn}(t_1, t_2)$, approximating the function $f(t_1, t_2)$ in the domain $[-1, 1] \times [-1, 1]$, can be constructed analogously. Calculate the integral Tf by the formula

$$Tf = \int_{-1}^{1}\int_{-1}^{1} \frac{f_{nn}(\tau_1, \tau_2)d\tau_1 d\tau_2}{((\tau_1 - t_1)^2 + (\tau_2 - t_2)^2)^\lambda} + R_{nn}(f). \tag{B.43}$$

Theorem B.16 *Let $\Psi = W^{r,r}(1), r = 1, 2, \ldots,$ and calculate the integral Tf by formula* (B.3) *with $\rho_1 = \rho_2 = r-1$, and $n_1 = n_2 = n$. Then cubature formula* (B.43)*, which has the error*

$$R_{nn}(\Psi) \leq (1 + o(1))\frac{2R_{r1}(1)}{(r+1)!(n - 1 + [R_{r1}(1)]^{1/r})^r}\int_{-1}^{1}\int_{-1}^{1}\frac{d\tau_1 d\tau_2}{(\tau_1^2 + \tau_2^2)^\lambda},$$

is optimal with respect to order. Here $R_{rq}(t)$ is a polynomial of degree r, least deviating from zero in $L_q([-1, 1])$.

As in the proof of the Theorem B.11 one gets the following estimate

$$R_{nn}(\Psi) \leq (1 + o(1))\frac{2R_{r1}(1)}{(r+1)!(n - 1 + [R_{r1}(1)]^{1/r})^r}\int_{-1}^{1}\int_{-1}^{1}\frac{d\tau_1 d\tau_2}{(\tau_1^2 + \tau_2^2)^\lambda}.$$

Comparing this estimate with the estimate of $\zeta_{nn}[W^{r,r}(1)]$ from Theorem B.14 one finishes the proof.

B.7 Calculation of Weakly Singular Integrals on Non-Smooth Surfaces

In Sections 5 and 6 asymptotically optimal methods for calculating weakly singular integrals defined on the squares $[0, 2\pi]^2$ or $[-1, 1]^2$ were constructed.

It is of interest to study optimal methods for calculating weakly singular integrals on piecewise-Lyapunov surfaces.

Consider the integral

$$Jf = \iint_G \frac{f(\tau_1, \tau_2, \tau_3)dS}{\left((\tau_1 - t_1)^2 + (\tau_2 - t_2)^2 + (\tau_3 - t_3)^2\right)^\lambda}, \quad t_1, t_2, t_3 \in G, \tag{B.44}$$

where G is a Lyapunov surface of class $L_s(B, \alpha)$.

We show that the results derived in Sections 5 and 6 can be partially generalized to the integrals (B.44).

Calculate integrals (B.44) by the formula:

$$Jf = \sum_{k=1}^{n} \sum_{|v|=0}^{\rho} p_{kv} f^{(v)}(M_k) + R_n(f, G, M_k, p_{kv}, t), \tag{B.45}$$

where $t = (t_1, t_2, t_3)$, $v = (v_1, v_2, v_3)$, $|v| = v_1 + v_2 + v_3$, $f^{(v)}(t_1, t_2, t_3) = \frac{\partial^{|v|} f}{\partial t_1^{v_1} \partial t_2^{v_2} \partial t_3^{v_3}}$.

The error of formula (B.45) is:

$$R_n(f, G, M_k, p_{kv}) = \sup_{t \in G} |R_n(f, G, M_k, p_{kv}, t)|.$$

Assume $f \in \Psi_1$, and $G \in \Psi_2$. Then the error of formula (B.45) on the classes Ψ_1 and Ψ_2 is:

$$R_n(\Psi_1, \Psi_2, M_k, P_{kv}) = \sup_{f \in \Psi_1, G \in \Psi_2} R_n(f, G, M_k, p_{kv}).$$

Let

$$\zeta_n[\Psi_1, \Psi_2] := \inf_{M_k, p_{kv}} R_n(\Psi_1, \Psi_2, M_k, p_{kv}).$$

A cubature formula with nodes M_k^* and weights p_{kv}^* is called optimal, asymptotically optimal, optimal with respect to order on the class of functions Ψ_1 and surfaces Ψ_2, if

$$\frac{R_n(\Psi_1, \Psi_2, M_k^*, p_{kv}^*)}{\zeta_n[\Psi_1, \Psi_2]} = 1, \sim 1, \asymp 1,$$

respectively.

Let $\Psi_1 = H_\alpha(1)$, $0 < \alpha \leqslant 1$, and $\Psi_2 = L_1(B, \beta)$ $0 < \beta \leqslant 1$. Let us construct an optimal with respect to order method for calculating integrals (B.44) on the classes of functions Ψ_1 and surfaces Ψ_2. Let $S(G)$ be a "square" of the surface G. Divide the surface G into n parts g_k, $k = 1, 2, \ldots, n$, so that a "square" of each of the domains g_k has the area of order $|S(G)|/n$, where $|S(G)|$ is the area of $S(G)$. We take a point M_k in each of domains g_k at the center of the domain g_k.

Calculate integral (B.44) by the formula

$$Jf = \sum_{k=1}^{n} f(M_k) \iint_{g_k} \frac{dS}{\left((\tau_1 - t_1)^2 + (\tau_2 - t_2)^2 + (\tau_3 - t_3)^2\right)^{\lambda}} + R_n(f, G). \tag{B.46}$$

Theorem B.17 *Formula* (B.46), *has the error*

$$R_n(\Psi_1, \Psi_2) \asymp n^{-\alpha/2},$$

and is optimal with respect to order on the classes $\Psi_1 = H_\alpha$, $0 < \alpha \leqslant 1$, *and* $\Psi_2 = L_1(B, \beta)$, $0 < \beta \leqslant 1$, *among all formulas* (B.45) *with* $\rho = 0$.

Proof. Assume for simplicity that the surface G is given by the equation $z = \varphi(x, y)$, $(x, y) \in G_0$, $\varphi(x, y) \geqslant 0$. Let $\varphi_x(x, y) := p$, $\varphi_y(x, y) := q$. Write the integral Jf as

$$Jf = \iint_{G_0} \frac{f(\tau_1, \tau_2, \varphi(\tau_1, \tau_2))\sqrt{1 + p^2(\tau_1, \tau_2) + q^2(\tau_1, \tau_2)}d\tau_1 d\tau_2}{\left[(\tau_1 - t_1)^2 + (\tau_2 - t_2)^2 + (\varphi(\tau_1, \tau_2) - \varphi(t_1, t_2))^2\right]^{\lambda}}. \tag{B.47}$$

The function $f(\tau_1, \tau_2, \varphi(\tau_1, \tau_2))$ belongs to the Hölder class H_α over G_0, and the function $\frac{\sqrt{1+p^2+q^2}}{[(\tau_1-t_1)^2+(\tau_2-t_2)^2+(\varphi(\tau_1,\tau_2)-\varphi(t_1,t_2))^2]^\lambda}$ is positive.

Let $M_k = (m_1^k, m_2^k, m_3^k)$ be the nodes of cubature formula (B.45). Let $\psi(\tau) := (d(\tau, \{M_k\}))^\alpha$, where $d(\tau, \{M_k\})$ is the distance between the point τ and the set of the nodes $\{M_k\}$, where the distance is measured along the geodesics of the surface G. This distance satisfies the Hölder condition $H_\alpha(1)$. Hence the function $\psi^*(\tau_1, \tau_2) = \psi(\tau_1, \tau_2, \varphi(\tau_1, \tau_2))$ belongs to the Hölder class $H_\alpha(A)$ and vanishes at the nodes (m_1^k, m_2^k), $k = 1, 2, \ldots, n$. Thus,

$$\begin{aligned}
\zeta_n(\Psi_1, \Psi_2) &\geqslant \frac{1}{S(G_0)} \\
&\times \iint_{G_0} \iint_{G_0} \frac{\psi(\tau_1, \tau_2, \varphi(\tau_1, \tau_2))\sqrt{1 + p^2 + q^2}d\tau_1 d\tau_2 dt_1 dt_2}{\left[(\tau_1 - t_1)^2 + (\tau_2 - t_2)^2 + (\varphi(\tau_1, \tau_2) - \varphi(t_1, t_2))^2\right]^{\lambda}} \\
&\geqslant \frac{1}{S(G_0)} \iint_{G_0} \psi(\tau_1, \tau_2, \phi(\tau_1, \tau_2))d\tau_1 d\tau_2
\end{aligned}$$

$$\times \min_t \iint_{G_0} \frac{\sqrt{1+p^2+q^2}}{[(\tau_1-t_1)^2+(\tau_2-t_2)^2+(\varphi(\tau_1,\tau_2)-\varphi(t_1,t_2))^2]^\lambda} d\tau_1 d\tau_2$$
$$\geqslant \frac{A}{n^{\alpha/2}} \min_t \iint_G \frac{ds}{(r(t,\tau))^\lambda},$$

where $S(G_0)$ is the "square" of the surface G_0.

Therefore the error of formula (B.46) is estimated by the inequality $R_n \leqslant \frac{A}{n^{\alpha/2}}$.

Theorem B.17 is proved. □

Remark B.5 *The method of decomposition of the domain G into smaller parts g_k, $k = 1, 2, \ldots, n$, described below, is optimal with respect to order for classes of functions $\Psi_1 = H_\alpha$, $0 < \alpha \leqslant 1$, and of surfaces $\Psi_2 = L_0(B, \beta)$, $0 < \beta \leqslant 1$ for $\alpha \leqslant \beta$.*

Remark B.6 *From formula* (B.47) *it follows that if the function $f \in W^{r,r}(1)$ and the surface $G \in L_s(B, \alpha)$, then the function $f(\tau_1, \tau_2, \varphi(\tau_1, \tau_2)) \in W^{v,v}(A)$, where $v = \min(r, s)$. Therefore, repeating the above arguments, one proves that the accuracy of calculation of integral* (B.47) *by cubature formulas using n values of integrand function does not exceed $O(n^{-v/2})$.*

From this remark it follows that if the surface G consists of several parts, for example of surfaces G_1 and G_2 having common edge L, then it is necessary to calculate the integrals for the surface G_1 and the surface G_2 separately. If the surface G is divided into smaller parts g_k, $k = 1, 2, \ldots, n$, the domains g_k, the curve L passes inside of these domains, should be associated with the class of surfaces $L_0(B, 1)$. In these domains the accuracy of calculation of the integral does not exceed than $O(n_k^{-1})$, where n_k is the number of nodes of the cubature formula used in the domain g_k.

For this reason the cusps and the nodes, in which three or more domains G_k, which are parts of the domain G touch each other, must belong to the boundaries of the covering domains g_k, $k = 1, 2, \ldots, n$.

The universal code for computing capacitances, described in Section B.9, is based on optimal with respect to order cubature formulas for calculating integrals on the classes of functions H_α, $0 < \alpha \leqslant 1$, on surfaces of the class $L_0(B, \beta)$, $B = const$, $\alpha \leqslant \beta$, $\beta \leqslant 1$.

The algorithm constructed in Section B.9 is optimal on this class of surfaces and does not require special treatment of edges and conical points of the surface.

When one studies cubature formulas on the classes $W^{r,r}(A)$, $r > 1$, and $L_s(B,\beta)$, $s \geqslant 1$, $0 \leqslant \beta \leqslant 1$, one has to develop a method to compute accurately the integrals in a neighborhood of the above singular points of the surface.

B.8 Calculation of Weights of Cubature Formulas

In calculating weakly singular integrals by cubature formulas (B.35) it is necessary to calculate integrals of the form of

$$J_{kl}(t_1,t_2) = \int\limits_{\Delta_{kl}} \frac{d\tau_1 d\tau_2}{((\tau_1 - t_1)^2 + (\tau_2 - t_2)^2)^\lambda}$$

for different values $(t_1,t_2) \in [-1,1]^2$.

Let $(t_1,t_2) \in \Delta_{ij}$. Let us consider two possibilities:

(1) the square Δ_{kl} and the square Δ_{ij} have nonempty intersection;
(2) the square Δ_{kl} is does not have common points with the square Δ_{ij}.

First consider the second case, when the function

$$\varphi(\tau_1,\tau_2) = \frac{1}{((\tau_1 - t_1)^2 + (\tau_2 - t_2)^2)^\lambda},$$

is smooth. Here $(\tau_1,\tau_2) \in \Delta_{kl}$, and $(t_1,t_2) \in \Delta_{ij}$.

In this case one has

$$\left|\frac{\partial^r \varphi(\tau_1,\tau_2)}{\partial \tau_1^r}\right| \leq \frac{r!2^{2r}}{((\tau_1 - t_1)^2 + (\tau_2 - t_2)^2)^{\lambda + r/2}}$$

and, if the squares Δ_{kl} and Δ_{ij} do not have common points, one gets

$$\left|\frac{\partial^r \varphi(\tau_1,\tau_2)}{\partial \tau_1^r}\right| \leq \frac{2^r r! n^{2\lambda + r}}{2^\lambda}.$$

Similar estimates holds for partial derivative with respect to τ_2.

Calculate the integral $J_{kl}(t_1,t_2)$ by the Gauss cubature formula

$$J_{kl}(t_1,t_2) = \int\limits_{\Delta_{kl}} P_{mm}\left[\frac{1}{((\tau_1 - t_1)^2 + (\tau_2 - t_2)^2)^\lambda}\right] d\tau_1 d\tau_2 + R_{mm}(\Delta_{kl}),$$

where $P_{mm} = P_m^{\tau_1} P_m^{\tau_2}$, $P_m^{\tau_i}$ $(i = 1,2)$ is the projection operator onto the set of interpolation polynomials of degree m with nodes at the zeros of the

Legendre polynomial, which maps the segment $[-1,1]$ onto the segment $[x_k, x_{k+1}]$ for $i=1$, and onto the segment $[x_l, x_{l+1}]$ for $i=2$.

An integer m is chosen so that $|R_{mm}| \le n^{-2-\alpha}$ for cubature formulas on the Hölder class $H_{\alpha\alpha}$, and $|R_{mm}| \le n^{-r-\alpha}$ for cubature formulas on the class W^{rr}.

This requirement is made because the error of calculation of the coefficients $J_{kl}(t_1,t_2)$ must not exceed the error of formula (B.32).

Using r derivatives of the integrand in the error $R_{mm}(\Delta_{kl})$, one gets:

$$|R_{mm}(\Delta_{kl})| \le \frac{B_r 2^r r!}{m^{r-1}} \left(\frac{2}{n}\right)^{2-2\lambda},$$

where B_r is the constant appearing in Jackson's theorems. It is known that the constants B_r are bounded by a constant, denoted b, uniformly with respect to r. In the case of periodic functions $b=1$ ([51]), and in the general case b is apparently unknown.

If $r=2$ and $m = B_r 2^r r! n^{2\lambda}$, then one gets the error estimate given for cubature formula (B.32).

Now, consider a method for calculating the integrals $J_{kl}(t_1,t_2)$ when the square Δ_{kl} has nonempty intersection with the square Δ_{ij}. For definiteness we consider the calculation of the integral $J_{ij}(t_1,t_2)$ by the formula:

$$J_{ij}(t_1,t_2) = \int\limits_{\Delta_{ij}} P_{mm}\left[\frac{1}{((\tau_1-t_1)^2+(\tau_2-t_2)^2)^{\lambda}+h}\right] d\tau_1 d\tau_2 + R_{mm}(\Delta_{ij}),$$

where $h = const > 0$ will be specified below.

One has:

$$|R_{mm}(\Delta_{ij})| \le h \int\limits_{\Delta_{ij}} \frac{d\tau_1 d\tau_2}{((\tau_1-t_1)^2+(\tau_2-t_2)^2)^{\lambda}(((\tau_1-t_1)^2+(\tau_2-t_2)^2)^{\lambda}+h)}$$

$$+ \int\limits_{\Delta_{ij}} D_{mm}\left[\frac{1}{((\tau_1-t_1)^2+(\tau_2-t_2)^2)^{\lambda}+h}\right] d\tau_1 d\tau_2 = r_1 + r_2,$$

where $D_{mm} = I - P_{mm}$, and I is an identity operator, and

$$r_1 \le h \int\limits_{\Delta_{ij}} \frac{d\tau_1 d\tau_2}{((\tau_1-t_1)^2+(\tau_2-t_2)^2)^{(\lambda+1)/2}(((\tau_1-t_1)^2+(\tau_2-t_2)^2)^{\lambda}+h)^{(1+\lambda)/2}}$$

$$\le \frac{2\pi}{1-\lambda} h^{(1-\lambda)/2} \left(\frac{2\pi}{n}\right)^{1-2\lambda}. \quad \text{(B.48)}$$

The function $\frac{1}{((\tau_1-t_1)^2+(\tau_2-t_2)^2)^\lambda+h}$ is infinitely smooth. Using bounds for its first derivatives for $\lambda \geq 1/2$, one gets:

$$r_2 \leq \frac{8\lambda B_1}{n^4 h^2 m}. \tag{B.49}$$

From inequality (B.48) it follows that for getting accuracy $O(n^{-1-\alpha})$ one has to have $h = n^{-2(2\lambda+\alpha)/(1-\lambda)}$ and from inequality (B.49) it follows that one has to have $m = max([n^{(8\lambda+4\alpha)/(1-\lambda)+\alpha-3}], 1)$.

B.9 Iterative Methods for Calculating Electrical Capacitancies of Conductors of Arbitrary Shapes

Numerical methods for solving electrostatic problems, in particular, calculating capacitancies of conductors of arbitrary shapes, are of practical interest in many applications. There exists a vast literature on calculation of the capacitances of perfect conductors. In [43] there is a reference section which gives the capacitance of the conductors of certain shapes (more than 800 shapes are considered in [43]). In Chapter 3 iterative methods for solving interior and exterior boundary value problems in electrostatics are proposed and mathematically justified. Upper and lower estimates for some functionals of electrostatic fields are obtained in Chapter 3 as well. Such functionals are the capacitances of perfect conductors and the polarizability tensors of bodies of arbitrary shape. These bodies are described by their dielectric permittivity, magnetic permeability and conductivity. They can be homogeneous or flaky. The main point is: these bodies have arbitrary geometrical shapes.

The methods, developed in Chapter 3, yield analytical formulas for calculation of the capacitances and polarizability tensors of bodies of arbitrary shapes with any given accuracy. Error estimates for these formulas are obtained in Chapter 3. Recall the formulas for calculating the capacitances of the conductors of arbitrary shapes (see Chapter 3):

$$C^{(n)} = 4\pi\varepsilon_0 S^2 \left\{ \frac{(-1)^n}{(2\pi)^n} \int_\Gamma \int_\Gamma \frac{ds\,dt}{r_{st}} \underbrace{\int_\Gamma \cdots \int_\Gamma}_{n \text{ times}} \psi(t,t_1)\cdots\psi(t_{n-1},t_n)dt_1\cdots dt_n \right\}^{-1},$$

where S is the surface area of the surface Γ of the conductor, ε_0 is the dielectric constant of the medium, $r_{st} := |s-t|$, and $\psi(t,s) := \frac{\partial}{\partial N_t}\frac{1}{r_{st}}$,

$$C^{(0)} = \frac{4\pi\varepsilon_0 S^2}{J} \le C, \quad J \equiv \int_\Gamma\int_\Gamma \frac{ds\,dt}{r_{st}}, \quad S = \operatorname{meas}\Gamma.$$

It is proved in Chapter 3 that

$$\left|C - C^{(n)}\right| \le Aq^n, \quad 0 < q < 1,$$

where A and q are constants which depend only on the geometry of Γ.

We use these formulas are used to construct the computer code for calculating the capacitances of the conductors of arbitrary shapes.

It is proved in Chapter 3 that

$$C^{(n)} = 4\pi\varepsilon_0 S^2 \left(\int_\Gamma\int_\Gamma r_{st}^{-1}\sigma_n(t)dtds\right)^{-1}, \tag{B.50}$$

where σ_n is defined by the iterative process:

$$\sigma_{n+1} = -A\sigma_n, \ \sigma_0 = 1, \int_\Gamma \sigma_n dt = S, \tag{B.51}$$

and A is defined by the formula:

$$A\sigma := \int_\Gamma \sigma(t)\frac{\partial}{\partial N_s}\frac{1}{2\pi r_{st}}dt,$$

where N_s is the outer unit normal to Γ at the point s.

To use iterative process (B.51), one has to calculate the weakly singular integral

$$\frac{1}{2\pi}\int_\Gamma \sigma(t)\frac{\partial}{\partial N_s}\frac{1}{r_{st}}dt. \tag{B.52}$$

Let us describe the construction of a cubature formula for calculating integral (B.52), assuming for simplicity that the domain G, bounded by the surface Γ, is convex. This asumption can be removed.

Let $\mathbb{S}$ be the inscribed in the conductor sphere of maximal radius r^*, centered at the origin. Introduce the spherical coordinates system (r,ϕ,θ), and the set of the nodes (r^*,ϕ_k,θ_l), where $\phi_k = 2k\pi/n$, $k = 0,1,\dots,n$, $\theta_l = \pi l/m$, $l = 0,1,\dots,m$. Assume that m is even, and cover the sphere $\mathbb{S}$ with the spherical triangles Δ_k, $k = 1,2,\dots,N$, $N = 2n(m-1)$.

Let us describe the construction of the spherical triangles. For $0 \leq \Theta \leq \pi/m$ the triangles $\Delta_k, k = 1, 2, \ldots, n$ have vertices $(r^*, 0, 0)$, $(r^*, \phi_{k-1}, \theta_1)$, (r^*, ϕ_k, θ_1), $k = 1, 2, \ldots, n$.

For $\theta_l \leq \theta \leq \theta_{l+1}$, $l = 1, 2, \ldots, m/2-1$, the triangles $\Delta_k, k = n+2n(l-1)+j$, $1 \leq j \leq 2n$ are constructed as follows. The rectangle $[0, 2\pi; \theta_l, \theta_{l+1}]$ is covered with the squares $\Delta_{kl} = [\phi_k, \phi_{k+1}; \theta_l, \theta_{l+1}]$, $k = 0, 1, \ldots, n-1$. Each of the squares Δ_{kl} is divided into two equal triangles Δ_{kl}^1 and Δ_{kl}^2, $k = 0, 1, \ldots, n-1$, $l = 1, 2, \ldots, m/2-1$. The spherical triangles Δ_{kl}^1 and Δ_{kl}^2, $k = 0, 1, \ldots, n-1$, $l = 1, 2, \ldots, m/2-1$, are images of triangles Δ_{kl}^1 and Δ_{kl}^2 on the sphere $\mathbb{S}$

As a result of these constructions the sphere $\mathbb{S}$ is covered with triangles Δ_k, $k = 1, 2, \ldots, N$.

We draw the straight lines through the origin and vertices of the triangle Δ_k, $k = 1, 2, \ldots, N$. The points of intersection of these lines with the surface Γ are vertices of the triangle $\overline{\Delta}_k$, $k = 1, 2, \ldots, N$. As a result of these constructions the surface Γ is approximated by the surface Γ_N consisting of triangle $\overline{\Delta}_k$, $k = 1, 2, \ldots, N$, and integral (B.52) is approximated by the integral

$$U(s) = \frac{1}{2\pi} \int\limits_{\Gamma_N} \sigma(t) \frac{\partial}{\partial N_s} \frac{1}{r_{st}} dt. \tag{B.53}$$

We fix each triangle $\overline{\Delta}_k$, $k = 1, 2, \ldots, N$, and associate with it a point $\tau_k \in \overline{\Delta}_k$, $k = 1, 2, \ldots, N$, equidistant from the vertices of the triangle $\overline{\Delta}_k$, $k = 1, 2, \ldots, N$. We calculate integral (B.53) at the points τ_k, $k = 1, 2, \ldots, N$, by the cubature formulas constructed in paragraphs 5-7 for the Hölder classes. After calculating the values $U(\tau_k)$, $k = 1, 2, \ldots, N$ by these cubature formulas, the integral

$$\tilde{C}^{(1)} = -4\pi\varepsilon_0 S_N^2 \left(\int\limits_{\Gamma_N} \int\limits_{\Gamma_N} r_{st}^{-1} \tilde{U}(t) dt ds \right)^{-1}$$

is calculated, where $\tilde{U}(t) = U(\tau_k)$ for $t \in \overline{\Delta}_k$, $k = 1, 2, \ldots, N$, S_N is area of the surface Γ_N, $\tilde{C}^{(1)}$ is approximation to the value of $C^{(1)}$. The successive iterations are calculated analogously.

B.10 Numerical Examples

In this section the numerical results are given. As an example we calculated the capacitances of various ellipsoids, because for ellipsoids one knows the analytical formula for the capacitance, which makes it possible to evaluate the accuracy of the numerical results. Consider the ellipsoid:

$$\frac{x^2}{a^2} + \frac{y^2}{b^2} + \frac{z^2}{c^2} = 1.$$

It is known [43] that the exact value of the capacitance of ellipsoid with $a = b$ is:

$$C = \frac{4\pi\varepsilon_0\sqrt{a^2 - c^2}}{\arccos(c/a)}.$$

Let $a = b = 1$, and $\varepsilon_0 = 1$. We have calculate the capacitance C for different values of the semiaxis c. The results of the calculations are given in Table B.10.

It is known (see Chapter 3), that the capacitance of a metallic disc of radius a is $C = 8a\varepsilon_0$, and one can see from Table 1, that asymptotically, as $c \to 0$, this formula can be used practically for the ellipsoids with $c \leq 0.001$ with the error approximately equal to 0.005.

C	n	m	N	Exact value	Error	Relative error	Calculation time
0.9	40	30	2320	12.144630	−0.221200	0.018212	25 sec
0.5	40	30	2320	10.392304	−0.222042	0.021366	25 sec
0.1	40	30	2320	8.5020638	−0.301189	0.035425	25 sec
0.01	40	30	2320	8.050854	0.072132	0.008959	25 sec
0.001	40	30	2320	8.005092	−0.821528	0.106374	25 sec
0.0001	40	30	2320	8.000509	−1.068178	0.133513	25 sec
0.9	50	40	3900	12.144630	−0.180510	0.014801	1 min 15 sec
0.5	50	40	3900	10.392304	−0.185642	0.017860	1 min 15 sec
0.1	50	40	3900	8.5020638	−0.288628	0.033947	1 min 15 sec
0.01	50	40	3900	8.050854	−0.372047	0.046212	1 min 15 sec
0.001	50	40	3900	8.005092	−0.586733	0.073295	1 min 15 sec
0.0001	50	40	3900	8.000509	−0.933288	0.116653	1 min 15 sec
0.9	60	50	5880	12.144630	−0.152009	0.012516	4 min
0.5	60	50	5880	10.392304	−0.160023	0.015391	4 min
0.1	60	50	5880	8.5020638	−0.283364	0.033328	4 min
0.01	60	50	5880	8.050854	0.532250	0.061110	4 min
0.001	60	50	5880	8.005092	−0.391755	0.048939	4 min
0.0001	60	50	5880	8.000509	−0.880394	0.110042	4 min

Problems

1. Write a computer program for calculating the capacitance and the polarizability tensors for a body of arbitrary shape. According to formulas (5.8)–(5.13) the program is to calculate multiple integrals over the surface of the body. The integrands are functions with weak singularities, e.g.,

$$\int_\Gamma \int_\Gamma r_{st}^{-1} ds\, dt, \quad \int_\Gamma \int_\Gamma r_{st}^{-1} N_i(s) N_j(s) ds\, dt,$$

$$\int_\Gamma \int_\Gamma r_{st}^{-1} \left(\int_\Gamma \frac{\partial r_{tt_1}^{-1}}{\partial N_t} dt_1 \right) ds\, dt,$$

where $r_{st} := |s - t|$ is the Euclidean distance between points s and t, amd $N_j(s)$ is the j–th component of the exterior uni normal to the surface Γ at the point s.

Finding good algorithms for calculating multiple integrals of functions with moving weak singularities is a problem of general interest. This problem is discussed in the Appendix, but it is of great interest to develop efficient computer codes for calculationg such integrals.

2. Carry out a numerical study of the dependence of the scattering amplitude on: (1) the shape of the body, (2) on the boundary conditions, (3) on the coating of the body (e.g., a flaky-homogeneous body with two layers of which the exterior layer is thin).

3. Carry out a numerical study of the many-body problem using formulas (7.59), (7.63), (7.67), (7.71), (7.72), (7.75), (7.79), and (7.80).

4. Study the inverse problem of finding the properties of the medium consisting of many small particles from the scattering data.

5. Develop a theory of elastic wave scattering by small bodies of arbitrary shape similar to the theory of acoustic and electromagnetic wave scattering given in Chapter 7.

Bibliographical Notes

Some of the results mentioned in the introduction are presented in the books [13], [15], [43], [58], [60], [68], [71], [74], [73]. Variational principles and two-sided estimates of various functionals of static fields are given in [79], [76]. Low-frequency scattering, first studied by Rayleigh (1871), is studied in [147], [151], [58], [40], [49], [50], [42], [39]. Scattering from small holes is studied in [9], [61], [25]. Potential theory for domains with smooth boundaries is given in [38]. The theory for domains with non-smooth boundaries is presented in [15], [52], and for Lipschitz domains in [22], [23], [70], [157], and in many other references. Boundary-value problems and scattering problems in domains with rough boundaries, much rougher than the Lipschitz ones, were studied in [107], [104], [31], see also [144], where the domains are of finite perimeter. In [65], [30] and [31] the embedding theorems for rough domains are given.

Reference material and an extensive bibliography on electrical capacitance can be found in [43]. There is an extensive literature on scattering by a system of many bodies and wave propagation in random media, topics outside of the scope of this book. Among many contributors to this field are [27], [59]. The effective field in random media and in media consisting of small particles has been studied in [162], [163], [26], [41], [152], [64], [159], [146], [113], to name only a few references.

In [69] there is a discussion of wave scattering by small non-spherical particles with many applications. The formulas for the scattering matrix for acoustic and electromagnetic wave scattering by particles of arbitrary shapes, derived in [113] apparently are not familiar to the authors of [69]. In many applications these formulas are important.

Integral equations of the first kind were used in electrostatics [155]. In [24] an integral equation of the second kind was derived for screens (non-closed surfaces). A numerical approach to problem (6.103), different from the one given in Section 6.3, was given in [161]. In [18] the possibility of calculating the cardiac electric potential of a human body from the potential, measured on the surface of the body, is discussed. A computer program for calculating the elements of the polarizability tensor of rotationally symmetric metallic bodies was given in [150]. In [4] some methods for finding small subsurface inhomogeneities from the measurements on the surface are discussed.

The main results presented in this book were obtained by the author in [113], [133], [107], in the author's papers and joint papers mentioned in the bibliography and in the book.

These results include:

(1) approximate analytical formulas for polarizability tensors and capacitances for bodies of arbitrary shapes,
(2) two-sided estimates of the polarizability tensors,
(3) approximate analytical formulas for the scattering amplitude and scattering matrix in the problem of wave scattering by a small body of an arbitrary shape and by a system of such bodies,
(4) investigation of the influence of the boundary conditions on the scattering amplitude,
(5) methods for a study of obstacle scattering problems in rough domains,
(6) methods for obtaining low-frequency asymptotics of the solutions to boundary-value problems,
(7) methods for finding small subsurface inhomogeneities from the scattering data measured on the surface or in the far-field region,
(8) MRC (Modified Rayleigh Conjecture) method for solving obstacle scattering problems and static problems,
(9) optimal methods for calculating multidimensional integrals with weak singularities,
(10) construction of convergent iterative schemes for solving integral equations for interior and exterior boundary-value problems,
(11) equations for the self-consistent field in a medium consisting of many small particles.

Bibliography

1. Agranovich, M., Katzenelenbaum, B., Sivov, A., Voitovich, N., *Generalized Method of Eigenoscillations in Diffraction Theory*, Wiley-VCH, Berlin, 1999.
2. Albeverio, S., Gesztesy, F., Hoegh-Krohn, R., Holden, H., *Solvable Models in Quantum Mechanics*, Springer, Berlin, 1988.
3. Ambrosio L., Kirchenheim B., Rectifiable sets in metric and Banach spaces, Math.Ann., 318, 2000, pp. 527–555.
4. Ammari, H., Kang, H., *Reconstruction of small inhomogeneities from boundary measurements*, Lecture Notes in Math, 1846, Springer, Berlin, 2004.
5. Barantsev R, *Concerning the Rayleigh hypothesis in the problem of scattering from finite bodies of arbitrary shapes,* Vestnik Lenungrad Univ., Math., Mech., Astron., **7**, (1971), 56–62.
6. Bakhvalov, N., *Theoretical foundation and construction of numerical algorithms for problems of mathyematical physics*, Nauka, Moscow, 1979 (editor K.Babenko).
7. Bakhvalov, N., *Properties of Optimal Methods of Solution of Problem of Mathematical Physics,* Journ. Comp. Math. and Math. Phys., 10, N3, (1970), 555–588.
8. Bakhvalov, N., *Optimal linear methods of approximation operators on convex classes of functions,* Journ. Comp. Math. and Math. Phys., 11, N4, (1971), 1014–1018.
9. Bethe, H., *Theory of diffraction by small holes*, Phys. Rev., 66, (1944), 163–182.
10. Boikov, I., Ramm, A. G., *Optimal with respect to accuracy algorithms for calculation of multidimensional weakly singular integrals and applications to calculation of capacitances of conductors of arbitrary shapes*, Acta Applicandae Math, 79, N3, (2003), 281–326.
11. Boikov, I. V., N.F.Dobrunina and L.N.Domnin, *Approximate Methods of Calculation of Hadamard Integrals and Solution of Hypersingular Integral Equations,* Penza Technical State Univ. Press, Penza, 1996. (in Russian).

12. Boikov, I. V., *Optimal with Respect to Accuracy Algorithms of Approximate Calculation of Singular Integrals,* Saratov State University Press, Saratov, 1983. (in Russian).
13. Buhgolz, G., *Calculating Electric and Magnetic Fields*, IL, Moscow, 1961.
14. Bukhshtab, A. A., *Theory of numbers,* Izdat. Prosv., Moscow, 1960.
15. Burago, Yu. and Mazya, V., *Potential Theory and Function Theory for Irregular Regions*, Consult. Bureau, New York, 1969.
16. Christiansen, S. and R.E. Kleinman, *On a misconception involving point collocation and the Rayleigh hypothesis*, IEEE Trans.Anten.Prop., 44,10, (1996), 1309–1316.
17. Colton D., Kress R., *Inverse Acoustic and Electromagnetic Scattering Theory,* Springer-Verlag, New York, 1992.
18. Colli Franzone, P. and Magenes, E., *On the inverse potential problem of electrocardiology*, Calcolo, 4, (1980), 459–538.
19. Dassios, G., Kleinman, R., *Low frequency scattering*, Clarendon Press, Oxford 2000.
20. Dahlberg, B., Kenig, C., *Hardy spaces and the Neumann problem in* L^p *for Laplace's equation in Lipschitz domains*, Ann. Math., 125, (1987), 437–465.
21. Davis P. J. and P. Rabinovitz, *Methods of numerical integration* 2nd ed., Academic Press, New York, 1984.
22. Fabes, E. B., Jodeit, M. Jr., and Riviere,N., Potential techniques for boundary value problem on C^1 domains, Acta Math., 141, (1978), 165–186.
23. Fabes, E., Mendez, O., Mitrea, M., *Boundary layers on Sobolev-Besov spaces and Poisson's equation for the Laplacian in Lipschitz domains*, J. Funct. Anal., 159, (1998), 323–368.
24. Fel'd, Ya. and Suharevskij, I., *Reduction of diffraction problems an nonclosed surfaces to integral equations of the second kind*, Radio Engrg. Electron. Phys., 11, (1966), 1017–1024.
25. Fihmanas, R. and Fridberg, P., *Two-sided estimates for the coefficients of polarizability in diffraction from small holes*, Soviet Phys. Dokl., 189, (1969), 969–972.
26. Finkelberg, V., *Wave propagation in random media*, JETP, 53, (1967), 401–415.
27. Foldy, L., The multiple scattering of waves, Phys. Rev., 67, (1945), 107–119.
28. Gesztesy, F., Ramm, A. G., *An inverse problem for point inhomogeneities, Methods of Functional Analysis and Topology*, 6, N2, (2000), 1–12.
29. Gilbarg, D., Trudinger, N., *Elliptic partial differential equations of second order*, Springer, Berlin, 2001.
30. Gol'dshtein V., Ramm A.G., *Embedding operators for rough domains*, Math. Ineq. and Applic., 4, N1, (2001), 127–141.
31. Gol'dshtein V., Ramm A.G., *Embedding operators and boundary-value problems for rough domains*, Internat. Journ. Appl Math. Mech., 2, (2005).
32. Gol'dshtein,V., Reshetnyak, Yu. G., *Quasiconformal Mappings and Sobolev Spaces*, Kluwer Academic Publishers, Dordrecht, 1990.
33. Grinberg, G. A., *Selected Topics in the Mathematical Theory of Electrical and Magnetical Phenomena*, Acad. of Sci. USSR, Leningrad, 1948 (in Russian).

34. Gutman S., Ramm A. G., *Numerical implementation of the MRC method for obstacle scattering problems,* J. Phys. A: Math. Gen. **35**, (2002), 8065–8074.
35. Gutman S., Ramm A. G., *Support Function Method for inverse scattering problems,* In the book "Acoustics, mechanics and related topics of mathematical analysis", (ed. A.Wirgin), World Scientific, New Jersey, 2003, pp. 178–184.
36. Gutman, S., Ramm, A. G. *Application of the hybrid stochastic-deterministic minimization method to a surface data inverse scattering problem*, in the book "Operator Theory and its Applications", Amer. Math. Soc., Fields Institute Communications vol.25, Providence, RI, 2000, pp. 293–304. (editors A.G.Ramm, P.N.Shivakumar, A.V.Strauss).
37. Gutman, S., Ramm, A. G., *Modified Rayleigh Conjecture with optimally placed sources*, J. Appl. Funct. Anal., 1, N2, (2006), 223–236.
38. Günter, N., *Potential Theory and Its Applications to Basic Problems of Mathematical Physics*, Ungar, New York, 1967.
39. Hönl, G., Maue, A. and Westpfahl, K., *Theorie der Beugung, Springer-Verlag*, Berlin, 1961.
40. Hulst, Van de., *Light Scattering by Small Particles*, Dover, New York, 1961.
41. Ishimaru, A., *Wave propagation and scattering in random media*, Acad. Press, New York, 1978.
42. Jones, D. S., *Low frequency electromagnetic radiation*, J. Inst. Math. Appl., 23, (1979), 421–427.
43. Jossel, Yu., Kochanov, E. and Strunskij, M., *Calculation of Electrical Capacity, Energija*, Leningrad, 1969 (in Russian).
44. Kantorovich, L., Akilov, G., *Functional analysis*, Pergamon Press, New York, 1982.
45. Kato, T., *Perturbation theory for linear operators, Springer Verlag*, New York, 1984.
46. Katsevich, A., Ramm, A. G., *Approximate inverse geophysical scattering on a small body*, SIAM J.Appl.Math., 56, N1, (1996), 192–218.
47. Kazandjian L., *Rayleigh-Fourier and extinction theorem methods applied to scattering and transmission at a rough solid-solid interface*, J. Acoust. Soc. Am., 92, 1679–1691, 1992.
48. Kazandjian L., Comments on *"Reflection from a corrugated surface revisited"*, [J. Acoust. Soc. Am., 96, 1116–1129 (1994)] J. Acoust. Soc. Am., 98, (1995), 1813–1814.
49. Keller, J., et al. *Dipole moments in Rayleigh scattering*, J. Inst. Appl. Math., 9, (1972), 14–22.
50. Kleinman, R., *Low frequency electromagnetic scattering*, in the book Electromagnetic Scattering, ed. P. Uslenghi, Acad. Press, New York, 1978.
51. Kornejchuk, N. P., *Exact Constants,* Nauka, Moscow, 1990 (in Russian).
52. Kral, I., *Integral Operators in Potential Theory, Springer-Verlag*, New York, 1980.
53. Krasnoselskij, M., Vainikko, G., Zabreiko, P., Rutickij, Ja., and Stecenko, V., *Approximate Solution of Nonlinear Equations*, Wolters-Noordhoff, Groningen, 1972.

54. Krylov, V. and Shulgina, L., *Reference Book in Numerical Integration*, Nauka, Moscow, 1968 (in Russian).
55. Krylov, V. I., *Approximate calculation of integrals*, Nauka, Moscow, 1967 (in Russian).
56. Kumano-go, H., *Pseudodifferential operators*, MIT Press, Cambridge, 1974.
57. Ladyzhenskaya, O. A., Ural'tseva, N., *Linear and quasilinear elliptic equations*, Academic Press, New York, 1968.
58. Landau, L., Lifschitz, E., *Electrodynamics of Continuous Media*, Pergamon Press, New York, 1960.
59. Lax, M., *Multiple scattering of waves*, Rev. Mod. Phys., 23, (1951), 287–310.
60. Lebedev, N., Skalskaya, I., Uflyand, Ya., *Worked Problems in Applied mathematics*, Dover, New York, 1965.
61. Levine, H., Schwinger, J., *On the theory of electromagnetic wave diffraction by an aperture in an infinite plane conducting screen*, Comm. Pure Appl. Math., 3, (1950), 355.
62. Lions, J. L., Magenes, E., *Non-homogeneous boundary value problems and applications*, Springer Verlag, New York, 1972.
63. Lorentz, G. G., *Approximation of function*, Chelsea Publishing Company, New York, 1986.
64. Marchenko, V. Hruslov, E., *Boundary value problems in domains with granular boundary*, Naukova Dumka, Kiev, 1974.
65. Mazja, V., *Sobolev spaces, Springer, Berlin*, 1985.
66. Mikhlin, S., *Variational Methods in Mathematical Physics, Macmillan*, New York, 1964.
67. Millar R., *The Rayleigh hypothesis and a related least-squares solution to the scattering problems for periodic surfaces and other scatterers*, Radio Sci., **8**, (1973), 785–796.
68. Miroljubov, N., *Methods of Calculating of Electrostatic fields*, High School, Moscow, 1963. (in Russian).
69. Mishchenko, M., Hovenier, J., Travis, L., (Editors) *Light Scattering by Nonspherical Particles: Theory, Measurements, and Applications*, Academic Press, San Diego, 2000.
70. Mitrea, D., Mitrea, M., Taylor, M., *Layer potentials, the Hodge Laplacian, and global boundary problems in nonsmooth Riemannian manifolds*, Mem. Amer. Math. Soc. 150, (2001), no. 713.
71. Morse, P. and Feshbach, H., *Methods of Theoretical Physics*, Vols. 1 and 2, McGraw-Hill, New York, 1953.
72. Muskhelishvili, N., *Singular Integral Equations*, Noordhoff Int., Leiden, 1972.
73. Smythe, W., *Static and Dynamic Electricity*, McGraw-Hill, New York, 1939.
74. Newton, R., *Scattering of Waves and Particles*, McGraw-Hill, New York, 1966.
75. Nikolskii, S. M., *Quadrature Rules*, Nauka, Moscow, 1979 (in Russian).
76. Noble, B., *Wiener-Hopf Methods for Solution of Partial Differential Equations*, Pergamon Press, New York, 1958.

77. Noether, F., *Über eine Klasse singulärer Integralgleichungen*, Math. Ann., 82, (1921), 42–63.
78. Payne, L., *Isoperimetric inequalities and their application*, SIAM Rev., 9, (1967), 453–488.
79. Polya G., Szego, G., *Isoperimetric Inequalities in Mathematical Physics,* Princeton Univ. Press, Princeton, 1951.
80. Press W. H., Teukolsky S.A., Vetterling W.T., Flannery B.P., *Numerical Recepies in FORTRAN,* Second Ed., Cambridge University Press, 1992.
81. Ramm, A. G., *A characterization of unbounded Fredholm operators*, Cubo a Mathem. Journ., 5, N3, (2003), 91–95.
82. Ramm, A. G., *A geometrical inverse problem*, Inverse problems, 2, (1986), L19–21.
83. Ramm, A. G., *Approximate formulas for polarizability tensor and capacitance for bodies of an arbitrary shape*, Radiofisika, 14, (1971), 613–620.
84. Ramm, A. G., *Approximate formulas for tensor polarizability and capacitance of bodies of arbitrary shape and its applications*, Doklady Acad Sci. USSR, 195, (1970), 1303–1306; English translation, 15, (1971), 1108–1111.
85. Ramm, A. G., *Approximate formulas for polarizability tensors and capacitances of bodies of arbitrary shapes and applications*, Doklady Acad. Sci. USSR, 195, (1970), 1303–1306. English translation, 15, (1971), 1108–1111.
86. Ramm, A. G., *A remark on integral equations theory*, Differential Equations, 8, (1972), 1517–1520; English translation, 1177–1180.
87. Ramm, A. G., *Exterior boundary value problems as limits of interface problems*, J. Math. Anal. Appl. 84, (1981), 256–263.
88. Ramm, A. G., *A simple proof of the Fredholm alternative and a characterization of the Fredholm operators*, Amer. Math. Monthly, 108, N9, (2001), 855–860.
89. Ramm, A. G., Pang, P., Yan, G., *A uniqueness result for the inverse transmission problem*, Internat. Jour. of Appl. Math., 2, N5, (2000), 625–634.
90. Ramm A. G., *Behavior of solutions to exterior boundary value problems at low frequencies*, J. Math. Anal. Appl., 117, (1986), 561–569.
91. Ramm, A. G., *Boundary value problem with discontinuous boundary condition,* Differential Equations, 13, (1976), 931–933.
92. Ramm, A. G., *Calculation of the initial field from scattering amplitude*, Radio Engrg. Electron. Phys., 16, (1971), 554–556.
93. Ramm, A. G., Frolov, N., *Calculation of the magnetization of thin films*, Microelectronics, 6, (1971), 65–68.
94. Ramm, A. G., *Calculation of the scattering amplitude of electromagnetic waves by small bodies of an arbitrary shape* II, Radiofisika, 14, (1971), 1458–1460.
95. Ramm, A. G., Golubkova,M., Usoskin, V., *Calculation of the capacitance of a parallelepiped*, Electricity, 5, (1972), 90–91.
96. Ramm, A. G., *Calculation of the capacitance of a conductor placed in anisotropic inhomogeneous dielectric*, Radiofisika, 15, (1972), 1268–1270.

97. Ramm, A. G., *Continuous dependence of the scattering amplitude on the surface of an obstacle*, Math. Methods in the Appl. Sci., 18, (1995), 121–126.
98. Ramm, A. G., *Electromagnetic wave scattering by small bodies of an arbitrary shape*, Proc. of 5th All-Union Sympos. On Wave Diffraction, Trudy Mat. Inst. Steklov, Leningrad, 1971, 176–186.
99. Ramm, A. G., *Electromagnetic wave scattering by small bodies of an arbitrary shape and relative topics*, Proc. Intern. Sympos. URSI, Moscow, 1971, 536–540.
100. Ramm, A. G., *Electromagnetic wave scattering by small bodies of arbitrary shapes,* in the book: "Acoustic, electromagnetic and elastic scattering-Focus on T-matrix approach" Pergamon Press, New York 1980. 537–546. (editors V. Varadan and V. Varadan).
101. Ramm, A. G., Somersalo, E., *Electromagnetic inverse problem with surface measurements at low frequencies*, Inverse Probl., 5, (1989), 1107–1116.
102. Ramm, A. G., *Estimates of some functionals in quasistatic electrodynamics*, Ukrain. Phys. Journ., 5, (1975), 534–543.
103. Ramm, A. G., Ruiz, A., *Existence and uniqueness of scattering solutions in non-smooth domains*, J. Math. Anal. Appl., 201, (1996), 329–338.
104. Ramm, A. G., Sammartino, M., *Existence and uniqueness of the scattering solutions in the exterior of rough domains*, in the book "Operator theory and its applications", Amer. Math. Soc., Fields Institute Communications, vol. 25, pp. 457–472, Providence, RI, 2000 (Editors A.G. Ramm, P. Shivakumar, A. Strauss).
105. Ramm, A. G., *Finding small inhomogeneities from surface scattering data*, Jour. of Inverse and Ill-Posed Problems, 8, N2, (2000), 205–210.
106. Ramm, A. G., *Numerically efficient version of T-matrix method*, Applic. Anal., 80, N3, (2002), 385–393.
107. Ramm, A. G., *Inverse problems*, Springer, New York, 2005.
108. Ramm, A. G., *Inverse scattering for geophysical problems.* Inverse problems, 1, N2, (1985) 133–172.
109. Ramm, A. G., *Investigation of the scattering problem in some domains with infinite boundaries* I, II, Vestnik Leningrad Univ, ser. math., mech. and astronomy, 7, (1963), 45–66; 19, (1963), 67–76.
110. Ramm, A. G., *Iterative solution of the integral equation in potential theory*, Doklady Acad. Sci. USSR, 186, (1969), 62–65.
111. Ramm, A. G., *Iterative methods for solving some heat transfer problems*, Eng. Phys. Jour., 20, (1971), 936–937.
112. Ramm A. G., *Materials with a desired refraction coefficient*, Annals of Communications in Mathematics, 8, N1, (2025), 38–42. (open access Journal).
113. Ramm A.G., *Iterative methods for calculating static fields and wave scattering by small bodies*, Springer Verlag, New York, 1982.
114. Ramm, A. G., *Iterative process to solve the third boundary value problem*, Differential equations, 9, (1973), 2075–2079.
115. Ramm, A. G., *Light scattering matrix for small particles of an arbitrary shape*, Optics and Spectroscopy, 37, (1974), 125–129.

116. Ramm A. G., *Modified Rayleigh Conjecture and Applications,* J. Phys. A: Math. Gen., 35, (2002), L357–L361.
117. Ramm A. G., *Modified Rayleigh Conjecture for static problems*, Appl. Math. Lett., 18, N12, (2005), 1396–1399.
118. Ramm A. G., Gutman S., *Modified Rayleigh Conjecture for scattering by periodic structures,* Internat. Jour. of Appl. Math. Sci., 1, N1, (2004), 55–66.
119. Ramm A. G., Gutman, S., *Modified Rayleigh Conjecture method for multidimensional obstacle scattering problems*, Numer. Func. Anal. Optim., 26, N2, (2005), 69–80.
120. Ramm, A. G., *Multidimensional inverse scattering problems*, Longman/Wiley, New York, 1992, pp. 1–385.
121. Ramm, A. G., *Multidimensional Inverse Scattering Problems,* Mir, Moscow, 1994. (expanded Russian edition of [120]).
122. Ramm, A. G., *New methods of calculating the static and quasistatic electromagnetic waves*, Proc. 5th Intern. Sympos. "Radioelectronics 74", Sofia, 3, (1974), 1–8 (report 12).
123. Ramm, A. G., *Numerically efficient version of the T-matrix method*, Applic. Anal., 80, N3, (2002), 385–393.
124. Ramm, A. G., *Numerical method for solving inverse scattering problems*, Doklady of Russian Acad. of Sci., 337, N1, (1994), 20–22
125. Ramm, A. G., Chen, W., *Numerical method for solving obstacle scattering problems by an algorithm based on the Modified Rayleigh Conjecture*, Internat. Jour. Appl. Math. Sci., 2, N1, (2005), 11–21.
126. Ramm, A. G., *On a property of the set of radiation patterns*, J. Math. Anal. Appl. 98, (1984), 92–98.
127. Ramm, A. G., Dolph, C., *On the quasistatic boundary value problem of electrodynamics*, J.Math. Anal. Appl., 75, (1980), 300–305.
128. Ramm, A. G., *On the skin-effect theory*, J. Tech. Phys., 42, (1972), 1316–1317.
129. Ramm, A. G., Gutman, S., *Optimization methods in direct and inverse scattering*, In the book *Optimization Methods and Applications*, editor A. Rubinov, Springer, New York, 2005.
130. Ramm, A. G., *Reconstruction of the domain shape from the scattering amplitude*, Radiotech. Electron., 11, (1965), 2068–2070.
131. Ramm, A. G., Makrakis, G., *Scattering by obstacles in acoustic waveguides*, In the book *Spectral and scattering theory*, editor A.G.Ramm, Plenum publishers, New York, 1998, pp. 89–110.
132. Ramm, A. G., *Scalar scattering by the set of small bodies of arbitrary shape*, Radiofisika, 17, (1974), 1062–1068.
133. Ramm, A. G., *Scattering by obstacles*, Reidel, Dordrecht, 1986.
134. Ramm, A. G., Weaver, O., Weck, N. and Witsch, K., *Dissipative Maxwell's equations at low frequencies,* Math. Meth. in the Appl. Sci. 13, (1990), 305–322.
135. Ramm, A. G., *Scattering amplitude as a function of the obstacle*, Appl.Math.Lett., 6, N5, (1993), 85–87.

136. Ramm, A. G., *Singularities of the inverses of Fredholm operators*, Proc. of Roy. Soc. Edinburgh, 102A, (1986), 117–121.
137. Ramm, A. G., *Some theorems on equations with parameters in Banach space*, Doklady Acad. of Sci. Azerb. SSR, 22, (1966), 3–6.
138. Ramm A. G., *Continuity of solutions to operator equations with respect to a parameter*, Internat. Journ of Pure and Appl. Math. Sci., 1, N1, (2004), 1–5.
139. Ramm A. G., *Stability of solutions to inverse scattering problems with fixed-energy data*, Milan Journ of Math., 70, (2002), 97–161.
140. Ramm, A. G., *Stability of the solution to inverse obstacle scattering problem*, J. Inverse and Ill-Posed Problems, 2, N3, (1994), 269–275.
141. Ramm, A. G., *Stability estimates for obstacle scattering*, J. Math. Anal. Appl. 188, N3, (1994), 743–751.
142. Ramm A. G., *Two-sided estimates of the scattering amplitude at low energies*, J.Math. Phys., 21, (1980), 308–310.
143. Ramm, A. G., *Theory and Applications of Some New Classes of Integral Equations*, Springer-Verlag, New York, 1980.
144. Ramm, A. G., *Uniqueness theorems for inverse obstacle scattering problems in Lipschitz domains*, Applic. Analysis, 59, (1995), 377–383.
145. Ramm, A. G., *Wave scattering by small particles, Optics and Spectrocopy,* 43, (1977), 523–532.
146. Ramm, A. G., *Equations for the self-consistent field in random medium*, Phys. Lett. A, 312, N3-4, (2003), 256–261.
147. Rayleigh, J., Scientific Papers, Cambridge, 1922 (in particular, papers from Phil. Mag., vols. 35, 41, 44).
148. Resvykh, K., *Calculating the Electrostatic Fields*, Energy, Moscow, 1967 (in Russian).
149. Scotti, T., Wirgin, A., *Shape reconstruction of an impenetrable body via the Rayleigh hypothesis*, Inverse probl., 12, (1996), 1027–1055.
150. Senior, T., Ahlgren, D., *Rayleigh scattering*, IEEE Trans., AP-21, (1971), 134.
151. Stevenson, A., *Solution of electromagnetic scattering problems as power series in the ratio (dimension of scatterer/wavelength)*, J. Appl. Phys., 24, (1953), 1134–1142.
152. Tatarskij, V., *Wave propagation in a turbulent atmosphere*, Nauka, M., 1967.
153. Tosoni, O., *Calculation of Electromagnetic Fields on Computers,* Technika, Kiev, 1967. (in Russian).
154. Traub, I., Wozniakowski, H., *A General Theory of Optimal Algorithms,* Academic Press, New York, 1980.
155. Tsyrlin, L., *On a method of solving of integral equations of the first kind in potential theory problems*, J. Vycisl. Math. and Math. Phys., 9, (1969), 235–238.
156. Vainberg, M., Trenogin, V., *Theory of branching of solutions of nonlinear equations*, Noordhoff, 1974.

157. Verchota, G., *Layer potentials and regularity for the Dirichlet problem for Laplace's equation in Lipschitz domains*, J. Func. Anal., 39, (1984), 572–611.
158. Wainstein, L., *Static problems for circular hollow cylinder of finite length,* J. Tech. Phys, 32 (1962) 1162–1173; 37 (1967), 1181–1188.
159. Waterman, P., Korringa, J., Ström, S., Varadan, V., in *Acoustic, Electromagnetic and Elastic Wave Scatterinq-Focus on the T-matrix Approach,* Pergamon Press, New York, 1980.
160. Weck, N., Witsch, K., *Exterior Dirichlet problem for the reduced wave equation: asymptotic analysis of low frequencies*, Comm. PDE, 16, (1991), 173–195.
161. Wendland, W. et al. *On the integral equation method for the plane mixed boundary value problem of the Laplacian,* Math. Meth. in the Appl. Sci., 1, (1979), 265–321.
162. Kravtsov, Yu., Rytov, S., Tatarskij, V., *Statistical problems in diffraction theory,* Soviet Phys. Uspekhi, 18, (1975), 118–130.
163. Barabanenkov, Yu., Kravtsov, Yu., Rytov, S., Tatarskij, V., *State of the theory of wave propagation in randomly nonhomogeneous medium,* Soviet Phys. Uspekhi, 13, (1971), 551.
164. Zabreiko, P., Koshelev, A., Krasnoselskij, M., Mihlin, S., Rakovzcik, L. and Stecenko, V., *Integral Equations,* Noordhoff Int., Leiden, 1975.
165. Ziemer, W., *Weakly Differentiable Functions*, Springer Verlag, 1989.

List of Symbols

$A \times B = [A, B]$-vector product

$A \cdot B = (A, B)$-scalar product

Δ-Laplacian

∇ - gradient, $\hat{\nabla}$ - surface gradient

tr - trace

$\alpha_{ij}(\gamma)$ - polarizability tensor	Section 5.1
β_{ij} - magnetic polarizability tensor	Section 5.1
$\tilde{\alpha}_{ij}$, $\tilde{\beta}_{ij}$ - polarizability tensors for screens and films	Section 5.4
P_i - electric dipole moment	Section 5.1
M_i - magnetic dipole moment	Section 5.4

V - volume of the domain D

$r_{st} = |s - t|$

$\psi(t, x) = \frac{\partial}{\partial N_t} \frac{1}{r_{st}}$

$\frac{\partial}{\partial N_{i(e)}}$ - the limiting value of the normal derivative from the interior (exterior)

Γ, S - closed surfaces

F - screen (unclosed surface)

L - the edge of F

$D' = D_e = \Omega$ - exterior domains

f, $A(\alpha', \alpha)$- scattering amplitudes

C_{ij} - electrical inductance coefficients Section 3.5

$C_{ij}^{(-1)}$ - potential coefficients Section 3.5

$Y_{\ell,m}$ - the spherical harmonics

ε, μ - dielectric and magnetic constants of the scatterer

$\varepsilon_e, \mu_e,$ - dielectric and magnetic constants of the medium

$\gamma = (\varepsilon - \varepsilon_e)(\varepsilon + \varepsilon_e)^{-1}$

$G^{\perp}$ - orthogonal complement to the subspace G Section 7.1

$R(A)$ - range of the linear operator A Section 7.1

$N(A) = \{f\ : Af = 0\}$ - the null space of the operator A

st - stationary value

$H_q = W_2^q$, $W^{\ell,p}$ - the Sobolev spaces Section 6.3

$\in$ - member of

For some symbols we do not give the page numbers because these symbols are standard.

Curriculum Vitae of Alexander G. Ramm

NAME:	Alexander G. Ramm, Professor
CITIZENSHIP:	U.S. Citizen
ADDRESS:	Department of Mathematics Kansas State University Manhattan, KS 66506-2602, USA.
PHONE:	510-922-9388
E-MAIL:	ramm@ksu.edu
HOME PAGE:	http://www.math.ksu.edu/~ramm

FIELD:	Differential and integral equations Operator theory, ill-posed and inverse problems Mathematical Physics (scattering theory, inverse scattering, wave propagation) Functional analysis and spectral theory Applied mathematics Theoretical numerical analysis Theoretical electrical engineering, signal estimation, tomography
DEGREES:	B.S., Leningrad State University, 1959 M.S., Leningrad State University, 1961 Ph.D., Moscow State University, 1964 Dr.Sci., Mathematics Institute Academy of Science, Minsk, 1972

EXPERIENCE:

1. Academic

Instructor, Leningrad Institute of Precision Mechanics and Optics, 1962–63
Assistant Professor, Leningrad Institute of Precision Mechanics and Optics, 1964–65
Associate Professor, Leningrad Institute of Precision Mechanics and Optics, 1965–78
Visiting Professor and Research Scientist, University of Michigan, 1979–81
Professor, Kansas State University, 1981-
Visiting Professor: University of Vienna, Goeteborg, Stuttgart, Bonn, Heidelberg, Manchester, London, Uppsala, Royal Inst. of Technology, Stockholm, Acad. Sinica, Taipei, Indian Institute of Science Bangalore, Concordia Univ., Montreal, Institute of Mathematics Ac.Sci USSR, Novosibirsk, Univ of Stockholm, Technion, Israel, Univ. of Cagliari and Milan, Wright Patterson Air Force Base, Univ. of Madrid, Univ. of Grenoble, Politecnico Milan, Univ of Giessen, Univ. of Singapore, Tokyo Metropolitan Univ., Univ. of Palermo, Hebrew Univ., IMPA-Brazil, LMA/CNRS-France, KAIST, Univ. of Leicester, IMPAN, Fulbright Research Professor, Univ. of Lviv (2015 May-Aug, 2016, May-Aug.)

2. Industry, Consulting

Senior Research Scientist, Institute of Precision Mechanics & Optics Research Institute, 1964–78 (Electrical Engineering: Antennas and Propagation, Signal Estimation; Optics: Wave Scattering, Resolution Ability, Inverse Problems; Systems Theory, Ill-posed Problems, Network Theory).
Visiting Scientist, Schlumberger-Doll Research, (1983, August).
Consultant: Dikewood Corporation, Standard Oil Production Co., Los Alamos National Laboratory.

PROFESSIONAL RECOGNITION AND HONORS:

Distinguished Visiting Professor supported by Royal Acad. of Engineers, Sep.10-Oct.10, 2009, Univ. of Leicester.
Invited plenary one-hour speaker at the 7th PACOM. (Pan African Congress of Mathematicians), Aug. 3–8, 2009, Ivory Coast, Yamoussoukro.
Plenary speaker at the International Conference Chaos 2009, June 1–5, Chania, Greece,
Mercator Professor, 2007, Germany, TU Darmstadt.

Distinguished speaker at HKSTAM, June 18, 2005.
London Math. Soc. lecturer, May 24-June 10, 2005.
Khwarizmi International Award, Feb., 2004.
Distinguished foreign professor at the University of Cairo, Amer. Univ. of Cairo, Al-Azhar Univ. of Cairo, (Nov. 2004, Dec. 2006), Academy of Science of Mexico (Oct. 1997)
Distinguished Graduate Faculty Award (1997).
Fulbright Research Professorship at the Technion 1991–92.
Elected member of the New York Academy of Sciences (1994); Elected member of the Electromagnetics Academy MIT (1990).
Founding Member of the Board of the International Society for Analysis, its Applications and Computation (ISAAC) (1994).
Organizer of the special sessions at the A.M.S. meetings 1980, April, 1982 Jan., 1988 Jan., 1990 March, 1998 March
Certificate of Appreciation, Univ. of Michigan, 1981, for technologically promising ideas.
Faculty research award, KSU, 1982
Senior Visiting Fellowships: SERC of Great Britain, June-Aug 1984; University of Bonn, 1984, 1985, 1989; Univ. of Heidelberg 1987; NSERC of Canada, Concordia Univ., May-Aug 1990.
NATO grant for joint research with Prof. R. Burge and Dr. M. Fiddy (King's College), 1985–1986.
AFOSR, NSF and ONR travel and ordinary grants.
Certificate of Recognition, 11th world IMACS Congress, (1985).
Research Professor at WPAFB (summer 1993).
Research Professor at the Univ. of Cagliari and Milan (summer 1994).
Research Professor, Complutense Univ., Madrid, Univ. of Grenoble, Politecnico Milan, Technion, Haifa, 1995.
Research Professor, Univ. of Bremen, Inst. for Appl. and Comput. Math, Heraklion, Crete, 1996 (summer).
Research Professor, INRIA, Univ. of Grenoble, 1997 (summer).
DAAD Research Professor, Institute of theoretical physics, Univ. of Giessen, 1998 (summer)
Research Professor, Hebrew Univ and Ben-Gurion Univ., May-June 1999.
Research Professor, Univ. of Singapore, July 1999.
Research Professor, Univ. of Milano and Palermo, May-July, 2000.
Research Professor, IMPA, Jan 3–27, 2001,
DAAD Research Professor, May 20-Aug 20, 2001, Univ. of Giessen
CNRS Research Professor, Feb. 1, 2002-Jan. 31, 2003.

KAIST, May 20-June 19, 2003
MFO, June 23-Aug 19, 2003
IAS Ben Gurion Univ., Hebrew Univ., May-June 2004
London Math. Soc. lecturer at the Univ. of Belfast, Bath, Edinburgh, Queen Mary Univ. and King's College, May 24-June 10, 2005
Distinguished speaker at HKSTAM, June 18, 2005, and City Univ. of Hong Kong, June 15–24, 2005.
IAS Ben Gurion Univ., May-June 2006.
Univ. of Leicester, May-June 2008.
ETHZ, May-June 2009, Univ. of Leicester, Sep. 10-Oct 10, 2009.
IMPAN, May 27-July 1, 2010.
MPI-Leipzig, May 16-Aug 13, 2011.
Beijing Institute of Technology, May 16-June 14, 2013
Fudan University, Dec 18, 2013-Jan 9, 2014
Fulbright Research Professor, Lviv's National University, May-Aug. 2015
Fulbright Research Professor, Lviv's National University, May-Aug. 2016

EDITORIAL WORK:

Associate Editor of the Journals: Research and Applications of Physics,(RAP), Global Journ Math. Anal., (GJMA), Mathematics, Jour. of Inequalities in Pure and Appl. Math. (JIPAM), International J. Comp. Sci and Math (IJCSM), Australian Jour. of Math Anal and Appl (AJMAA), International J. of Appl. Math. Sci., Jour. of Basic and Appl. Phys. (JBAP), International J. of Tomography and Statistics, Jour. of Functional Analysis and Approximation Theory (JFAAT), J. of Comput. Anal. Appl (JCAA), Advances in Nonlinear Analysis and Applications (ANAA), PanAmerican Math. Journal, Math. Sci. Research Journal, Internat. Journ. of Appl. Math., Internat. Journ. of Diff. Equations and Applications, Nonlinear Functional Analysis and Applications (NFAA), Cubo a Math. Journal, Journ. of Egypt. Math. Soc.

New York Academy of Science,
Electromagnetic Academy MIT,
International Society for Analysis, Applications and Computing,
American Mathematical Society,
International Assoc. of Math. Physics.

COLLOQUIUM TALKS:

Princeton, Cornell, Brown, Madison-Wisconsin, Rice, Univ. of Delaware, Iowa St. Univ., Wayne St. Univ. Detroit, Univ. Of Utah, Salt Lake City, Texas A & M Univ., Kansas Univ. at Lawrence, Univ. of Alberta, Canada, Univ. of Bonn, Univ. of Stuttgart, Univ. of Mich., Argonne Nat. Lab., Naval Res. Lab., Math. Res. Center in Madison-Wisconsin, UCSD, UC Berkeley, Stanford, Mich. State Univ., Univ. of Pittsburgh, General Motors Res. Labs., Shell Res. Center, Houston, Univ. of W. VA., Univ. of Arizona, Washington St. Univ., Case Western Reserve Univ., Kansas St. Univ., Univ. of Madrid, Univ. of Florence, Univ. of Goteborg, Royal Institute of Technology, Stockholm, Univ. of Uppsala, Vrije Univ., Amsterdam, Nat. Bureau of Standards, Howard Univ., Univ. of Vienna, Tech. Univ. of Vienna, Schlumberger-Doll Research Lab., Ridgefield, CT., Univ. of Gottingen, Sohio Petroleum Research Lab., Dallas, Univ. of Minnesota, Minneapolis, Schlumberger Well Services, Houston, Univ. of Manchester, Univ. of London, Univ. of Tubingen, Univ. of Minnesota, Univ. of Marceille (Luminy), Univ. of Helsinki, Univ. of Muenchen, Royal Radar and Signal Establishment, Great Malvern. Catholic Univ. Washington D.C., ERIM, Ann Arbor, MI, Univ. of Nevada, Las Vegas, Acad. Sinica, Taipei, Indiana University, Bloomington, Wichita State University, University of Heidelberg, Bonn, Regensburg, Karlsruhe, Frankfurt, Bielefeld, Indian Institute of Science, Bangalore, Institute for basic research, Istanbul, IIT Delhi, Cornell University, Univ. of Munster, Univ. of Lund, Univ. of Linkoping, Courant Inst. NYU., RPI, Troy, Yale Univ., Boston Univ., Univ. of MD., Univ. of Lowell, Univ. of Paris Süd (Orsay), Univ. of Pierre and Marie Curie, Paris, Florida Atl. Univ., Boca Raton, Tech. Univ. of Berlin, Freie Univ. of Berlin, Univ. of Essen, Univ. of Düsseldorf, NASA Langley Research Center, Virginia Tech. Blacksburg, Institute of Mathematics, Bulgarian Acad. of Sci., Sofia, Univ. of Southwest Louisiana, Nat. Inst. of Standards and Technology, Boulder, Univ. of Aachen, FRG., Univ. of Montreal, Univ. of Kyoto, Univ. of Oklahoma, Univ. of TN., Inst. of Math. Acad Sci., Novosibirsk, Inst. of Math., Acad. of Sci., Kiev., Univ. of Stockholm, Univ. of Paris-Sud., Polytech. of Turin, Univ. of Pavia, Univ. of Pisa, Politechnico of Milan Ecole Super. Electr., Gif sur Yvette, Inst. for Problems of Mechanics, Moscow, Univ of Stockholm, Univ. of Uppsala, Inst. of Appl. Math., Moscow, Weizmann Institute, Hebrew Univ., Tel-Aviv Univ., Technion, Bar-Ilan Univ., Univ of Beer Sheva, Univ. of Haifa, Tübitak Istanbul, TH Darmstadt, Univ. of Siegen, Univ. of Karlsruhe

Geophys. Inst., Inst. of App. Math. Acad. Sinica, Beijing Univ., Fudan Univ., Univ. of Sci.& Tech., Hefei, Graduate School, Acad. Sinica, Univ. of Sci. & Tech., Hong Kong, Tokyo Metropol. Univ., Kyoto Univ., Univ. of Muenster, Georgia Tech., Univ. of Cincinnati, CWRU, Univ. of Dayton, Wright-Patterson AFB at Dayton, Novosibirsk Univ., Inst. of Math. Bulg. Acad. Sci., Los Alamos Nat. Lab., Univ of New Mexico, Sandia Nat. Lab., Univ. of Cagliari, Univ. of Rome, Autonoma Univ. Madrid, Complutense Univ. Madrid, Univ. of Bilbao, Univ. of Barcelona, Supelec Paris, Univ. of Tenerife, IMA Univ. of Minnesota, Univ. of Carlos III, Madrid, Univ. of Grenoble, CNRS Marceille, Univ. of Paris 13, Univ. of Milan, Univ. of Naples, Univ. of Liege, INRIA-Rocquencourt, Univ. of Rome, IAC Rome, Univ of Lyon, Univ of Beer Sheva, Technion, Univ of Milan, TH Darmstadt, Univ. of Erlangen, Univ. of Bremen vision center, Univ. of Athens, Tech. Univ. of Athens, Inst. for Appl and Comput. Math., Heraklion, Crete, UNAM Mexico City, INRIA-Grenoble, CNRS-Marceille, Univ de Franche Comte-Besancon, IMAG-Grenoble, Univ of Manitoba-Winnipeg, Autonoma Univ, Mexico City, Univ. of Tennessee at Knoxville, Oak Ridge National Lab., Univ of Giessen, Chemnitz, Freiberg, Dresden, Bonn, Darmstadt, TU Vienna, Univ. of Vienna, Forschungszentrum Karlsruhe, Complutense Univ Madrid, Autonoma Univ. Madrid, SDR Research, Univ of Missouri, Univ. of Nagoya, Kyoto, Tokyo, UCLA, Caltech, Univ. of California at Irwine, at San Diego, Technion, Haifa Univ., Hebrew Univ., Ben-Gurion Univ., Weizmann Inst., Univ of Palermo, Univ. of Singapore, Inst. of Math. of Nat. Acad. of Sci. of Ukraine, Kiev, Univ. of Lvov, Univ. of MO, Univ. of Milano, Bologna, Roma (La Sapienza), Palermo, Darmstadt, Stuttgart, Giessen, Pusan University, Korea, FAU (Frontiers of Science lecture), UCF, Orlando, Institute of mathematics, Yerevan, Los Alamos CNLS, Univ. Fed. Rio de Janeiro, IMPA, Univ. of Bologna, Pavia, Univ. of Muenster, Bonn, Giessen, Ecole Polytechnique, CNRS-LMS, Frankfurt, Stuttgart, Linz, Chemnitz, Autonoma Univ. Mexico City, Tokyo Metropolitan Univ., Nihon Univ., Meiji Univ., Univ. of Tokyo, Univ of Kyoto, Polytech. Univ. of Torino, LMA/CNRS-Marseille, INRIA-Antibes, Univ. of Nice, Luminy-theor. physics, Weizmann Inst, Rehovot, Technion, Univ. of Haifa, Ben-Gurion Univ., Univ. of Uppsala, Univ. of Guanajuato at Salamanca, Univ of Marseille-CMI, INRIA-Rockencourt, Yonsei Univ., SNU (Seoul Nat. Univ), KAIST, Univ. of Dresden, Dresden Math. Seminar, Ben-Gurion Univ., Hebrew Univ., Technion, Queen's Univ of Belfast, Univ. of Bath, Univ of Edinburgh, Univ. of Queen Mary, King's College, City Univ. of Hong Kong, Hong Kong Soc. of Theor. and Appl.

Mechanics, Oklahoma Univ., LANL, Univ. of Giessen, Dresden, TU Darmstadt, GSI (Gesellshaft für Schwere Ions) Physics Institute, Karlsruhe, Konstanz, Lyon, Charles Univ. Prague, Math. Inst., Prague, TU Delft, Univ. of Barcelona, TU Darmstadt-EM division, IMDEA-Madrid, Bergacadem.-Freiberg, ETH-Zuerich, Univ. of Manchester, Leicester, Birmingham, UCL-Univ. Coll. London, UCIrvine, Oklahoma Univ., UCF Orlando, ETH, Univ. of Konstanz, Univ. of Lausanne, ETH Zurich, Univ. of Innsbruck, Univ. of Leicester, Birmingham, Loughborough, UCL-Univ. Coll., Imperial College-London, Autonoma Univ. Madrid, Univ. of Granada, Military Tech. Univ. of Warsaw, Inst Probl. Tech. Acad. of Sci. of Poland, IMPAN, WIAS, Humboldt Univ., Univ. of Chemnitz, Univ. of Jena, Univ. of Crete, BIT (Beijing Inst. of Technology), Fudan Univ. (Mathematics Center), Fudan Univ. (Physics Institute), CCNU Wuhan, Univ of Lisbon (Math. Department), Univ of Lisbon (Phys. Department), Univ. of Lviv, Inst. for problems of mechanics and mathematics of Ukrainian Acad. of Sci., Univ. of Lviv, IMPAN, IFPAN

MONOGRAPHS and BOOKS:

1. Theory and applications of some new classes of integral equations. Springer Verlag, New York, 1980, pp. 1–356;
   isbn 0-387-90540-5.
2. Iterative methods for calculating the static fields and wave scattering by small bodies. Springer Verlag, New York, 1982, pp. 1–130;
   isbn 0-387-90682-7
3. Scattering by obstacles. Reidel, Dordrecht, 1986, pp. 1–442;
   isbn 90-277-2103-3
4. Random fields estimation theory. Longman Scientific & Wiley, New York, 1990, pp. 1–281;
   isbn 0-582-03768-9
5. Random fields estimation theory, Expanded Russian edition, Mir, Moscow, 1996, pp. 1–352;
   isbn 5-03-003031-X
6. Multidimensional inverse scattering problems, Longman Scientific & Wiley, New York. 1992, pp. 1–385;
   isbn 0-582-05665-9;
7. Multidimensional inverse scattering problems, Expanded Russian edition, Mir, Moscow, 1994, pp. 1–496;
   isbn 5-03-002939-7

8. The Radon transform and local tomography. CRC Press, Boca Raton, 1996, pp. 1–503 (with A. Katsevich);
   isbn 0-8493-9492-9.
9. Spectral and scattering theory, Plenum publishers, New York, 1998 (editor A.G. Ramm)
   isbn 0-306-45829-2
10. Inverse problems, tomography and image processing, Plenum publishers, New York, 1998 (editor A.G. Ramm)
    isbn 0-306-45828-4
11. Operator Theory and Applications, Amer. Math. Soc., Fields Institute Communications, Providence RI, 2000 (editors A.G.Ramm, P.N. Shivakumar, A.V. Strauss).
    isbn 0-8218-1990-9
12. Inverse problems, Springer, New York, 2005. isbn 0-387-23195-1
13. Wave scattering by small bodies of arbitrary shapes, World Sci. Publishers, Singapore, 2005.
    isbn 981-256-186-2
14. Random fields estimation, World Sci. Publishers, Singapore, 2005.
    isbn 981-256-536-1
15. Dynamical Systems Method for solving operator equations, Elsevier, Amsterdam, 2007.
    isbn 0-444-52795-8
16. **Dynamical Systems Method and Applications. Theoretical Developments and Numerical Examples.** Wiley and Sons, Hoboken, 2012. isbn 978-1-118-02428-7 (with N.S. Hoang)
17. **Scattering of Acoustic and Electromagnetic Waves by Small Bodies of Arbitrary Shapes. Applications to Creating New Engineered Materials**,
    Momentum Press, New York, 2013.
    isbn-13: 978-1-60650-621-9
18. **Scattering by obstacles and potentials**, World Sci. Publ., Singapore, 2017.
    isbn 9789813220966
19. **Creating materials with a desired refraction coefficient,** IOP Concise Physics, Morgan & Claypool Publishers, San Rafael, CA, USA, 2017.
    isbn 978-1-6817-4709-5
20. **Symmetry Problems. The Navier-Stokes Problem,** Morgan & Claypool Publishers, San Rafael, CA, USA, 2019.
    isbn 9781681735054

21. **Inverse obstacle scattering with non-over-determined scattering data**, Morgan & Claypool Publishers, San Rafael, CA, 2019. isbn: 9781681735887
22. **Creating materials with a desired refraction coefficient,** IOP Publishers, Bristol, UK, 2020. Second edition.
23. **The Navier-Stokes problem**, Morgan & Claypool Publishers, 2021. isbn 978163639243
24. **Analysis of the Navier-Stokes problem. Solution of a Millennium Problem**, Springer, 2023. ISBN 978-3-031-30722-5 ISBN 978-3-031-30723-2 (eBook)
25. **Wave scattering by small bodies. Creating materials with a desired refraction coefficient and other applications**, World Sci. Publishers, Singapore, 2023.

US PATENTS:

1. Pseudolocal tomography (with A. Katsevich), number 5,539,800 issued July 23, 1996.
2. Enhanced local tomography (with A. Katsevich), number 5,550,892 issued Aug.27, 1996.

INVITED ADDRESSES:

1. All-union symposium on wave diffraction; Tbilisi, 1964
2. All-union conference on numerical mathematics, Moscow, 1965
3. International URSI symposium, Stressa/Italy, 1968
4. Third all-union meeting on theoretical and applied mechanics, Moscow, 1968
5. Third all-union conference on heat and mass transfer, Minsk, 1968
6. All-union symposium on wave diffraction, Leningrad, 1970
7. International URSI symposium on electromagnetic waves, Tbilisi, 1971
8. Fourth all-union conference on heat and mass transfer, Minsk, 1972
9. Conference on technical cybernetics, Moscow, 1972
10. International symposium on radioelectronics, Varna, 1974
11. International congress on acoustics, London, 1974
12. All-union seminar on the atom and atomic spectra theory, Tashkent, 1974
13. All-union symposium on the interaction of cosmic dust with the atmosphere, Ashhabad, 1974

14. International symposium on nonlinear networks, Split, Yugoslavia, 1975
15. International conference CVUT, Prague, 1975
16. All-union conference on differential equations, Rjazan, 1976
17. International Conference on Computer-Aided Design of Electromagnetic and Microwave Circuits and Systems, Hull, England 1977
18. International symposium on approximation theory, Campinas, Brazil 1977
19. All-union winter mathematical school, Voronezh, 1977
20. International Congress on applied mathematics, Weimar, DDR, 1978
21. All-union 10 symposium on the representation and analysis of random fields and processes, Suhumi, 1978
22. GAMM Tagung, Wiesbaden, BRD, April 1979
23. International symposium–workshop on wave scattering, Columbus, Ohio, June 1979
24. International symposium on ill-posed problems, Newark, Delaware, October 1979
25. Solutions of some inverse and ill-posed problems, Nav. Res. Lab., Wash. D.C., Oct. 9, 1979
26. A.M.S. Meeting, Kent, Ohio, Nov. 1979
27. A.M.S. Meeting, Boulder, Colorado, March 1980
28. A.M.S. Meeting, Bloomington, Indiana, April 1980
29. International symposium on nonlinear phenomena, Arlington, Texas, June 1980
30. Symposium on real analysis, Mich. State University, June 1980
31. Symposium on scattering theory, Oberwolfach, FRG, August 1980
32. A.M.S. 1980 summer meeting, August 1980 Ann Arbor, MI
33. Conference on integral equations, Oberwolfach, FRG, Dec. 1980
34. Mathematical foundations of the singularity and eigenmode expansion methods, Meeting at the University of Kentucky, Lexington, KY, Nov. 1980
35. International conference on spectral theory of differential operators, Birmingham, Alabama, March 26–28, 1981
36. IEEE International symposium on circuits and systems, Chicago, IL, April 27–29, 1981
37. A.M.S. Annual meeting, January 1981
38. 7th International Dundee conference on ordinary and partial differential equations, 3-29-82 to 4-3-82

39. IEEE International symposium on information theory, Les Arcs. France, 6-21-25-1982
40. IEEE International symposium on antennas and propagation, Univ. of New Mexico, Albuquerque, May 1982.
41. AMS annual meeting, Denver, Jan. 1983, Monotone operators and nonlinear passive systems (special session, invited talk)
42. 1983 International symposium on the mathematical theory of networks and systems, June 20–24, 1983, Ben Gurion Univ., Beer Sheva, Israel
43. 1983 International IEEE symposium CAS, Newport Beach, California, May 2–4, 1983
44. Conference on scattering theory, Oberwolfach, July 1983, FRG
45. NATO advanced research workshop on inverse scattering, Bad-Windsheim, FRG, Sept. 1983
46. AMS annual meeting, January 1984, special session PDE
47. International conference on P. D. E., Dundee, June 1984
48. Conference on PDE, Oberwolfach, March, 1985
49. 11th world IMACS congress, Oslo, August 1985, plenary talk
50. Finnish mathematical society meeting, May, 1985
51. Conference of the Chinese mathematicians, Taiwan, July 1986, plenary talk
52. International conference on operator theory, Oct. 1986
53. Conference on inverse problems, Montpellier, Dec. 1986
54. International conference on mathematical geophysics, West Berlin, Feb. 1987
55. AMS meetings March, April 1987
56. European Congress on Simulation, ECS-87, Sep. 1987, Plenary talk
57. Conference on numerical integration, Nov. 1987, Oberwolfach
58. AMS annual meeting, January 1988
59. Annual GAMM meeting, Vienna, April 1988, plenary talk
60. Workshop on inverse problems, Univ. of MD, March 1988
61. The first Woodward conference on Wave phenomena, June 1988, plenary talk
62. International conference on inverse problems, Montpellier, France, Dec. 1988 (2 one-hour lectures)
63. NSF conference on nonlinear wave equations Jan. 1989
64. Oberwolfach conference on differential equations (March 1989)
65. NSF workshop on inverse problems (July 29-August 4, 1989)

66. International conference on inverse problems (Bulgaria, Sep. 1989)
67. Oberwolfach conference on solitons, Jan. 1990
68. NSF conference on inverse scattering, June 1990
69. SIAM annual meeting, July 1990, minisymposium on inverse scattering.
70. International conference "Inverse problems in science and engineering", Osaka, Aug. 1990
71. International Congress of Mathematicians, Kyoto, August, 1990
72. Oberwolfach conference on statistical estimation, Nov. 1990
73. South Eastern conference on differential equations, Blacksburg, VA, Nov. 1990
74. International conference on mathematical modeling, key-note speaker, Univ. of MD, Apr. 1991
75. International conference on signal processing, Cetraro, Italy, plenary speaker, May 1991
76. International conference on ill-posed problems, plenary speaker, Moscow, Aug. 1991
77. International Workshop on inverse problems, invited speaker, Novosibirsk, Aug. 1991
78. US-Israel NSF workshop on operator theory, Beer Sheva, Feb.1992, invited speaker.
79. International Conference of Computational Engineering Science, ICES-92, invited speaker, Dec. 1992, Hong Kong.
80. Third Midwest conference on geometry, Columbia, Apr. 1993.
81. International conference on quantum inversion, Bad Honnef, FRG, May 1993, plenary speaker.
82. International conference on dynamical systems, May 1993, Atlanta, plenary speaker.
83. International symposium on computerized tomography, Aug. 1993, Novosibirsk, Russia, plenary speaker.
84. International symposium on numerical methods, Aug. 1993, Plovdiv, plenary speaker.
85. International symposium on differential equations, Aug. 1993, Plovdiv, plenary speaker.
86. International symposium on Inverse problems, Sept. 1993, Potsdam, FRG, plenary speaker.
87. Oberwolfach conference on pseudodifferential operators, Jan. 1994, invited speaker.
88. Oberwolfach conference on tomography, Sep.1994, invited speaker.

89. 26th Midwest conference on differential equations, invited speaker, Oct 7–8, 1994.
90. International ASME conference, Chicago, Nov.7–11, 1994, invited speaker, special session on inverse problems in mechanics.
91. AMS-SIAM workshop on inverse problems, March 1995, invited speaker.
92. Oberwolfach conference on inverse problems, Feb. 1996, invited speaker.
93. World Congress of Nonlinear Analysts, WCNA-96,Jul.10–17, 1996, Athens, plenary speaker
94. International conference on inverse scattering, Sep.3–7, 1996, Lake Balaton-96, plenary speaker
95. International conference on inverse and ill-posed problems, IIPP-96, Sep.9–14, Moscow, plenary speaker.
96. Mexican math. soc. meeting, Oct.7–11, 1996, invited speaker.
97. ISAAC International Congress, June 2–7, 1997, plenary speaker.
98. The mathematics of life sciences, Jan. 28–31, 1998, Texas Tech. Univ., one-hour invited speaker.
99. International conference MTCP-98, modern trends in comput. physics, Joint Instit. for Nuclear Research, Dubna, June 15–20, 1998, plenary speaker.
100. Oberwolfach conference on tomography, Aug. 2–8, 1998, invited speaker.
101. International conference "Operator theory and applications" Winnipeg, Oct. 7–11, 1998, plenary speaker
102. Workshop on the Radon transform, Univ. of Nagoya, Nov. 1998, key-note speaker.
103. Braude College PDE days, main speaker, May 18–20, 1999,
104. Israel Math. Union annual meeting, invited speaker, May 26, 1999.
105. Internat. workshop on inverse problems and wave scattering, Lvov, Sep. 20–23, 1999, plenary speaker.
106. Internat. conference PDE 2000, Clausthal, Germany, July 24–28, 2000.
107. Internat. conference on nonlinear analysis, Korea, Pusan, Aug. 31–Sep. 5, 2000, plenary speaker.
108. Internat. conference on dynamical systems and chaos, Armenia, Sep. 11–18, 2000, plenary speaker.
109. Mathematics and medical imaging, Frontiers of Science Lecture, FAU, Oct. 11, 2000

110. Dynamical systems and linear and nonlinear ill-posed problems, Los Alamos Nat. Lab, CNLS colloquium, Dec. 6, 2000
111. Inverse and direct problems and applications, Gargnano, Apr. 2–6, 2001, main lecturer
112. Dynamical systems and linear and nonlinear ill-posed problems, lectures at the Auton. Univ., Mexico City, Sep.17–21, 2001.
113. AMRTMA conference on acoustic, mechanics and related topics of mathematical analysis, June 2002, France.
114. Oberwolfach conference on tomography, Aug. 11–17, 2002
115. Conference on mathematical modelling of wave phenomena, Vaxjo University, Sweden, Nov. 3–8, 2002, plenary speaker
116. Internat. workshop on random fields, Guanajuato, Nov 27–30, 2002, plenary speaker
117. International seminar on nonlinear analysis and spectral problems, Complutense Univ., Madrid, June 14–16, 2004, invited speaker
118. Workshop on PDE, Hebrew Univ., June 2004, invited speaker.
119. IPAM conference "Mathematics of the Ear and Sound Signal Processing", January 31–February 2, 2005
120. Midwest Geometry Conference, Apr 28-May 1, 2005, Ohio St. Univ
121. LMS lectures, May 24-June 10, 2005
122. HKSTAM, June 18, 2005, distinguished invited speaker.
123. 5icipe, Cambridge, July 9–16, 2005.
124. ICAM05-Internat. Conference on Appl. Math., Bandung, Aug. 22–26, 2005, plenary speaker
125. ICMAA06-Internat. Conference on Math. Anal. and Appl., Assiut, Egypt, Jan 3–6, 2006, plenary speaker.
126. Midwest geometry conference, Univ. of Oklahoma, May 5–7, 2006.
127. ETOPIM7, Sydney, July 9–14, 2006, plenary speaker.
128. IPDO-2007, Miami, Apr. 16–18, 2007, key-note speaker.
129. World Congress of Engineering and Applied Mathematics, London, July 2–4, 2007, key-note speaker.
130. International Conference on Inverse Quantum Scattering Theory, Aug. 27–31, 2007, Hungary, Lake Balaton-Siofok, plenary speaker.
131. Workshop on PDE, Darmstadt, Sep. 24–26, 2007, invited speaker
132. Analysis of Multiphase Problems, Prague, Oct. 8–12, 2007, special lecture.
133. IMDEA-Madrid, Nov 29, 2007, invited talk.
134. Oberwolfach workshop, Material Theories, Dec. 16–21, 2007, invited speaker

135. International Conference Chaos-2008, Chaotic modeling and simulation, June 3–6, 2008, Chania, Crete, Greece, plenary speaker.
136. World Congress of Nonlinear Analysts, WCNA-2008, Orlando, Florida, July2–9, 2009, key-note speaker.
137. International Conference Chaos-2009, Chaotic modeling and simulation, June 1–5, 2009, Chania, Crete, Greece, plenary speaker,
138. PanAfrican Congress of Mathematicians, PACOM7, Aug. 3–8, 2009, plenary one-hour speaker.
139. International Workshop, DIPED2009, Lvov, Sep.21–24, 2009, plenary one-hour speaker.
140. International Conference Chaos-2010, June 1–5, 2010, Chania, Crete, Greece, plenary speaker,
141. International Conference Computational methods in Applied Mathematics, CMAM-4, June 20–26, 2010, Bedlewo, Poland,
142. International Conference Chaos-2011, May 31-June 3, 2011, Agios Nikolaos, Crete, Greece, plenary speaker,
143. International Conference on Differential and Difference Equations and Applications, July 4–8, 2011, Ponta Delgada, Univ of Azores, plenary speaker.
144. ACEX-13, Plenary speaker, Madrid, July 1–5, 2013.
145. International Conference Chaos-2014, Plenary speaker, Lisbon, June 7–10, 2014
146. International Conference Chaos-2015, Plenary speaker, Paris, May 26–29, 2015
147. International Conference MMET-2016, invited speaker, Lviv, July 5–7, 2016
148. Intern. Conference Materials 2017, plenary speaker, Nov. 13–15, Las Vegas, Renaissance Hotel.
149. Intern. Conference HSA-2020, Hungary, June 8–11, 2020, invited speaker
150. Prague, Sep. 20–24, 2020, Internat. Conference on PDE, plenary speaker.
151. Materials Science Conference, plenary speaker, Nov. 16–17, 2020, Istanbul
152. How to create materials with a desired refraction coefficient? 2nd Virtual Congress on Materials Science and Engineering, March 29–31, 2021, key-note speaker
153. Bremen Zoom Workshop on Light Scattering 2021, 22–23. March 2021, plenary speaker

154. First Western Balkan Conference in Mathematics and Applications, June 10–12, 2021, Analysis of the Navier-Stokes problem, plenary speaker.
155. Third global webinar on Appl. Science, Engineering and Technology, Feb. 19–20, 2022, How to create materials with a desired refraction coefficient? Plenary talk.
156. Global Magnus group conference, March 28–30, Keynote talk, Analysis of the Navier-Stokes problem.
157. "Solution of the millennium problem concerning the Navier-Stokes equations", Keynote talk at the World Conference on Physics and Mathematics, May 22–23, 2023, Berlin, Germany

LIST OF COURSES TAUGHT:

Undergraduate courses:

Calculus and analytic geometry, advanced calculus, differential equations, technical calculus, linear algebra, elementary PDE, integral equations, special functions and their applications, mathematics:its form and impact.

Graduate courses:

Ordinary differential equations, PDE, functional analysis and its applications, spectral and scattering theory for differential equations, singular integral equations, complex analysis, theoretical numerical analysis, ill-posed problems, integral transforms, asymptotic methods, iterative solution of the static problems, mathematical methods for engineers, mathematics of wave propagation, electrodynamics, quantum mechanics, integral equations and applications, inverse scattering theory, inverse problems in analysis and PDE, nonlinear functional analysis, theory of passive networks, entire functions in antenna synthesis and optics, approximation theory, potential theory, calculus of variations, distribution theory, probability theory, random fields estimation theory.

Ph.D students

T. Miller, A. Zade-Chavoshi, Peiqing Li, A. Katsevich, Yan Chuntao, R. Hayrapetyan, A. Smirnova, N.S. Hoang, S. Indratno, N. Tran, Cong Van.

Articles for mathematical encyclopedia, Kluwer, Dordrecht, 2001, Supplement volume 3.

1. Ordinary differential equations, property C for, pp. 295–296.
2. Local Tomography, pp. 241–242.
3. Partial Differential Equations, Property C for, pp. 298–299.
4. Inverse Scattering, half-axis case, pp. 209–211.
5. Inverse scattering, full line case, pp. 207–208.
6. Obstacle scattering, pp. 284–286.
7. Inverse scattering: multidimensional case, pp. 211–212.
8. Pseudolocal tomography, pp. 310–311.
9. Reproducing kernel, pp. 328–329.
10. Reproducing kernel Hilbert Space, pp. 329–331.

PROFESSIONAL PUBLICATIONS:

[1.] On the Kotelnikow's theorem. Electrocommunication, 10, (1962), 71–72.

[2.] A necessary and sufficient condition for compactness of embedding, Vestnik Leningr. Univ., N1, (1963), 150–151. (Math. Rev. 27 #1808)

[3.] Investigation of the scattering problem in some domains with infinite boundaries I, II, Vestnik 7, (1963), 45–66; 19, (1963), 67–76. 27 #483, 23 #374.

[4.] Spectral properties of the Schrödinger operator in some domains with infinite boundaries, Doklady Acad of Sci. USSR, 152, (1963) 282–285. 27 #3930.

[5.] Absence of the discrete positive spectrum of the Dirichlet Laplacian in some infinite domains. Vestnik 13, (1964), 153–156; N 1, (1966), 176. 30 #1295.

[6.] On the analytic continuation of the solution of the Schrödinger equation in the spectral parameter and the behavior of the solution to the nonstationary problem as $t \to \infty$, Uspechi Mat. Nauk, 19, (1964), 192–194.

[7.] Statement of the diffraction problem in domains with infinite boundaries. Proc. 3 all-union wave diffraction symp., Nauka, Moscow, (1964), 28–31.

[8.] Reconstruction of the shape of a domain from the scattering amplitude. Proc. 3 all-union wave diffraction symp., Nauka, Moscow, (1964), 143–144.

[9.] Conditions under which the scattering matrix is analytic, Doklady Acad. of Sci. of USSR, 157, (1964), 1073–1075. 32 #2049.

[10.] Analytic continuation of the resolvent kernel of the Schrödinger operator in the spectral parameter and limiting amplitude principle in some infinite domains, Doklady Acad. of Sci. Azerb. SSR, 21, (1965), 3–7.

[11.] Spectral properties of the Schrödinger operator in some infinite domains, Mat. Sbor. 66, (1965), 321–343. 30 #3297, 34 #7994.

[12.] On wave diffusion, Mathematics. Izvestija vuzov, 2, (1965), 136–138. 32 #1451.

[13.] On a method of solving the Dirichlet problem in some infinite domains, Mathematics. Izvestija vuzov, 5, (1965), 124–127. 32 #7993.

[14.] On the conditions under which integral operators are nuclear and existence of the S-matrix in the problem of scalar scattering on a potential and surface. Ukrain. Math. Jour. 17, (1965), 92–98. 34 #1887.

[15.] Necessary and sufficient conditions for the validity of the limiting amplitude principle. Doklady Acad of Sci. USSR, 163, (1965), 584–586. 33 #7673.

[16.] Reconstruction of the domain shape from the scattering amplitude, Radiotech. i Electron., 11, (1965), 2068–2070.

[17.] Reconstruction of a signal from its values on a discrete sequence of time moments, Radiotech. i Electron., 11, (1965), 1957–1959.

[18.] Behavior of the solution to a nonstationary problem as $t \to \infty$, Mathematics, Izvestija vuzov, 1, (1966), 124–138. 33 #7674.

[19.] Domain free from the resonances in the three-dimensional scattering problem, Doklady Acad of Sci. USSR, 166, (1966), 1319–1322. 34 #3902.

[20.] Spectrum of Schrödinger operators with spin-orbit potential, Doklady Acad of Sci. USSR, 169, (1966), 799–802. 34 #7993.

[21.] Antenna synthesis with the prescribed pattern. 22 sci. session dedicated the day of radio, Moscow, 1966, section of antennas, 9–13.

[22.] Statement and numerical solution of inverse ionospheric problem, 22 sci. session dedicated the day of radio, Moscow, 1966, section wave propagation, 3–6.

[23.] Reconstruction of the potential and domain boundary from the scattering amplitude, 22 sci. session dedicated the day of radio, Moscow, 1966, section wave propagation, 7–10.

[24.] Estimates of the temperature field for bodies of complicated shape. In collection of papers Investigation of nonstationary heat and mass transfer, Minsk, 1966, 64–70.

[25.] Some theorems on equations with parameters in Banach space, Doklady Acad. of Sci. Azerb. SSR, 22, (1966), 3–6. 33 #7963.

[26.] Some inverse problems of wave propagation. Proc. of the 4-th all-union wave diffraction symp., Moscow, 1967, 7–11.

[27.] Asymptotic behavior of eigenvalues in the case when the potential depends on parameter, Math. Zametki, 1, (1967), 599–608. (with Levitan B. M.). 37 #1817.

[28.] About estimates of the thermoresistances for bodies of complicated shape, Eng. Phys. Journ., 13, 1967, 914–920.

[29.] On the limiting amplitude principle, Diff. eq., 4, (1968), 714–720. 37 #1759.

[30.] Estimates of the thermoresistances, Proc. of the third all-union conference on heat and mass transfer, Minsk, 1968, 12–17.

[31.] Optimal solution of the antenna synthesis problem, Doklady Acad. of Sci. USSR, 180, (1968), 1071–1074.

[32.] On numerical differentiation, Mathem., Izvestija vuzov, 11, (1968), 131–135. 40 #5130.

[33.] On equations of the first kind, Diff. eq. 4, (1968), 2056–2060. 40 #817; English transl., 1062–1064.

[34.] Asymptotic distribution of the Schrödinger operator eigenvalues when the potential tends to infinity and the boundary is infinite, Doklady Acad. of Sci. USSR, 183, (1968), 780–783. 40 #1827.

[35.] Some theorems on analytic continuation of the Schrödinger operator resolvent kernel in the spectral parameter, Izvestiya Acad. Nauk Armyan. SSR, Mathematics, 3, (1968), 443–464. 42 # 5563.

[36.] Random fields filtering in optical instruments in the case of finite entrance pupil size, Optics and Spectroscopy, 26, N3, (1969), 421–428.

[37.] Perturbation by damping of the eigenfrequencies of small oscillations, Appl. Math. and Mechanics, 33, (1969), 328–330. 40 #8318.

[38.] Filtering of nonstationary random fields in optical systems, Opt. and Spectroscopy, 26, N5, (1969), 832–836.

[39.] On antenna synthesis theory, Collection "Antennas" N 5, (1969), 35–46, Moscow, Izd. Svjaz.
[40.] Iterative solution of the integral equation in potential theory, Doklady Acad. Sci. USSR, 186, (1969), 62–65. 41 #9462.
[41.] Nonlinear antenna synthesis problems, Doklady Acad. Sci. USSR, 186, (1969), 1277–1280. 41 #4904.
[42.] Apodization theory, Optics and Spectroscopy, 27, N3, (1969), 508–514.
[43.] Filtering of nonhomogeneous random fields, Optics and Spectroscopy, 27, N6, (1969), 881–887.
[44.] Green's function study for differential equation of the second order in domains with infinite boundaries, Diff. eq. 5, (1969), 1509–1516. 40 #6078; E.t. 1111–1116.
[45.] Calculation of the scattering amplitude for the wave scattering from small bodies of an arbitrary shape, Radiofisika, 12, (1969), 1185–1197. 43 #7131.
[46.] Optimal solution of the linear antenna synthesis problem, Radiofisika, 12, (1969), 1842–1848. 43 #8223.
[47.] Asymptotic behavior of the eigenvalues and eigenfunction expansions for the Schrödinger operator with increasing potential in the domains with infinite boundary. Izvestija Ac. Nauk. Arm. SSR, Mathematics, 4, (1969), 442–467. 42 #3451.
[48.] Nonlinear problems of antenna synthesis, Radiotech. i Electron., 15, (1970), 21–28. Rad. Eng. El. Phys. 15, (1970), 15–22.
[49.] Nonlinear problem of plane antenna synthesis, Radiotech. i Electron., 15, (1970), 591–593. E. T. 500–503.
[50.] Eigenfunction expansion for nonselfadjoint Schrödinger operator, Doklady Acad. Sci. USSR, 191, (1970), 50–53. 42 #703.
[51.] On some integral operators, Diff. eq., 6, (1970), 1439–1451. Et. 1096–1106. 42 #8339.
[52.] Exponential decay of solution of the hyperbolic equation, Diff. eq., 6, (1970), 2099–2100. 44 #631. E.t. 1598–1599.
[53.] Apodization theory II, Opt. and Spectroscopy, 29, N2, (1970), 390–394.
[54.] Increasing of the resolution ability of the optical instruments by means of apodization, Opt. and Spectroscopy, 29,N3, (1970), 594–599.
[55.] On resolution ability of optical systems, Opt. and Spectroscopy, 29, N4, (1970), 794–798.

[56.] Reconstruction of the shape of a reflecting body from the scattering amplitude, Radiofisika, 13, (1970), 727–732.

[57.] Approximate formulas for polarizability tensors and capacitances of bodies of arbitrary shapes and applications, Doklady Acad. Sci. USSR, 195, (1970), 1303–1306. MR 55 #1947; E.t. 15, (1971), 1108–1111.

[58.] Filtering and extrapolation of some nonstationary random processes, Radiotechnika i Electronika, 16, (1971), 80–87. E.t. 68–75.

[59.] Calculation of the initial field from the scattering amplitude, Radiotechnika i Electronika, 16, (1971), 554–556.

[60.] Necessary and sufficient conditions for the validity of the limiting amplitude principle, Dif. eq., 7, (1971), 366–367. 45 #5523; E.t. 284–285.

[61.] Eigenfunction expansions for exterior boundary problems, Dif. eq., 7, (1971), 737–742. 44 # 2094; E.t. 565–569.

[62.] Approximate formulas for polarizability tensor and capacitances for bodies of an arbitrary shape. Radiofisika, 14, (1971), 613–620. 47 #1386.

[63.] Iterative methods for solving some heat transfer problems, Engin. Phys. Jour., 20, (1971), 936–937.

[64.] Electromagnetic wave scattering by small bodies of an arbitrary shape, Proc. of the 5-th all-union sympos. on wave diffraction, Trudy. Math. Inst. Steklova, Leningrad, 1971, 176–186.

[65.] On multidimensional integral equation with convolution kernel, Diff. eq. 7, (1971), 2234- 2239. 44 #7235; E.t. 1683–1687.

[66.] Calculation of the magnetization of thin films, Microelectronics 6, (1971), 65–68. (with Frolov).

[67.] Calculation of the scattering amplitude for electromagnetic wave scattering by small bodies of arbitrary shapes II. Radiofisika, 14, (1971), 1458- 1460.

[68.] Electromagnetic wave scattering by small bodies of an arbitrary shape and related topics, Proc. Intern. Sympos. URSI, Moscow, 1971, 536–540.

[69.] Calculation of thermal fields by means of iterative processes, Proc. 4-th all-union conference on heat and mass transfer. Minsk, 1972, 133–137.

[70.] Calculation of the capacitance of a parallelepiped, Electricity, 5, (1972), 90–91 (with M. Golubkova, V. Usoskin).

[71.] Simplified optimal differentiators, Radiotech.i Electron. 17, (1972), 1325–1328; E.t.1034–1037.

[72.] On exterior diffraction problems, Radiotech.i Electron, 7, (1972), 1362–1365. 51 #4864; e.t. 1064–1067.
[73.] On the skin-effect theory. J. of Techn. Phys., 42, (1972), 1316–1317.
[74.] Calculation of the quasistationary states in quantum mechanics, Doklady Acad. Sci. USSR, 204, (1972), 1071–1074. 56 #14326.
[75.] Calculation of the capacitance of a conductor placed in anisotropic inhomogeneous dielectric, Radiofisika, 15, (1972), 1268–1270. 47 #2284.
[76.] Remark on integral equations theory, Diff. Uravneniya, 8, (1972), 1517–1520. 47 #2284; E.t. 1177–1180.
[77.] On some class of integral equations, Diff. Uravneniya, 9, (1973), 931–941. 49 #5749; E.t. 706–713.
[78.] Iterative process to solve the third boundary problem, Diff. Uravneniya, 9, (1973), 2075–2079. 48 #6861.
[79.] Optimal harmonic synthesis of generalized Fourier series and integrals with randomly perturbed coefficients, Radiotechnika, 28, (1973), 44–49.
[80.] Eigenfunction expansion corresponding the discrete spectrum, Radiotech. i Electron., 18, (1973), 496–501. 50 #1641 E.t. 364–369.
[81.] Discrimination of the random fields in noises, Problems of information transmission, 9, N3, (1973), 22–35. 48 #13439.
[82.] Light scattering matrix for small particles of an arbitrary shape, Opt. and spectroscopy, 37, N1, (1974), 125–129.
[83.] Scalar scattering by the set of small bodies of an arbitrary shape, Radiofisika, 17, (1974), 1062–1068.
[84.] New methods of calculation of the static and quasistatic electromagnetic waves, Proc. of the Fifth Intern. sympos. "Radioelectronics-74" Sofia, 3, (1974), 1–8 (report 12).
[85.] Influence of the shape of cosmic particles on the scattering amplitude, Proc. all-union sympos. Interaction of cosmic dust with the atmosphere. Ashkhabad, 1974, 11–12.
[86.] Approximate solution of some integral equations of the first kind, Diff. eq. 11, (1975), 582–586. 51 #13613; E.t. 440–443.
[87.] Estimates of some functionals in quasistatic electrodynamics, Ukrain. Phys. Jour., 5, (1975), 534–543. 56 #14165.
[88.] Filtering and signal detection for random fields and vectorial processes, Proc. internat. confer. Prague, Sept. 1975, 45–59.
[89.] Existence uniqueness and stability of the periodic regimes in some nonlinear networks, Proc. of the third intern. sympos. on network theory, Split, Yugoslavia, Sept. 1975, 623–628.

[90.] Diffraction losses in open resonators. Opt. and spectroscopy, 40, N1, (1976), 160–163.

[91.] Boundary value problem with discontinuous boundary conditions, Diff. eq. 13, (1976), 931–933. 54 #10830.

[92.] Investigation of a class of integral equations, Doklady Acad. Sci. USSR, 230, (1976), 283–286. 54 #3341, 55 #1016.

[93.] Iterative process for calculation of the periodic and almost periodic oscillations in some nonlinear systems, Radiotech. i Electron., 21, (1976), 2429–2433.

[94.] Optimization of the resolution ability, Opt. and Spectroscopy, 42, N3, (1977), 540–545 (with Rodionov).

[95.] Wave scattering by small particles, Opt. and Spectroscopy, 43, N3, (1977), 523–531.

[96.] On simultaneous approximation of a function and its derivative by interpolation polynomials, Bull. Lond. Math. Soc. 9, (1977), 283–288.

[97.] New method of calculation of the stationary regimes in some nonlinear networks, Proc. of Intern. Conf. on computer-aided design of electron. and microwave systems. Hull, July, 1977.

[98.] A new class of nonstationary processes and fields and its applications, Proc. 10 all-union sympos. "Methods of representation and analysis of random processes and fields" Leningrad, 3, 1978, 40–43.

[99.] Existence of the periodic solutions to some nonlinear problems, Diff. Eq., 13, (1977), 1701–1708; E.t. 1186–1191, 57 #10530.

[100.] On eigenvalues of some integral equations, Diff. Eq., 15, (1978), 932–934; 58 #1528. E.t. 665–667.

[101.] On stability of control systems, Diff. Eq., 15, (1978), 1670–1677. E.t. 1188–1193.

[102.] Necessary and sufficient conditions for the validity of the limiting amplitude principle, Mathematics, Izv. vuzov, 5, (1978), 96–102.

[103.] Perturbations, preserving asymptotics of spectrum, Atti Ac. Naz. Lincei, ser. 8, vol. 64, fasc 1, Jan. (1978), p.30–31.

[104.] Investigation of a class of systems of integral equations, Proc. Intern. Congr. on appl. math., Weimar, DDR, 1978, 345–351.

[105.] Existence, uniqueness and stability of solutions to some nonlinear problems, Proc. Intern. Congr. on appl. math., Weimar, DDR, 1978, 352–356.

[106.] Existence in the large of solution of systems of nonlinear differential equations, Bull. Acad. Pol. Sci., 9–10, (1978), pp. 795–797.

[107.] Investigation of some classes of integral equations and their application, In collection "Abel inversion and its generalizations", edited by N. Preobrazhensky, Siberian Dep. of Acad. Sci. USSR, Novosibirsk 1978, pp. 120–179.

[108.] A uniqueness theorem for the Dirichlet problem, Sibir. Math. J. N6, (1978), 1421–1423.

[109.] Approximation by entire functions, Mathematics, Izv. vusov, 10, (1978), 72–76.

[110.] On an integral equation, Comptes Rendus Acad. Sci. Bulg. 32, N6, (1979), 715–717.

[111.] Iterative process for calculating periodic and almost periodic oscillations in some nonlinear systems, Radiotech. i Electronica, 24, N1, (1979), 190–191.

[112.] On uniqueness of harmonic coordinate systems in general relativity, Uspehi Math. Nauk. 34, N1, (1979), 239–240 (with Mishnaevskii P.).

[113.] On a class of integral equations, Math. Nachr., 92, (1979), 21–24.

[114.] On nonlinear equations with unbounded operators, Math. Nachr., 92, (1979), 13–20.

[115.] Linear filtering of some vectorial nonstationary random processes, Math. Nachr., 91,(1979), 269–280.

[116.] On spectrum of operator Schrödinger equation, Rev. Roum. Math. Pure et Appl. 25, (1980), 789–795.

[117.] Electromagnetic wave scattering by small bodies of arbitrary shapes, in the book: "Acoustic, electromagnetic and elastic scattering-Focus on T-matrix approach" Pergamon Press, N. Y. 1980. 537–546. (ed. V. Varadan).

[118.] **Theory and applications of some new classes of integral equations, Springer-Verlag, New York, 1980.**

[119.] Investigation of a class of systems of integral equations, J. Math. Anal. Appl., 76,(1980), 303–308.

[120.] Perturbation preserving asymptotics of spectrum of linear operators, J. Math. Anal. Appl., 76,(1980), 10–17.

[121.] Uniqueness theorem of use in general relativity, J. Math. Anal. Appl., 75,(1980), 58–65. (with Mishnaevskii P.)

[122.] On the quasistatic boundary value problem of electrodynamics, J. Math. Anal. Appl., 75, (1980), 300–305. (with C.L. Dolph)

[123.] Two sided estimates of the scattering amplitude at low energies, J. Math. Phys., 21, (1980), 308–310.

[124.] Analytical results in random fields filtering theory, Zeitschr. Angew. Math. Mech., 60, (1980), T 361-T 363.

[125.] Nonselfadjoint operators in diffraction and scattering, Math. Methods in Appl. Sci., 2, (1980), 327–346.

[126.] A new proof of absence of positive discrete spectrum of the Schrödinger operator, J. Math. Phys. 21, (1980), 2395–2397. (with B. A. Taylor)

[127.] Eigenfunction expansion for nonselfadjoint operators, Rev. Roum. Math. Pure Appl., 25, (1980), 797–809.

[128.] Theoretical and practical aspects of singularity and eigenmode expansion methods, IEEE A-P, 28, N6, (1980), 897–901.

[129.] Stationary regimes in passive nonlinear networks, in "Nonlinear Electromagnetics", Ed. P.L.E. Uslenghi, Acad. Press, N. Y., 1980, pp. 263–302.

[130.] A variational principle for resonances. J. Math. Phys. 21, (1980), 2052–53.

[131.] Existence uniqueness and stability of solutions to some nonlinear problems. Appl. Analysis, 11, (1981), 223–232.

[132.] Exterior boundary value problems as limits of interface problems, J. Math. Anal. Appl. 84, (1981), 256–263.

[133.] Stable solutions of some ill-posed problems, Math. Meth. in the appl. Sci. 3, (1981), 336–363.

[134.] Variational principles for spectrum of compact nonselfadjoint operators, J. Math. Anal. Appl. 80, (198l), 291–293.

[135.] On the basis property for the root vectors of some nonselfadjoint operators, Jour. Math. Anal. Appl. 80, (1981), 57–66.

[136.] On some properties of solutions of Helmholtz equation, J. Math. Phys., 22, (1981), 275–276.

[137.] Spectral properties of some nonselfadjoint operators, Bull, Am. Math. Soc., 5, N3, (1981), 313–315.

[138.] Existence, uniqueness, stability and calculation of the stationary regimes in some nonlinear systems, Proc. of the AMS Special session in Math. Physics, Boulder, CO, March 1980. Plenum Publish. Co., 1981, 337–342.

[139.] Nonselfadjoint operators in diffraction and scattering, Proc. of the AMS Special session in Math. Physics, Boulder, CO, March 1980. Plenum Publish. Co., 1981, 179–182. (with C. L. Dolph)

[140.] Stability in the large and calculation of the stationary regimes in a feedback nonlinear system. Proc. IEEE Intern. Sympos. on circuits and systems, (1981), 955–956. (with G.S. Ramm)

[141.] Spectral properties of some nonselfadjoint operators and some applications, in "Spectral theory of differential operators", Math. Studies, North Holland, Amsterdam, 1981, ed. I. Knowles and R. Lewis, 349–354.

[142.] On the singularity and eigenmode expansion methods, Electromagnetics, 1, N4, (1981), 385–394.

[143.] Electromagnetic wave scattering by small bodies, Nonlinear Vibr. Probl., 20, (1981), 109–120.

[144.] **Iterative methods for calculating static fields and wave scattering by small bodies, Springer Verlag, New York, 1982.**

[145.] Mathematical foundations of the singularity and eigenmode expansion methods. J. Math. Anal. Appl., 86, (1982), 562–591.

[146.] Asymptotic of resonant states, J. Math. Anal. Appl., 87, (1982), 323–331. (with P.A. Mishnaevskii)

[147.] Perturbations preserving asymptotic of spectrum with a remainder. Proc. Amer. Math. Soc., 85, N2, (1982), 209–212.

[148.] Variational principles for resonances II, J. Math. Phys., 23, N6, (1982), 1112–1114.

[149.] Perturbation of resonances. J. Math. Anal. Appl. 88, (1982), 1–7.

[150.] Convergence of the T-matrix approach to scattering theory, J. Math. Phys. 23, N6, (1982), 1123–1125.

[151.] Topics in scattering and spectral theory, Notices AMS, 29, (1982), 327–329.

[152.] Justification of the T-matrix approach, Proc. of the internat. IEEE sympos. on Antennas and Propagation, Albuquerque 1982, vol. 1, p. 13–14, 1982.

[153.] Convergence of the T-matrix approach in the potential scattering, J. Math. Phys. 23, (1982), 2408–2409.

[154.] Investigation of a class of nonlinear integral equations and calculation of passive nonlinear networks, Nonlin. Vibr., 21, (1983), 19–37.

[155.] Variational principles for eigenvalues of compact nonselfadjoint operators II, J. Math. Anal. Appl., 91, (1983), 30–38.

[156.] Eigenfunction expansions for some nonselfadjoint operators and the transport equation, J. Math. Anal. Appl., 92, (1983), 564–580.

[157.] Convergence of the T-matrix approach in scattering theory II, J. Math. Phys., 24, N 11, (1983), 2619- 2631. (with G. Kristensson and S. Ström)

[158.] Convergence of the T-matrix scheme, Proc. of the IEEE intern. symposium on antennas and propagation, May 1983, Houston.

[159.] Inverse scattering for geophysical problems, Phys. Letters, 99A, (1983), 258–260.
[160.] An inversion formula in scattering theory. Phys, Lett., 99A, (1983), 201–204.
[161.] On a property of the set of radiation patterns, J. Math. Anal. Appl., 98, (1984), 92–98.
[162.] A uniqueness theorem in scattering theory. Phys, Rev. Lett., 52, N1, (1984), 13.
[163.] Scattering by a penetrable body, J. Math. Phys., 25, N3, (1984), 469–471.
[164.] Representations of solutions to Helmholtz's equation, J. Math. Phys., 25, N4, (1984), 807–809.
[165.] Existence of infinitely many purely imaginary resonances in the problem of potential scattering, Phys. Lett., 101A, N4, (1984), 187–188.
[166.] A uniqueness theorem in scattering theory, Phys. Rev., 102A, N5-6, (1984), 218–219.
[167.] Description of the degree of nonuniqueness in inverse source problem, J. Math. Phys., 25, N6, (1984), 1791–1793.
[168.] Estimates of the derivatives of random functions, J. Math. Anal. Appl., 102, (1984), 244–250.
[169.] Remarks about inverse diffraction problem, J. Math. Phys., 25, N11, (1984), 2672–2674.
[170.] Inverse scattering for geophysical problems II, Inversion of acoustical data, J. Math. Phys., 25, N11, (1984), 3231–3234. (with A. Weglein)
[171.] Analytic theory of random fields estimation and filtering, Proc. of the intern sympos. on Mathematics in systems theory (Beer Sheva, 1983), Lecture notes in control and inform. sci. N58, Springer Verlag, 1984, 764–773.
[172.] On inverse diffraction problem, J. Math. Anal. Appl., 103, (1984), 139–147.
[173.] Inverse diffraction problem, Inverse methods in electromagnetic imaging, Reidel, Dordrecht, 1985, pp. 231–249. (ed. W. Boerner)
[174.] Limit of the spectra of the interior Neumann problems when a solid domain shrinks to a plane one, J. Math. Anal. Appl., 108, (1985), 107–112.
[175.] Extraction of resonances from transient fields, IEEE AP Trans., 33, (1985), 223–226.

[176.] Inverse scattering for geophysical problems III, On the velocity inversion problems of acoustics, Proc. Roy. Soc. Lond., A 399, (1985), 153–166. (with P. Martin)

[177.] Scattering amplitude and algorithm for solving the inverse scattering problem for a class of non-convex obstacles, Phys. Lett. A, 108A, (1985), 238–240. (with H. Alber)

[178.] Calculating resonances (natural frquencies) and extracting them from transient fields, J. Math. Phys., 26, N5, (1985) 1012–1020.

[179.] Inverse scattering for geophysical problems, Inverse Problems, 1, N2, (1985) 133–172.

[180.] Numerical solution of integral equations in a space of distributions, J. Math. Anal. Appl. 110, (1985), 384–390.

[181.] Estimates of the derivatives of random functions II, J. Math. Anal. Appl. 110, (1985), 429–435. (with T. Miller)

[182.] Wave scattering by small bodies, Reports Math. Phys. , 21, (1985), 69–77.

[183.] On a uniqueness theorem in inverse scattering by an infinite surface with a finite inhomogeneity, Opt. Comm., 59, (1985), 8–10. (with M. Fiddy).

[184.] On the limit amplitude principle for a layer, Jour. für die reine und angew. Math., 360, (1985), 19–46. (with P. Werner)

[185.] Offset measurements on a sphere at a fixed frequency do not determine the inhomogeneity uniquely, Inverse problems, 1, (1985), L35–L37.

[186.] Adjusting migration data for incompleteness: notes on Doeve's method, XX General conference, International Union for the scientific study of population, Florence, Italy, June 1985, pp. 1–44. Population index 52, N4, (1986). (with W. Doeve).

[187.] Inversion of the back scattering data and a problem of integral geometry. Phys. Lett. 113A, (1985), 172–176.

[188.] Inverse scattering in an absorptive medium. Inverse problems, 2, (1986), L5–L7.

[189.] Singularities of the inverses of Fredholm operators. Proc. of Roy. Soc. Edinburgh, 102A, (1986), 117–121.

[190.] **Scattering by obstacles, D. Reidel, Dordrecht, 1986, pp. 1–442.**

[191.] Inverse scattering for geophysical problems IV. Inversion of the induction logging measurements. Geophysical prospecting, 34, N3, (1986), 293–301.

[192.] Behavior of solutions to exterior boundary value problems at low frequencies, J. Math. Anal. Appl., 117, (1986), 561–569.
[193.] Scattering amplitude and algorithm for solving the inverse scattering problem for a class of non-convex obstacles, J. Math. Anal. Appl., 117, (1986), 570–597. (with H.D. Alber)
[194.] A geometrical inverse problem. Inverse problems, 2, (1986), L19–21.
[195.] Nonuniqueness of the solution to an inverse geophysical problem, Inverse problems, 2, (1986), L23–25.
[196.] On completeness of the products of harmonic functions, Proc. A.M.S., 99, (1986), 253–256.
[197.] Inverse scattering: asymptotic analysis, Inverse problems, 2, (1986), L43–46. (with H.D. Alber)
[198.] Inversion of the Laplace transform from the real axis, Inverse problems, 2, (1986), L55–59.
[199.] Inverse scattering for geophysical problems when the background is variable, J. Math. Phys., 27, (1986), 2687–2689.
[200.] An inverse problem for the Helmholtz equation in a semi-infinite medium, Inverse problems, 3, (1987), L19–22.
[201.] A method for solving inverse diffraction problems, Inverse problems, 3, (1987), L23–25.
[202.] Characterization of the low-frequency scattering data in the inverse problems of geophysics, Inverse problems, 3, (1987), L33–35.
[203.] Sufficient conditions for zero not to be an eigenvalue of the Schrödinger operator, J. Math Phys., 28, (1987), 1341–1343.
[204.] Optimal estimation from limited noisy data, J. Math. Anal. Appl., 125 (1987), 258–266.
[205.] Signal estimation from incomplete data, J. Math. Anal. Appl., 125 (1987), 267–271.
[206.] Analytic and numerical results in random fields estimation theory, Math. Reports of the Acad. of Sci., Canada, 9, (1987), 69–74.
[207.] Recovery of the potential from I-function, Math. Reports of the Acad. of Sci., Canada, 9, (1987), 177–182.
[208.] Example of a potential in one-dimensional scattering problem for which there are infinitely many purely imaginary resonances, Phys. Lett. A. 124, (1987), 313–319. (with B.A. Taylor)
[209.] Characterization of the scattering data in multidimensional inverse scattering problem, in the book: "Inverse Problems: An Interdisciplinary Study." Acad. Press, NY, 1987, 153–167. (Ed. P. Sabatier).

[210.] A characterization of the scattering data in 3D inverse scattering problem, Inverse problems, 3, (1987), L49–52. (with O. Weaver)

[211.] Necessary and sufficient conditions for a function to be the scattering amplitude corresponding to a reflecting obstacle, Inverse problems, 3, (1987), L53–57.

[212.] An inverse problem for Helmholtz's equation II, Inverse Problems, 3, (1987), L59–61.

[213.] Necessary and sufficient conditions on the scattering data for the potential to be in L^2, Inverse Problems, 3, (1987), L71–L76

[214.] Completeness of the products of solutions to PDE and uniqueness theorems in inverse scattering, Inverse problems, 3, (1987), L77–L82

[215.] Inverse problem for Helmholtz's equation, Intern J. of Math and Math Sci., 10, (1987), 825–27

[216.] Equisummability for linear operators in Banach spaces, Proc. of Roy. Soc. Edinburgh, 106A, (1987), 315–325. (with M. Kon, L. Raphael)

[217.] A uniqueness theorem for two-parameter inversion, Inverse Probl., 4, (1988), L7–10.

[218.] A uniqueness theorem for a boundary inverse problem, Inverse Probl., 4, (1988), L1–5.

[219.] Inverse scattering on half-line, J. Math. Anal. App. 133, 2, (1988), 543–572.

[220.] Multidimensional inverse problems and completeness of the products of solutions to PDE, J. Math. Anal. Appl. 134, 1, (1988), 211–253; 139, (1989) 302.

[221.] An inverse problem for biharmonic equation, Int. J. of Math. and Math. Sci., 11, (1988), 413–415.

[222.] Characterization of the scattering data in three dimensional inverse scattering problems. Proc. of the fifth intern. seminar on model optimization in explorational geophysics, editor A. Vogel, F. Vieweg & Sohn, Braunschweig, 1988, pp. 39–45.

[223.] A uniqueness theorem for an inverse problem, Appl. Math. Lett 1, (1988), 185–187.

[224.] Estimates for Green's functions, Proc. Amer. Math. Soc., 103, N3, (1988), 875- 881. (with L. Li)

[225.] Inversion of the acoustic well to well data, Appl Math. Letters, 1, (1988), 127–131, (with J. Harris).

[226.] A criterion for completeness of the set of scattering amplitudes, Phys. LettA. 129, (1988), 191–194.

[227.] Conditions for zero not to be an eigenvalue of the Schrödinger operator, J. Math. Phys. 29, (1988), 1431–1432.

[228.] Recovery of the potential from fixed energy scattering data, Inverse Problems, 4, (1988), 877–886; 5, (1989) 255.

[229.] A simple proof of uniqueness theorem in impedance tomography, Appl. Math. Lett., 1, N3, (1988), 287–290.

[230.] Numerical method for solving 3D inverse scattering problems, Appl. Math. Lett., 1, N4, (1988), 381–384.

[231.] Application of operator approximation to Fourier analysis, in the book "Constructive theory of functions", Proc. of the intern. conference on constructive theory of functions, Varna, May 24–31, (1987). Publishing House of Bulgar. Acad. of Sciences, Sofia, 1988, pp. 276–282 (with M. Kon and L. Raphael).

[232.] Error estimate for a quadrature formula for H^2 functions, Proc. of the 1987 Oberwolfach conference on numerical integration in: Numerical Integration III. Birkhauser, Basel, 1988, p. 199–201. (ed. H. Brass and G. Hammerlin).

[233.] Uniqueness theorems for multidimensional inverse problems with unbounded coefficients, J. Math. Anal. Appl. 136, (1988), 568–574.

[234.] Numerical method for solving 3D inverse problems of geophysics, J. Math. Anal. Appl., 136, (1988), 352–356.

[235.] Multidimensional inverse problems: Uniqueness theorems, Appl. Math. Lett., 1, N4, (1988), 377–380.

[236.] Multidimensional inverse scattering problems and completeness of the products of solutions to homogeneous PDE, Zeitschr. f. angew. Math. u. Mech., 69, (1989) N4, T13–T22.

[237.] Numerical method for solving 3D inverse problems with complete and incomplete data, In the book: "Wave Phenomena", Springer-Verlag, New York 1989, (eds. L. Lam and H. Morris), 34–43.

[238.] Numerical recovery of the 3D potential from fixed energy incomplete scattering data, Appl. Math. Lett., 2, N1, (1989), 101–104.

[239.] An inverse problem for Helmholtz's equation III, Appl. Math. Lett. 2, N1, (1989) 105–108.

[240.] Numerical recovery of the layered medium, J. of Comput. and Appl. Math. 25, N3, (1989), 267–276. (with P. Li)

[241.] Necessary and sufficient condition for a scattering amplitude to correspond to a spherically symmetric scatterer, Appl. Math. Let. 2, (1989), 263–265.

[242.] An inverse problem for Maxwell's equations, Phys. Let. A 138(1989), 459–462.

[243.] Electromagnetic inverse problem with surface measurements at low frequencies, Inverse Probl., 5, (1989), 1107–1116. (with E. Somersalo)

[243a] Electromagnetic inverse problems at low frequencies, Inverse methods in action (Montpellier, 1989), pp. 201–206, Inverse Probl. Theoret. Imaging, Springer, Berlin, 1990. (with E. Somersalo)

[244.] Necessary and sufficient condition on fixed energy scattering data for the potential to be spherically symmetric, Inverse Probl.5, (1989), L45–47. (with O. Weaver)

[245.] Property C and uniqueness theorems for multidimensional inverse spectral problem, Appl. Math. Lett., 3, (1990), 57–60.

[246.] **Random fields estimation theory, Longman Scientific and Wiley, New York, 1990.**

[247.] Stability of the numerical method for solving the 3D inverse scattering problem with fixed energy data, Inverse problems 6, (1990), L7–12.

[248.] Algorithmically verifiable characterization of the class of scattering amplitudes for small potentials, Appl. Math. Lett, 3, (1990), 61–65.

[249.] Is the Born approximation good for solving the inverse problem when the potential is small? J. Math. Anal. Appl., 147, (1990), 480–485.

[250.] Random fields estimation theory, Math. and Comput. Modelling 13, (1990), 87–100.

[251.] Dissipative Maxwell's equations at low frequencies, Math. Meth. in the Appl. Sci. 13, (1990), 305–322. (with O. Weaver, N. Weck, and K. Witsch)

[252.] Completeness of the products of solutions of PDE and inverse problems, Inverse Probl.6, (1990), 643–664.

[253.] Calculating singular integrals as an ill-posed problem, Numer. Math., 57, (1990) 139–145. (with van der Sluis)

[254.] Uniqueness result for inverse problem of geophysics I, Inverse Probl. 6, (1990), 635–642.

[255.] Uniqueness result for inverse problem of geophysics II, Appl. Math. Lett., 3, (1990), 103–105. (with G. Xie)

[256.] Uniqueness theorems for geophysical problems with incomplete surface data. Appl. Math. Lett.3, (1990), N4, 41–44.

[257.] An inverse problem for the wave equation, Math. Zeitschr., 206, (1991) 119–130. (with J. Sjostrand).

[258.] Numerical solution of some inverse problems of geophysics, Computers and Mathematics with Applications, 21, (1991), 75–80 (with Q. Zou)

[259.] Asymptotics of the solution to a singularly perturbed integral equation (with E.I. Shifrin), Appl. Math. Lett., 4, (1991), 67–70.

[260.] Symmetry properties for scattering amplitudes and applications to inverse problems, J. Math. Anal. Appl., 156, (1991), 333–340.

[261.] Property C and an inverse problem for a hyperbolic equation, J. Math. Anal. Appl., 156, (1991), 209–219. (with Rakesh)

[262.] Necessary and sufficient condition for a PDE to have property C, J. Math. Anal. Appl.156, (1991), 505–509.

[263.] A singular perturbation result and its application to mathematical ecology, Proc. AMS, 111, (1991), 1043–1050. (with L. Li)

[264.] Stability of the numerical method for solving 3D inverse scattering problem with fixed energy data, J.f.die reine und angew. Math, 414, (1991), 1–21.

[265.] Numerical solution of some integral equations in distributions, Comput. & Math with Appl. 21, (1991), 1–11. (with Peiqing Li)

[266.] Finding conductivity from boundary measurements, Comp.& Math. with Appl., 21, N8, (1991), 85–91

[267.] Exact inversion of fixed-energy data, in the book Mathematical and Numerical Aspects of Wave Propagation Phenomena, SIAM, Philadelphia, (1991), pp. 481–486

[268.] Justification of Fabrikant's method for solving mixed problems of potential theory, Comp. and Math. with Appl., 22, N6, (1991), 97–104. (with V. Fabrikant).

[269.] Property C and inverse problems, ICM-90 Satellite Conference Proceedings, Inverse Problems in Engineering Sciences, Proc. of a conference held in Osaka, Japan, Aug. 1990, Springer Verlag, New York, 1991, pp. 139–144.

[270.] Can a constant be a scattering amplitude? Phys. Lett., 154A, (1991), 35–37

[271.] Inversion of limited-angle tomographic data, Comp. and Math. with Applic., 22, 4/5, (1991), 101–112.

[272.] On 3D inverse scattering, Comp. and Math. with Appl., 22, 4/5, (1991), 1–26 (with O. Weaver)

[273.] On a problem of integral geometry, Comp. and Math. with Appl., 22, 4/5, (1991), 113–118.

[274.] Solution of some integral equations arising in integral geometry, Appl. Math. Lett., 4, (1991), 177–181

[275.] An approximation problem, Appl. Math. Lett., 4, N5, (1991), 75–77.

[276.] Uniqueness theorem for a Goursat-Darboux type problem, Soviet Math. Doklady, 321, 1, (1991), 19–22. (with P. Mishnaevskii)

[277.] Inversion of the Radon transform with incomplete data, Math. Methods in the Appl. Sci., 15, N3, (1992), 159–166.

[278.] **Multidimensional inverse scattering problems, Longman/ Wiley, New York, 1992, pp. 1–385.**

[279.] Stability of the solution to inverse scattering problem with exact data, Appl. Math. Lett., 5, 1, (1992), 91–94

[280.] Inversion of incomplete Radon transform (with A. Katsevich) Appl. Math. Lett., 5, N2, (1992), 41–46.

[281.] Inversion of limited-angle tomographic data II, Appl. Math. Lett., 5, N2, (1992), 47–49.

[282.] Inverse scattering problem with fixed-energy data, Appl. Math. Lett., 5, N4, (1992), 63–67.

[283.] A multidimensional Ambartsumian's theorem, Appl. Math. Lett, 5, N5, (1992), 87–88. (with P. Stefanov)

[284.] Uniqueness of the solution to a Goursat problem, Appl. Math. Lett., 5, N6, (1992), 11–13. (with P. Mishnaevskii)

[285.] Stability estimates in inverse scattering, Acta Appl. Math., 28, N1, (1992), 1–42.

[286.] FBP method for inversion of incomplete tomographic data, Appl. Math. Lett., 5, N3, (1992), 77–80. (with A. Katsevich)

[287.] Numerical solution of 3D inverse scattering problems with noisy discrete fixed-energy data, Appl. Math. Lett., 5, N6, (1992), 15–18.

[288.] Inversion of incomplete cone-beam data, Appl. Math. Lett., 5, N4, (1992), 91–94. (with A. Zaslavsky)

[289.] Stability of the inversion of 3D fixed-frequency data, J. Math. Anal. Appl., 169, N2(1992), 329–349.

[290.] Stability of the solution to 3D fixed-energy inverse scattering problem, J. Math. Anal. Appl., 170, N1 (1992), 1–15.

[291.] Singularities of the Radon transform, Bull. Am. Math. Soc., 25, N1, (1993), 109–115. (with A. Zaslavsky)

[292.] Asymptotics of the solutions to singularly perturbed integral equations II., J. Math. Anal. Appl., 178, N2, (1993), 322–343. (with E. Shifrin).

[293.] Multidimensional algorithm for finding discontinuities of functions from noisy data. Math. Comp. Modelling, 18, N1, (1993), 89–108. (with A. Katsevich).

[294.] A criterion for property C, J. Math. Anal. Appl., 177, N2, (1993), 491–495. (with D. Markushevich)

[295.] Consistency of rank tests against some general alternatives, Math. & Comput. Modelling, 18, N12, (1993), 49–56. (with A. Katsevich)

[296.] Uniqueness and inversion of cone-beam data, Appl. Math. Lett., 6, N1 (1993), 35–38.

[297.] Property C with constraints and inverse spectral problems with incomplete data, J. Math. Anal. Appl., 180, N1, (1993), 239–244

[298.] An inverse problem for the heat equation, Proc. Roy. Soc. Edinburgh, 123, N6, (1993), 973–976.

[299.] Reconstructing singularities of a function given its Radon transform, Math. Comp. Modelling, 18, N1, (1993), 109–138, (with A. Zaslavsky).

[300.] Stable calculation of the Legendre transform of noisy data, J. Math. Anal. Appl. 178, N2, (1993), 592–602 (with A. Steinberg and A. Zaslavsky).

[301.] Approximation by the scattering solutions and applications to inverse scattering, Math. Comp. Modelling, 18, N1, (1993), 47–56.

[302.] Fixed-energy inverse scattering for non-compactly supported potentials, Math. Comp. Modelling, 18, N1, (1993), 57–64. (with P. Stefanov).

[303.] Asymptotics of the Fourier transform of piecewise-smooth functions, Comptes Rendus Acad. Sci. Paris, 316, ser. 1, (1993), 541–545. (with A. Zaslavsky).

[304.] Property C with constraints and inverse problems, J. of Inverse and Ill-Posed Problems, 1, N3 (1993), 227–230.

[305.] Scattering amplitude is not a finite-rank kernel, J. of Inverse and Ill-Posed Problems, 1, N4, (1993), 349–354. (with P. Stefanov)

[306.] Inverse scattering at fixed energy for exponentially decreasing potentials, Proc. of the Lapland conference on inverse problems (with P. Stefanov). Lecture notes in Phys, N422, Springer-Verlag, 1993, 189–192.

[307.] Algorithm for solving 3D inverse scattering problems with noisy discrete fixed-energy data, Proceedings of ICES-92 conference on inverse problems. In the book: Inverse Problems, Atlanta Technology Publications, Atlanta, Georgia, (1993), pp. 70–74.

[308.] Property C and applications. Math. Comp. Modelling, 18, N1, (1993), 1–4.

[309.] Scattering amplitude is not a finite rank kernel in the basis of spherical harmonics, Appl. Math. Lett., 6, N5, (1993), 89–92.

[310.] New method for proving uniqueness theorems for obstacle inverse scattering problems, Appl. Math. Lett., 6, N6, (1993), 19–22.

[311.] Scattering amplitude as a function of the obstacle, Appl. Math. Lett., 6, N5, (1993), 85–87.

[312.] Inverse geophysical problems for some non-compactly supported inhomogeneities, Appl. Math. Lett., 6, N6, (1993), 15–18. (with D. Ghosh Roy)

[313.] **Multidimensional inverse scattering problems, Mir Publishers, Moscow, 1994, pp. 1–496. (Russian translation of the expanded monograph 278).**

[314.] A numerical approach to 3D inverse scattering problems, Appl. Math. Lett.7, N2, (1994), 57–61.

[315.] X-ray transform, the Legendre transform and envelopes, J. Math. Anal. Appl., 183, N3, (1994), 528–546. (with A. Zaslavsky)

[316.] Nonparametric estimation of the singularities of a signal from noisy measurements, Proc. AMS, 120, N8, (1994), 1121–1134. (with A. Katsevich)

[317.] Numerical method for solving inverse scattering problems, Doklady of Russian Acad. of Sci., 337, N1, (1994), 20–22.

[318.] Low frequency inversion of surface data in a finite-depth ocean, Appl. Math. Lett, 7, N1, (1994), 11–14 (with L. Couchman & D. Ghosh Roy).

[319.] Multidimensional inverse scattering: solved and unsolved problems, Proc. Intern. Confer. on Dynamical Syst. and Applic., Vol. 1, Atlanta, (1994), pp. 287–296. (Eds. G. Ladde and M. Sambandham)

[320.] Inversion of fixed-frequency surface data for layered medium, J. of Inverse and Ill-Posed Problems, 2, N3, (1994), 263–268

[321.] Stability estimate in scattering theory and its application to mesoscopic systems and quantum chaos, Physics A, 27, N18, (1994), 6157–6166. (with G. Berman)

[322.] A method for finding discontinuities of functions from tomographic data, Proc. AMS-SIAM summer seminar on the mathematics of tomography, impedance imaging and integral geometry, Lectures in Appl. Math., Vol. 30 (1994), pp. 115–123. (with A.I. Katsevich)

[323.] Uniqueness theorems for Goursat-type problems, J. Diff. Eq., 112, N1, (1994), 250–255. (with P. Mishnaevskii)

[324.] Modeling of the ejection process, Math. and Comp. Modelling, 20, N1, (1994), 95–102. (with I. Kaleps)

[325.] Stability of the solution to inverse obstacle scattering problem, J. Inverse and Ill-Posed Problems, 2, N3, (1994), 269–275.

[326.] Optimal local tomography formulas, PanAmer. Math. Journ., 4, N4, (1994), 125–127.

[327.] Stability of the solution to 3D inverse scattering problems with fixed-energy data. Proc. ASME Nov.6–11, 1994, meeting. Inverse problems in mechanics, AMD-Vol 186, pp. 99–102.

[328.] Mathematical results in signal and image processing, Doklady of Russian Acad. Sci., 339, N1, (1994), 11–14. (with A. Katsevich)

[329.] Stability estimates for obstacle scattering, J. Math. Anal. Appl. 188, N3, (1994), 743–751.

[330.] Examples of nonuniqueness for an inverse problems of geophysics, Appl. Math. Lett., 8, N4, (1995), 87–90.

[331.] Finding discontinuities from tomographic data, Proc. Amer. Math. Soc., 123, N8, (1995), 2499–2505.

[332.] Property C with constraints, Compt. Rendus, Paris, ser 1,321, N 11, (1995), 1413–1417.

[333.] Asymptotics of the solutions to singularly perturbed multidimensional integral equations, J. Math. Anal. Appl., 190, N3, (1995), 667–677. (with E. Shifrin)

[334.] Continuous dependence of the scattering amplitude on the surface of an obstacle, Math. Methods in the Appl. Sci., 18, (1995), 121–126.

[335.] An inverse problem for multiple scattering of fast charged particles in mesoscopic medium, Phys. Rev. B, 51, N4, (1995), 2406–2409. (with G. Berman).

[336.] Asymptotics of PDO on discontinuous functions near singular support, Appl. Analysis, 58, N3–4, (1995), 383–390. (with A. Katsevich).

[337.] New methods for finding values of the jumps of a function from its local tomographic data, Inverse Problems, 11, N 5, (1995) 1005–1023. (with A. Katsevich)

[338.] The Radon transform is an isomorhism between $L^2(B)$ and $H_e(Z_a)$, Appl. Math. Lett., 8, N1, (1995), 25–29.

[339.] Finding singular support of a function from its tomographic data, Proc. Japan Acad., Math. Sci., 71, N3, (1995), 62–67. (with A. Katsevich)

[340.] A formula for inversion of boundary data, J. of Inverse and Ill-Posed Problems, 3, N5, (1995), 411–415.

[341.] Radon transform on distributions, Proc. Japan Acad., ser A, 71, N9, (1995), 202–206.

[342.] Inverse geophysical and potential scattering on a small body, in the book: Experimental and Numerical Methods for Solving Ill-Posed Inverse Problems: Medical and Nonmedical Applications, vol. SPIE-2570, (1995), 151–162. (with A. Katsevich)

[343.] Uniqueness theorems for inverse obstacle scattering problems in Lipschitz domains, Applic. Analysis, 59, (1995), 377–383.

[344.] Inversion of cone-beam data and helical tomography, Jour. of Inverse and Ill-Posed Probl. 3, N6, (1995), 429–445. (with V. Faber and A. Katsevich)

[345.] Consistency of rank test against general alternatives of change points (surfaces) and continuous trend, Acta Appl. Math., 42, N2, (1996), 105–137. (with A. Katsevich)

[346.] Finding jumps of a function using local tomography, PanAmer. Math. Jour., 6, N2, (1996), 1–21. (with A. Katsevich)

[347.] Approximate inverse geophysical scattering on a small body, SIAM J. Appl. Math., 56, N1, (1996), 192–218. (with A. Katsevich)

[348.] **The Radon Transform and Local Tomography, CRC Press, Boca Raton 1996, pp. 1–503.** (with A. Katsevich)

[349.] Estimates from below for Lebesgue constants, J. Fourier Anal. and Appl. 2, N3, (1996), 287–301. (with E. Liflyand and A. Zaslavsky)

[350.] Finding potential from the fixed-energy scattering data via D-N map, J. of Inverse and Ill-Posed Problems, 4, N2,(1996), 145–152.

[351.] Completeness and non-completeness results for the set of products of solutions to differential equations, Applicable Analysis, 60, (1996), 241–249. (with G. Porru)

[352.] The scattering problem analyzed by means of an integral equation of the first kind, J. Math. Anal. Appl., 201, (1996), 324–327.

[353.] Necessary and sufficient conditions for a PDO to be a local tomography operator, Comptes Rend Acad Sci, Paris, 332, N7, (1996), 613–618.

[354.] Inversion formulas for the backprojection operator in tomography, Proc. Amer. Math. Soc., 124, N2, (1996), 567–577.

[355.] Approximate solution to inverse scattering problem for potentials with small support, Math. Meth. in the Appl. Sci., 19, (1996), 1121–1134. (with A. Katsevich)

[356.] Pseudolocal tomography, SIAM J. Appl. Math., 56, N1, (1996), 167–191. (with A. Katsevich)

[357.] Existence and uniqueness of scattering solutions in non-smooth domains, J. Math. Anal. Appl., 201, (1996), 329–338. (with A. Ruiz)

[358.] Minimization of the total radiation from an obstacle by a control function on a part of the boundary, Jour. of Inverse and Ill-posed Prob., 4, N6, (1996), 531–534.

[359.] Property C with constraints and PDE, Proc. of the Japan Acad., 72A, N. 10, (1996), 235–237.

[360.] **Random fields estimation theory, MIR, Moscow, 1996, pp. 1–352** (expanded Russian edition of monograph 246).

[361.] Small body approximation and efficient algorithm for solving inverse problems, Advances in Optic. Imaging and Photon migration, Opt. Soc. of America, AWD4, (1996), pp. 1–4. (with A. Katsevich and J. George)

[362.] New methods for finding discontinuities of functions from local tomographic data, Jour. of Inverse and Ill-Posed Problems, 5, N2, (1997), 165–174.

[363.] The Pompeiu problem, Applicable Analysis, 64, N1–2, (1997), 19–26.

[364.] Finding small objects from tomographic data, Inverse problems, 13, (1997), 1239–1246 (with L. Desbat)

[365.] Multidimensional inverse scattering with fixed-energy data, in the book "Quantum Inversion", Lecture Notes in physics vol. 488, pp. 373–384, Springer Verlag, Berlin, 1997 (ed. B. Apagyi).

[366.] New approach to scattering in irregular waveguides, Math. Sci. Research Hot-Line, 1, N3, (1997), 1–2.

[367.] Theory of ground-penetrating radars, Jour. of Inverse and Ill-Posed Problems, 5, N4, (1997), 377–384. (with A. Shcheprov)

[368.] A method for finding small inhomogeneities from surface data, Math. Sci. Research Hot-Line, 1, N10, (1997), 40–42.

[369.] Formula for the radius of the support of the potential in terms of the scattering data, Jour. of Phys. A, 31, N1, (1998), L39–L44. (with J.H. Arredondo and B.G. Izquierdo)

[370.] Scattering by obstacles in acoustic waveguides, In the book: Spectral and scattering theory, Plenum publishers, New York, 1998 (ed. A.G. Ramm), pp. 89–110. (with G. Makrakis)

[371.] Fundamental solutions to elliptic equations with discontinuous senior coefficients and an inequality for these solutions. Math. Ineq. and Applic., 1, N1, (1998), 99–104.

[372.] Inverse acoustic scattering by layered obstacles (with C. Athanasiadis and I. Stratis), In the book: Inverse problems, tomography and image processing, Plenum Publishers, New York, 1998, pp. 1–8. (editor A.G. Ramm).

[373.] Recovery of compactly supported spherically symmetric potentials from the phase shift of s-wave, In the book: Spectral and scattering theory, Plenum publishers, New York, 1998 (ed. A.G. Ramm), pp. 111–130.

[374.] Inequalities for norms of some integral operators Math. Ineq. and Applic. 1, N2, (1998), 259–265.

[375.] Recovery of a quarkonium system from experimental data, Jour. of Phys. A, 31, N15, (1998), L295-L299.

[376.] Compactly supported spherically symmetric potentials are uniquely determined by the phase shift of s-wave, Phys. Lett. A, 242, N4–5, (1998), 215–219.

[377.] On Saitoh's characterization of the range of linear transforms, In the book: Inverse problems, tomography and image processing, Plenum publishers, New York, 1998, (ed. A.G. Ramm) pp. 125–128.

[378.] **Spectral and scattering theory, Plenum publishers, New York, 1998** (editor A.G. Ramm)

[379.] **Inverse problems, tomography and image processing, Plenum publishers, New York, 1998** (editor A.G. Ramm)

[380.] Theory of ground-penetrating radars II, Jour of Inverse and Ill-Posed Probl., 6, N6, (1998), 619–624.

[381.] On the theory of reproducing kernel Hilbert spaces, Jour. of Inverse and Ill-Posed Problems, 6, N5, (1998), 515–520.

[382.] Necessary and sufficient condition for a domain, which fails to have Pompeiu property, to be a ball, Jour of Inverse and Ill-Posed Probl., 6, N2, (1998), 165–171.

[383.] Inequality for the minimal eigenvalue of the Dirichlet Laplacian in an annulus, Math. Inequalities and Applic. 1, N4, (1998), 559–563. (with P.N. Shivakumar)

[384.] A new approach to the inverse scattering and spectral problems for the Sturm-Liouville equation, Ann. der Phys., 7, N4, (1998), 321–338.

[385.] Calculation of waves scattered in irregular waveguides, Proc. of the third international seminar on "Direct and inverse problems of electromagnetic and acoustic wave theory: DIPED-98 pp. 57–61. (with N. Voitovich, Yu. Topolyuk and N. Zdeoruk)

[386.] Continuous analog of Gauss-Newton method, Math. Models and Methods in Appl. Sci., 9, N3, (1999), 463–474. (with R. Airapetyan and A. Smirnova)

[387.] Property C for ODE and applications to inverse scattering, Zeit. fuer Angew. Analysis, 18, N2, (1999), 331–348.

[388.] A numerical method for solving nonlinear ill-posed problems, Numerical Funct. Anal. and Optimiz., 20, N3, (1999), 317–332. (with A.B. Smirnova)

[389.] A numerical method for some nonlinear problems, Math. Models and Meth. in Appl. Sci., 9, N2, (1999), 325–335.

[390.] Inequalities for brachistohrone, Math. Ineq. and Applic., 2, N1, (1999), 135–140.

[391.] Inverse problem for an inhomogeneous Schrödinger equation, Jour. Math. Phys, 40, N8, (1999), 3876–3880.

[392.] Example of two different potentials which have practically the same fixed-energy phase shifts, Phys. Lett A, 254, N3–4, (1999), 141–148. (with R. Airapetyan and A. Smirnova).

[393.] Inverse scattering problem with part of the fixed-energy phase shifts, Comm. Math. Phys. 207, N1, (1999), 231–247.

[394.] An approximate method for solving inverse scattering problem with fixed-energy data, Jour. of Inverse and Ill-Posed Problems, 7, N6, (1999), 561–571. (with W. Scheid)

[395.] Finding small inhomogeneities from scattering data, Jour. of Inverse and Ill-Posed Problems, 8, N2, (2000), 205–210.

[396.] An inverse problem for point inhomogeneities, Methods of Functional Analysis and Topology, 6, N2, (2000), 1–12. (with F. Gesztesy)

[397.] Inequalities for the derivatives, Math. Ineq. and Appl., 3, N1, (2000), 129–132.

[398.] The ground-penetrating radar problem III Jour. of Inverse and Ill-Posed Problems, 8, N1, (2000), 23–31.

[399.] Application of the hybrid stochastic-deterministic minimization method to a surface data inverse scattering problem, In the book "Operator Theory and its Applications", Amer. Math. Soc., Fields Institute Communications vol. 25, Providence, RI, 2000, pp. 293–304. (editors A.G. Ramm, P.N. Shivakumar, A.V. Strauss). (with S. Gutman)

[400.] A numerical method for solving the inverse scattering problem with fixed-energy phase shifts, Jour. of Inverse and Ill-Posed Problems, 8, N3, (2000), 307–322. (with A. Smirnova)

[401.] Dynamical systems and discrete methods for solving nonlinear ill-posed problems, Appl. Math. Reviews, vol. 1, Ed. G. Anastassiou, World Sci. Publishers, 2000, pp. 491–536. isbn 981–02-4339–1. (with R. Airapetyan)

[402.] Property C for ODE and applications to inverse problems, in the book "Operator Theory and Its Applications", Amer. Math. Soc., Fields Institute Communications vol. 25,(2000), pp. 15–75, Providence, RI. (editors A.G. Ramm, P.N. Shivakumar, A.V. Strauss).

[403.] Continuous methods for solving nonlinear ill-posed problems, In the book "Operator theory and applications", Amer. Math. Soc., Fields Institute Communications, Providence,RI, 2000, pp. 111–138. (editors A.G. Ramm, P.N. Shivakumar, A.V. Strauss). (with R. Airapetyan, A. Smirnova)

[404.] A new approach to inverse spectral theory III. Short range potentials, J. d'Analyse Math., 80, (2000), 319–334. (with B. Simon)

[405.] Krein's method in inverse scattering, in the book "Operator Theory and Its Applications", Amer. Math. Soc., Fields Institute Communications vol. 25, pp. 441–456, Providence, RI, 2000. (editors A.G. Ramm, P.N. Shivakumar, A.V. Strauss),

[406.] Justification of the limiting absorption principle in R^2, in the book "Operator Theory and Its Applications", Amer. Math. Soc., Fields Institute Communications vol. 25, pp. 433–440, Providence, RI, 2000. (editors A.G. Ramm, P.N. Shivakumar, A.V. Strauss).

[407.] **Operator Theory and Its Applications, Amer. Math. Soc., Fields Institute Communications vol. 25, Providence RI, 2000, pp. 1–600.** (editors A.G. Ramm, P.N. Shivakumar, A.V. Strauss).
isbn 0–8218-1990–9

[408.] Numerical inversion of the Laplace transform from the real axis, J. Math. Anal. Appl., 248, (2000), 572–587. (with R. Airapetyan)

[409.] Existence and uniqueness of the scattering solutions in the exterior of rough domains, in the book "Operator Theory and Its Applications", Amer. Math. Soc., Fields Institute Communications vol. 25, pp. 457–472, Providence, RI, 2000. (with M. Sammartino) (editors A.G. Ramm, P.N. Shivakumar, A.V. Strauss).

[410.] A uniqueness result for the inverse transmission problem, Internat. Jour. of Appl. Math., 2, N5, (2000), 625–634. (with P. Pang and G. Yan)

[411.] Numerical implementation of the cross section method for irregular waveguides, Radiophysics and radioastronomy, 5, N3, (2000), 274–283. (with N. Voitovich, O. Zamorska)

[412.] A non-overdetermined inverse problem of finding the potential from the spectral function, IJDEA (Intern. J. of Diff. Eq. and Appl.), 3, N1, (2001), 15–29.

[413.] An inverse problem of ocean acoustics, Jour. of Inverse and Ill-Posed Probl., 9, N1, (2001), 95–102.

[414.] Singularities of the Radon transform, Applic. Analysis., 79, N3–4, (2001), 351–379. (with R. Airapetyan)

[415.] On stable numerical differentiation, Mathem. of Computation, 70, (2001), 1131–1153. (with A. Smirnova)

[416.] Embedding operators for rough domains, Math. Ineq. and Applic., 4, N1, (2001), 127–141. (with V. Gol'dshtein)

[417.] Some identification problems for integro-differential operator equations, Nonlinear Functional Analysis and Applic., 6, N1, (2001), 107–123. (with A. Lorenzi)

[418.] Linear ill-posed problems and dynamical systems, Jour. Math. Anal. Appl., 258, N1, (2001), 448–456.

[419.] A simple proof of the Fredholm alternative and a characterization of the Fredholm operators, Amer. Math. Monthly, 108, N 9, (2001), 855–860.

[420.] New proof of Weyl's theorem, IJDEA (Intern. J of Diff Eq. and Appl), 3, N1, (2001), 31–37.

[421.] Piecewise-constant positive potentials with practically the same fixed-energy phase shifts, Applicable Analysis, 78, N1–2, (2001), 207–217. (with S. Gutman)

[422.] An inverse problem for the heat equation, Jour. of Math. Anal. Appl., 264, N2, (2001), 691–697.

[423.] An inverse problem for an abstract evolution equation, Applic. Analysis, 79, N3–4, (2001), 475–482. (with S. Koshkin)

[424.] Reconstruction of the potential from I-function, Jour. of Inverse and Ill-Posed Probl., 10, N4, (2002), 385–395.

[425.] Stability of solutions to inverse scattering problems with fixed-energy data, Milan Journ of Math., 70, (2002), 97–161.

[426.] Continuous regularized Gauss-Newton-type algorithm for nonlinear ill-posed equations with simultaneous updates of inverse derivative, Intern. Jour. of Pure and Appl Math., 2, N1, (2002), 23–34. (with A.B. Smirnova)

[427.] Stable identification of piecewise-constant potentials from fixed-energy phase shifts, Jour. of Inverse and Ill-Posed Probl., 10, N4, (2002), 345–360. (with S. Gutman)

[428.] Regularization of ill-posed problems with unbounded operators, J. Math. Anal. Appl., 271, (2002), 547–550.

[429.] Numerically efficient version of the T-matrix method, Applic. Analysis, 80, N3, (2002), 385–393.

[430.] Modified Rayleigh Conjecture and applications, Jour. Phys. A, 35, (2002), L357–361.

[431.] Analysis of the Newton-Sabatier scheme for inverting fixed-energy phase shifts, Applic. Analysis, 81, N4, (2002), 965–975.

[432.] Estimation of random fields, Theory of Probability and Math. Statistics, 66, (2002), 95–108. Translation: Theory Probab. Math. Statist. No. 66 (2003), 105–118.

[433.] A counterexample to a uniqueness result, Applic. Analysis, 81, N4, (2002), 833–836.

[434.] Recovery of small inhomogeneities from partial boundary measurements, Compt. Rendus Acad. Sci. Paris, ser IIb, 330, N3, (2002), 199–205. (erratum C.R. Mechanique, 330 (2002), 601). (with H. Ammari)

[435.] Inverse scattering by the stability index method, Jour. of Inverse and Ill-Posed Probl., 10, N5, (2002), 487–502. (with S. Gutman and W. Scheid)

[436.] Convergence rates of the continuous regularized Gauss-Newton method, Jour. Inv. Ill-Posed Probl., 10, N3, (2002), 261–280. (with B. Kaltenbacher and A. Neubauer)

[437.] Numerical implementation of the MRC method, J. Phys A., 35, (2002), 8065–8074. (with S. Gutman)

[438.] An inverse problem for the heat equation II, Applic. Analysis, 81, N4, (2002), 929–937.

[439.] Acceleration of convergence of a continuous analog of the Newton method, Applic. Analysis, 81, N4, (2002), 1001–1004.

[440.] Injectivity of the spherical means operator, Compt. Rend. Acad Sci., Paris, Ser I, 335, N12, (2002), 1033–1038.

[441.] Stable numerical differentiation: when is it possible? Jour. Korean SIAM, 7, N1, (2003), 47–61. (with A. Smirnova)

[442.] Analytical solution of a new class of integral equations, Diff. Integral Eqs, 16, N2, (2003), 231–240.

[443.] Continuous modified Newton's-type method for nonlinear operator equations, Ann. di Mat. Pure Appl, 182, N1, (2003), 37–52. (with A. Smirnova, A. Favini)

[444.] On a new notion of regularizer, J. Phys A, 36 (2003), 2191–2195.

[445.] Comments on the letter of P. Sabatier, http://arXiv.org/abs/math-ph/0308025, PaperId: math-ph/0308025.

[446.] Optimal with respect to accuracy algorithms for calculation of multidimensional weakly singular integrals and applications to calculation of capacitances of conductors of arbitrary shapes, Acta Applicandae Math, 79, N3, (2003), 281–326. (with I. Boikov)

[447.] On deconvolution methods, Internat. Jour. of Engin. Sci., 41, N1, (2003), 31–43. (with A. Galstian)

[448.] Inequalities for the transformation operators and applications, JIPAM (Jour. of Inequalities in Pure and Appl. Math.) 4, N4, (2003), pp. 1–9. (paper 69).

[449.] On the discrepancy principle, Nonlinear Functional Anal. and Applic., 8, N2, (2003), 307–312.

[450.] Equations for the self-consistent field in random medium, Phys. Lett. A, 312, N3-4, (2003), 256–261.

[451.] A characterization of unbounded Fredholm operators, Cubo a Mathem. Journ., 5, N3, (2003), 91–95.

[452.] Global convergence for ill-posed equations with monotone operators: the dynamical systems method, J. Phys A, 36, (2003), L249–L254.

[453.] Support function method for inverse obstacle scattering problems, In the book "Acoustics, mechanics and related topics of mathematical analysis", World Scientific, New Jersey, 2003, (ed. A. Wirgin), pp. 178–184. (with S. Gutman)

[454.] Dynamical systems method for solving nonlinear operator equations, International Jour. of Applied Math. Sci., 1, N1, (2004), 97–110.

[455.] Explanation of Feynman's paradox concerning low-pass filters, International Jour. of Applied Math. Sci., 1, N1, (2004), 111–116 (with O.L. Weaver)

[456.] Dynamical systems method for solving operator equations, Communic. in Nonlinear Sci. and Numer. Simulation, 9, N2, (2004), 383–402.

[457.] Inequalities for solutions to some nonlinear equations, Nonlinear Functional Anal. and Applic., 9, N2, (2004), 233–243.

[458.] Optimization methods in direct and inverse scattering, The optimization research bridge, No 13, 2004, pp. 1–4.

http://www.ballarat.edu.au/ard/itms/CIAO/ORBNewsletter/ (with S. Gutman)

[459.] Inverse scattering with fixed-energy data, Jour. of Indones. Math. Soc., 10, N1, (2004), 53–62.

[460.] One-dimensional inverse scattering and spectral problems, Cubo a Mathem. Journ., 6, N1, (2004), 313–426.

[461.] Modified Rayleigh Conjecture for scattering by periodic structures, International Jour. of Applied Math. Sci., 1, N1, (2004), 55–66. (with S. Gutman)

[462.] Continuity of solutions to operator equations with respect to a parameter, Internat. Jour. of Pure and Appl. Math. Sci., 1, N1, (2004), 1–5.

[463.] Numerical solution of obstacle scattering problems, Internat. Jour. of Appl. Math and Mech., 1, (2005), 1–32. (with S. Gutman)

[464.] An essay on some problems of approximation theory, In the book: Ten Mathematical Essays on Approximation in Analysis and Topology, Elsevier, Boston, 2005, pp. 245–262 (Eds J. Perrera, J. Lopez-Gomez, F. Ruiz del Portal), isbn 0–444-51861–4

[465.] On deconvolution problems: numerical aspects, Jour. Comp. Appl. Math., 176, N2, (2005), 445–460. (with A. B. Smirnova)

[466.] Dynamical systems method and surjectivity of nonlinear maps, Communic. in Nonlinear Sci. and Numer. Simulation, 10, N8, (2005), 931–934.

[467.] Analysis of a linear sampling method for identification of obstacles, Acta Appl. Math. Sinica, 21, N3, (2005), 399–404. (with S. Gutman)

[468.] Symmetry problems in the elasticity theory problem for plane cracks of normal rapture, Prikl. Math. Mech., 69, No 1., (2005), 146–154. English translation: Journ of Appl. Math. and Mechan., 69, (2005), 127–134. (with E. Shifrin)

[469.] DSM for ill-posed equations with monotone operators, Comm. in Nonlinear Sci. and Numer. Simulation, 10, N8, (2005), 935–940.

[470.] **Inverse problems, Springer, New York,** 2005.

[471.] Inequalities for the derivatives and stable differentiation of piecewise-smooth discontinuous functions, Math. Ineq and Applic., 8, N1, (2005), 169–172.

[472.] Necessary and sufficient conditions for compactness of the embedding operator, JIPAM (Jour. of Ineq. in Pure and Appl. Math), 6, N5, Article 130, (2005).

[473.] Determination of the shape of the ear channel, Math. Sci. Res. Journ., 9(6)(2005), 139–141.

[474.] Discrepancy principle for the dynamical systems method, Communic. in Nonlinear Sci. and Numer. Simulation, 10, N1, (2005), 95–101

[475.] Modified Rayleigh Conjecture method for multidimensional obstacle scattering problems, Numer. Funct. Anal. and Optimization, 26, N2, (2005), 69–80. (with S. Gutman)

[476.] **Wave scattering by small bodies of arbitrary shapes**, World Sci. Publishers, Singapore, 2005.

[477.] Embedding operators and boundary-value problems for rough domains, Intern. Jour. of Appl. Math. Mech., 1, (2005), 51–72. (with V. Gol'dshtein)

[478.] Integral operators basic in random fields estimation theory, Internat. Jour. of Pure and Appl. Math. (IJPAM), 20, No. 3, (2005), 405–427. (with A. Kozhevnikov)

[479.] Optimization methods in direct and inverse scattering, in the book: Continuous Optimization: Current Trends and Modern Applications, Springer, New York, 2005, pp. 51–110. (Editors: V. Jeyakumar and A. M. Rubinov). (with S. Gutman)

[480.] A new discrepancy principle, J. Math. Anal. Appl., 310, (2005), 342–345.

[481.] Modified Rayleigh Conjecture for static problems, Appl. Math. Lett., 18, N12, (2005), 1396–1399.

[482.] Numerical method for solving obstacle scattering problems by an algorithm based on the Modified Rayleigh Conjecture, Intern. Jour. Appl Math. Sci, 2, N1, (2005), 11–21. (with W. Chen)

[483.] Singular perturbation theory for a class of Fredholm integral equations arising in random fields estimation theory, Integral Equations and Operator Theory (IEOT), 53, N1, (2005), 107–126. (with E. Shifrin)

[484.] Inverse problems for parabolic equations, Austral. Jour. Math. Anal. Appl. (AJMAA), 2, N2, (2005), Article 10, pp. 1–5.

[485.] Dynamical systems method (DSM) and nonlinear problems, in the book: Spectral Theory and Nonlinear Analysis, World Scientific Publishers, Singapore, 2005, 201–228. (ed J. Lopez-Gomez).

[486.] **Random fields estimation,** World Sci. Publishers, Singapore, 2005.

[487.] Uniqueness of the solution to inverse obstacle scattering problem, Phys. Lett A, 347, N4–6, (2005), 157–159.

[488.] Dynamical systems method for nonlinear equations in Banach spaces, Communic. in Nonlinear Sci. and Numer. Simulation, 11, N3, (2006), 306–310.

[489.] Dynamical systems method and a homeomorphism theorem, Amer. Math. Monthly, 113, N10, (2006), 928–933.

[490.] A nonlinear singular perturbation problem, Asymptotic Analysis, 47, N1–2, (2006), 49–53.

[491.] Dynamical systems method (DSM) for unbounded operators, Proc. Amer. Math. Soc., 134, N4, (2006), 1059–1063.

[492.] Completeness of the set of scattering amplitudes, Phys. Lett. A, 360, N1, (2006), 22–25.

[493.] Modified Rayleigh conjecture method with optimally placed sources, Jour. of Appl. Functional Analysis, 1, N2, (2006), 223–236. (with S. Gutman)

[494.] A scheme for a stable numerical differentiation, Jour. Comp. Appl. Math, 186, N2, (2006), 325–334. (with U Jin Choi and Soyoung Ahn)

[495.] Existence of a solution to a nonlinear equation, Jour. Math. Anal. Appl., 316, (2006), 764–767.

[496.] The shape of the ear canal, Phys. Lett. A., 355, N4–5, (2006), 247–249.

[497.] Finding discontinuities of piecewise-smooth functions, JIPAM (Journ of Inequalities in Pure and Appl. Math.) 7, N2, Article 55, pp. 1–7 (2006).

[498.] Compactness of embeddings, Nonlinear Functional Analysis and Applications, 11, N4, (2006), 655–658.

[499.] **Dynamical systems method for solving operator equations**, Elsevier, Amsterdam, 2007.

[500.] Ill-posed problems with unbounded operators, Journ. Math. Anal. Appl., 325, (2007), 490–495.

[501.] Dynamical systems method (DSM) for selfadjoint operators, Jour. Math. Anal. Appl., 328, (2007), 1290–1296.

[502.] Two results on ill-posed problems, Internat. Journ. Appl. Math. and Statist., 11, N7, (2007), 136–139.

[503.] Inverse problems for parabolic equations 2, Communic. in Nonlinear Sci. and Numer. Simulation, 12, (2007), 865–868.

[504.] Iterative solution of linear equations with unbounded operators, Jour. Math. Anal. Appl., 330, N2, (2007), 1338–1346.

[505.] A Schrödinger singular perturbation problem, Communic. in Nonlinear Sci. and Numer. Simulation, 12, (2007), 1390–1394.

[506.] Electromagnetic wave scattering by many small particles, Phys. Lett. A, 360, N6, (2007), 735–741.
[507.] Materials with the desired refraction coefficients can be made by embedding small particles, Phys. Lett. A, 370, 5–6, (2007), 522–527.
[508.] Scattering by many small bodies and applications to condensed matter physics, Europ. Phys. Lett., 80, (2007), 44001.
[509.] Many-body wave scattering by small bodies and applications, J. Math. Phys., 48, N10, (2007), 103511.
[510.] Wave scattering by small particles in a medium, Phys. Lett. A 367, (2007), 156–161.
[511.] Wave scattering by small impedance particles in a medium, Phys. Lett. A 368, N1–2,(2007), 164–172.
[512.] A symmetry problem, Ann. Polon. Math., 92, (2007), 49–54.
[513.] Distribution of particles which produces a desired radiation pattern, Communic. in Nonlinear Sci. and Numer. Simulation, 12, N7, (2007), 1115–1119.
[514.] Computational method for acoustic wave focusing, Intern. Journ. Comp. Sci. and Math., 1, N1, (2007), 1–15. (with S. Gutman)
[515.] Distribution of particles which produces a "smart" material, Jour. Stat. Phys., 127, N5, (2007), 915–934.
[516.] Distribution of particles which produces a desired radiation pattern, Physica B, 394, N2, (2007), 253–255.
[517.] Invisible obstacles, Ann. Polon. Math., 90, N2, (2007), 145–148.
[518.] Many-body wave scattering by small bodies, J. Math. Phys., 48, N2, 023512, (2007).
[519.] Creating wave-focusing materials, Inverse Problems, Design and Optimization, IPDO-2007 Vol. II, (2007), pp. 687–690. (Ed. G. Dulikravich et al).
[520.] Distribution of particles creating "smart" material, International Journ. Tomog. Stat., 8, (2008), 25–31.
[521.] An inverse problem with data on the part of the boundary, Comm. Nonlin. Sci. and Numer. Simulation, 13, (2008), 534–538.
[522.] On unbounded operators and applications, Appl. Math. Lett., 21, (2008), 377–382.
[523.] Inverse scattering problem with data at fixed energy and fixed incident direction, Nonlinear Analysis: Theory, Methods and Applications, 69, N4, (2008), 1478–1484.
[524.] Creating wave-focusing materials, LAJSS (Latin-American Journ. of Solids and Structures), 5, (2008), 119–127.

[525.] Discrepancy principle for DSM II, Comm. Nonlin. Sci. and Numer. Simulation, 13, (2008), 1256–1263.

[526.] Solving ill-conditioned linear algebraic systems by the dynamical systems method (DSM), Inverse Problems in Sci. and Engineering, 16, N5, (2008), 617–630. (with N.S. Hoang)

[527.] Modified Rayleigh Conjecture method and its applications, Nonlinear Analysis: Theory, Methods and Appl., 68, (2008), 3884–3908. (with S. Gutman)

[528.] Electromagnetic wave scattering by many conducting small particles, J. Phys A, 41, (2008), 212001.

[529.] Dynamical systems method (DSM) for general nonlinear equations, Nonlinear Analysis: Theory, Methods and Appl., 69, N7, (2008), 1934–1940.

[530.] On stable numerical differentiation, Australian J. Math. Anal. Appl., 5, N1, (2008), Article 5, pp. 1–7. (with N.S. Hoang)

[531.] Fixed-energy inverse scattering, Nonlinear Analysis: Theory, Methods and Appl., 69, N3, (2008), 971–978.

[532.] Some results on inverse scattering, Modern Phys. Lett. B, 22, N23, (2008), 2217–2240.

[533.] A recipe for making materials with negative refraction in acoustics, Phys. Lett. A, 372/13, (2008), 2319–2321.

[534.] Does negative refraction make a perfect lens? Phys. Lett. A, 372, (2008), 6518–6520.

[535.] Electromagnetic wave scattering by small bodies, Phys. Lett. A, 372/23, (2008), 4298–4306.

[536.] Wave scattering by many small particles embedded in a medium, Phys. Lett. A, 372/17, (2008), 3064–3070.

[537.] Creating materials with desired properties, Mathem. Forschungsinst. Oberwolfach, report 58/2007, pp. 10–13. "Material Theories" Dec. 16–22, 2007.

[538.] A nonlinear inequality, Jour. Math. Ineq., 2, N4, (2008), 459–464. (with N.S. Hoang)

[539.] An iterative scheme for solving nonlinear equations with monotone operators, BIT Numer. Math. 48, N4, (2008), 725–741. (with N.S. Hoang)

[540.] Creating wave-focusing materials, Direct and Inverse Problems of Electromagnetic and Acoustic Wave Theory, 2008. DIPED 2008. 13th International Seminar/Workshop, pp. 31–37. ISBN: 978-966-02–4649-2

[541.] Identification of obstacles for parabolic problems, Internat. Jour. Tomog. Stat., 11, S09, (2009), 53–60. (with H. Heck)

[542.] Dynamical systems method for solving linear ill-posed problems, Ann. Polon. Math., 95, N3, (2009), 253–272.

[543.] Dynamical systems method for solving linear finite-rank operator equations, Ann. Polon. Math., 95, N1, (2009), 77–93. (with N.S. Hoang)

[544.] An inverse problem for a heat equation with piecewise-constant thermal conductivity, J. Math. Phys., 50, 063512 (2009). (with N.S. Hoang)

[545.] A DSM proof of surjectivity of monotone nonlinear mappings, Annal. Polon. Math., 95, N2, (2009), 135–139.

[546.] Finding the position of a small body in the presence of a large body from scattering data, Inverse Probl. in Sci. and Engineering (IPSE), 17, N5, (2009), 699–712. (with Y. Ol'shansky)

[547.] Some open problems in analysis, Austral Jour. Math. Anal. Appl. (AJMAA), (2009)

[548.] Asymptotics of the solution to Robin problem, J. Math. Anal. Appl., 349, N1, (2009), 156–164. (with H.-D. Alber)

[549.] Dynamical Systems Gradient method for solving nonlinear equations with monotone operators, Acta Appl. Math., 106, (2009), 473–499. (with N.S. Hoang)

[550.] A new version of the Dynamical Systems Method (DSM) for solving nonlinear equations with monotone operators, Diff. Eqns and Appl., 1, N1, (2009), 1–25. (with N.S. Hoang)

[551.] On the relation between the S-matrix and the spectrum of the interior Laplacian, Bull. Polish Acad of Sci. Mathem., 57, N2, (2009), 181–188.

[552.] Preparing materials with a desired refraction coefficient and applications, In the book "Topics in Chaotic Systems: Selected Papers from Chaos 2008 International Conference", Editors C. Skiadas, I. Dimotikalis, Char. Skiadas, World Sci. Publishing, 2009, pp. 265–273.

[553.] Preparing materials with a desired refraction coefficient, Nonlinear Analysis: Theory, Methods and Appl., 70, N12, (2009), e186-e190.

[554.] A discrepancy principle for equations with monotone continuous operators, Nonlinear Analysis: Theory, Methods and Appl., 70, (2009), 4307–4315. (with N.S. Hoang)

[555.] Attractors of strongly dissipative systems, Bull. Polish Acad of Sci. Mathem., 57, N1, (2009), 25–31.

[556.] Symmetry problems 2, Annal. Polon. Math., 96, N1, (2009), 61–64. (with N.S. Hoang)

[557.] Slow invariant manifolds for dissipative systems, J. Math. Phys., 50, N1, (2009), 042701.

[558.] A nonlinear inequality and applications, Nonlinear Analysis: Theory, Methods and Appl., 71, (2009), 2744–2752. (with N.S. Hoang)

[559.] Numerical solution of many-body wave scattering problem for small particles, Proc. DIPED-2009, Lviv, Ukraine, Sept. 21–24, (2009), pp. 77–81. (with M. Andriychuk)

[560.] A singular integral equation for electromagnetic wave scattering, Internat. Journ. Pure and Appl. Math., 55, N4, (2009), 7–11.

[561.] Boundary integral equation for electromagnetic wave scattering by a homogeneous body of arbitrary shape, Intern. Journ. Pure and Appl. Math., 55, N4, (2009), 13–16.

[562.] Inverse scattering with non-overdetermined data, Phys. Lett. A, 373, (2009), 2988–2991.

[563.] A collocation method for solving integral equations, Internat. Journ. Comp. Sci and Math., 3, N2, (2009), 222–228.

[564.] Creating desired potentials by embedding small inhomogeneities, J. Math. Phys., 50, N12, 123525, (2009).

[565.] Dynamical Systems Method for solving ill-conditioned linear algebraic systems, Internat. Journ. Comp. Sci. Math. (IJCSM), 2, N4, (2009), 308–333. (with S. Indratno)

[566.] Property C for ODE and applications to an inverse problem for a heat equation, Bull. Polish Acad of Sci. Mathem., 57, N3–4, (2009), 243–249.

[567.] An iterative method for solving Fredholm integral equations of the first kind, Internat. Journ. Comp. Sci. Math. (IJCSM), 2, N4, (2009), 354–379. (with S. Indratno)

[568.] Existence of solution to an evolution equation and a justification of the DSM for equations with monotone operators, Comm. Math. Sci., 7, N4, (2009), 1073–1079. (with N.S. Hoang)

[569.] Inversion of the Laplace transform from the real axis using an adaptive iterative method, Internat. Jour. Math. Math. Sci (IJMMS), Vol. 2009, Article 898195, 38 pages; doi:10.1155/2009/898195 (with S. Indratno)

[570.] Asymptotic stability of solutions to abstract differential equations, Journ. of Abstract Diff. Equations and Applications (JADEA), 1, N1, (2010), 27–34.

[571.] How to prepare materials with a desired refraction coefficient, Proceedings of ISCMII and EPMESCXII, AIP Conference Proceedings 1233, (2010), pp. 165–168.

[572.] Dynamical systems gradient method for solving ill-conditioned linear algebraic systems, Acta Applic. Math., 111, N2, (2010), 189–204. (with N.S. Hoang)

[573.] Collocation method for solving some integral equations of estimation theory, Internat. Journ. of Pure and Appl. Math., 62, N1, (2010), 57–65.

[574.] Dynamical systems method for solving nonlinear equations with monotone operators, Math. of Comput., 79, 269, (2010), 239–258. (with N.S. Hoang)

[575.] Implicit Function Theorem via the DSM, Nonlinear Analysis: Theory, Methods and Appl., 72, N3–4, (2010), 1916–1921.

[576.] Inverse problem for a heat equation with piecewise-constant conductivity, J. Appl. Math and Informatics (JAMI), 28, N3–4, (2010), 551–561. (with S. Gutman)

[577.] The Dynamical Systems Method for solving nonlinear equations with monotone operators, Asian Europ. Math. Journ., 3, N1, (2010), 57–105. (with N.S. Hoang)

[578.] Dynamical Systems Method (DSM) for solving equations with monotone operators without smoothness assumptions on $F'(u)$, J. Math. Anal. Appl., 367, N2, (2010), 508–515. (with N.S. Hoang)

[579.] Slow manifolds for dissipative dynamical systems J. Math. Anal. Appl., 363, (2010), 729–732.

[580.] Scattering by many small particles and creating materials with a desired refraction coefficient, International Journ. Comp. Sci. and Math. (IJCSM), 3, N1/2, (2010), 102–121. (with M. Andriychuk)

[581.] Electromagnetic wave scattering by many small particles and creating materials with a desired permeability, Progress in Electromag. Research, M, 14, (2010), 193–206.

[582.] On a new notion of the solution to an ill-posed problem, J. Comp. Appl. Math., 234, (2010), 3326–3331.

[583.] Creating materials with a desired refraction coefficient: numerical experiments, International Journ. Comp. Sci. and Math. (IJCSM), 3, N1/2, (2010), 76–101. (with S. Indratno)

[584.] Uniqueness theorem for inverse scattering problem with non-overdetermined data, J. Phys. A, FTC, 43, (2010), 112001.

[585.] DSM of Newton-type for solving operator equations $F(u) = f$ with minimal smoothness assumptions on F, International Journ. Comp. Sci. and Math. (IJCSM), 3, N1/2, (2010), 3–55. (with N. Hoang)

[586.] Electromagnetic wave scattering by many small bodies and creating materials with a desired refraction coefficient, Progress in Electromagnetic Research M (PIER M), 13, (2010), 203–215.

[587.] A nonlinear inequality and evolution problems, Journ, Ineq. and Special Funct., (JIASF), 1, N1, (2010), 1–9.

[588.] Numerical modeling in wave scattering problem for small particles, Proc. of MIKON-2010, 18-th Internat. Conf. on microwave radar and wireless communications, Geozandas Ltd, Vilnius, Lithuania, 2010, pp. 224–227. (with M. Andriychuk)

[589.] Uniqueness of the solution to inverse scattering problem with backscattering data, Eurasian Math. Journ (EMJ), 1, N3, (2010), 97–111.
open access Journal.

[590.] A method for creating materials with a desired refraction coefficient, Internat. Journ. Mod. Phys B, 24, 27, (2010), 5261–5268.

[591.] Materials with a desired refraction coefficient can be created by embedding small particles into a given material, International Journal of Structural Changes in Solids (IJSCS), 2, N2, (2010), 17–23.

[592.] Electromagnetic wave scattering by a thin layer in which many small particles are embedded, Progress in Electromagnetic Research Letters (PIER L), 19, (2010), 147–154.

[593.] A theorem on entire functions, Rev Roum Math Pure Appl., 55, N6, (2010), 515–519.

[594.] Justification of the Dynamical Systems Method (DSM) for global homeomorphisms, Eurasian Math. Journ (EMJ), 1, N4, (2010), 116–123.

[595.] Wave scattering by many small bodies and creating materials with a desired refraction coefficient, Afrika Matematika, 22, N1, (2011), 33–55.

[596.] Nonlinear differential inequality, Mathematical Inequalities and Applications (MIA), 14, N4, (2011), 967–976. (with N. Hoang)

[597.] Scattering by many small inhomogeneities and applications, In the book "Topics in Chaotic Systems: Selected Papers from Chaos 2010 International Conference", Editors C. Skiadas, I. Dimotikalis, Char. Skiadas, World Sci. Publishing, 2011. pp. 41–52.

[598.] Electromagnetic wave scattering by a small impedance particle of arbitrary shape, Optics Communications, 284, (2011), 3872–3877.

[599.] A collocation method for solving some integral equations in distributions, Journal of Computational and Applied Mathematics, 236, (2011), 1296–1313 (with S. Indratno)

[600.] Series that can be differentiated termwise m times if the function is m-smooth, Mathematica Aeterna, 1, N3, (2011), 137–148.

[601.] On the DSM version of Newton's method, Eurasian Math. Journ (EMJ), 2, N3, (2011), 91–99.

[602.] Wave scattering by many small bodies and applications, J. Math. Phys., 59, 023519, (2011) (with A. Rona)

[603.] Uniqueness of the solution to inverse scattering problem with scattering data at a fixed direction of the incident wave, J. Math. Phys., 52, 123506, (2011).

[604.] Some nonlinear inequalities and applications, Journ. of Abstract Diff. Equations and Applications, 2, N1, (2011), 84–101 (with N. Hoang)

[605.] Stability of solutions to some evolution problems, Chaotic Modeling and Simulation (CMSIM), 1, (2011), 17–27.

[606.] How large is the class of operator equations solvable by a DSM Newton-type method ? Appl. Math. Lett, 24, N6, (2011), 860–865.

[607.] Scattering of scalar waves by many small particles, AIP Advances, 1, 022135, (2011).

[608.] Uniqueness of the solution to inverse scattering problem with non-overdetermined data, Proceedings of the International Conference on Inverse Problems in Engineering, May 4–6, 2011, Orlando, Florida, USA, vol. 5, (2011), pp. 281–286.

[609.] Numerical solution of many-body wave scattering problem for small particles and creating materials with desired refraction coefficient, Chapter in the book:
"Numerical Simulations of Physical and Engineering Processes", InTech., Vienna, 2011, pp. 1–28. (edited by Jan Awrejcewicz)
ISBN 978-953-307-620-1 (with M. I. Andriychuk)
available online http://www.intechopen.com/articles/show/title/numerical-solution-of-many-body-wave-scattering-p

[610.] Scattering of electromagnetic waves by many thin cylinders, Results in Physics, 1, N1, (2011), 13–16.

[611.] On the DSM Newton-type method, J. Appl. Math. and Comp., (JAMC), 38, N1–2, (2012), 523–533.

[612.] **Dynamical Systems Method and Applications. Theoretical Developments and Numerical Examples.** Wiley, Hoboken, 2012, ISBN-13: 978-1-118-02428-7 (with N. S. Hoang)
[613.] Dynamical Systems Method (DSM) for solving nonlinear operator equations in Banach spaces, Eurasian Math. Journ (EMJ), 3, N1, (2012), 86–96.
[614.] Scattering of electromagnetic waves by many thin cylinders: theory and computational modeling, Optics Communications, 285, N20, (2012), 4019–4026. (with M. Andriychuk)
[615.] A variational principle and its applications, Internat. Journ. of Pure Appl. Math., 77, N3, (2012), 309–313.
[616.] DSM for general nonlinear equations, Appl. Math. Lett., 25, (2012), 2009–2014.
[617.] Stability of solutions to abstract evolution equations with delay, Journ. Math. Anal. Appl. (JMAA), 396, (2012), 523–527.
[618.] Electromagnetic wave scattering by a small impedance particle: theory and modeling, Optics Communications, 285, (2012), 1684–1691. (with M. Andriychuk and S. Indratno)
[619.] A problem in analysis, Analysis, 32, N2, (2012), 1001–1003.
[620.] Electromagnetic wave scattering by many small perfectly conducting particles of an arbitrary shape, Optics Communications, 285, N18, (2012), 3679–3683.
[621.] Large-time behavior of the weak solution to 3D Navier-Stokes equations, Appl. Math. Lett., 26, (2013), 252–257.
[622.] Convergence of time-dependent Turing structures to a stationary solution, Acta Appl. Math., 123, N1, (2013), 31–42. (with V. Volpert)
[623.] Inverse scattering problem for Maxwell equations, Math. Model. Nat. Phenom., 8, N1, (2013), 200–206.
[624.] Heat transfer in a medium in which many small particles are embedded, Math. Model. Nat. Phenom., 8, N1, (2013), 193–199.
[625.] Spectral properties of Schrödinger-type operators and large-time behavior of the solutions to the corresponding wave equation, Math. Model. Nat. Phenom., 8, N1, (2013), 207–214.
[626.] Symmetry problem, Proc. Amer. Math. Soc., 141, N2, (2013), 515–521.
[627.] Stability result for abstract evolution problems, Math. Meth. Appl. Sci., 36(4), (2013), 422–426.

[628.] Electromagnetic wave scattering by small impedance particles of an arbitrary shape, J. Appl. Math and Comput., (JAMC), 43, N1, (2013), 427–444.
DOI: 10.1007/s12190-013-0671-3

[629.] The Pompeiu problem, Global Journ. of Math. Analysis (GJMA), 1, N1, (2013), 1–10.
Open access Journal: http://www.sciencepubco.com/index.php/GJMA/issue/current

[630.] A variational principle and its application to estimating electrical capacitance of a perfect conductor, Amer. Math. Monthly, 120 (8), (2013), 747–751.

[631.] Stability of the solutions to evolution problems, Mathematics, 1, (2013), 46–64.
doi:10.3390/math1020046
Open access Journal: http://www.mdpi.com/journal/mathematics

[632.] Many-body wave scattering problems in the case of small scatterers, J. of Appl. Math and Comput., (JAMC), 41, N1, (2013), 473–500.

[633.] Wave scattering by many small bodies: transmission boundary conditions, Reports on Math. Physics, 71, N3, (2013), 279–290.

[634.] Scattering of electromagnetic waves by many nano-wires, Mathematics, 1, (2013), 89–99.
doi: 10.3390/math1030089.
Open access Journal: http://www.mdpi.com/journal/mathematics

[635.] **Scattering of Acoustic and Electromagnetic Waves by Small Bodies of Arbitrary Shapes. Applications to Creating New Engineered Materials**, Momentum Press, New York, 2013.

[636.] Asymptotic of some integral, Analysis, 33, (2013), 377–382.

[637.] Electromagnetic wave scattering by a small impedance body of an arbitrary shape, Proceedings of the XIX-th International seminar/workshop on direct and inverse problems of electromagnetic and acoustic wave theory (DIPED-2014), Tbilisi, Georgia, Sept. 22–25, 2014, plenary talk, IEEE, pp. 9–11.

[638.] Calculation of electromagnetic wave scattering by a small impedance particle of an arbitrary shape, Math. Meth. in Natur. Phenomena (MMNP), 9, N5, (2014), 254–269. (with M. Andriychuk)

[639.] Inverse scattering with under-determined scattering data, Math. Meth. in Natur. Phenomena, (MMNP), 9, N5, (2014), 244–253.

[640.] Recovery of the potential from I-function, Reports on Math. Phys., (ROMP), 74, N2, (2014), 135–143.

[641.] Application of the asymptotic solution to EM wave scattering problem to creating medium with a prescribed permeability, Journ Appl. Math. and Computing, (JAMC), 45, (2014), 461–485. (With M. Andriychuk)
doi: 10.1007/s12190-013-0732-7

[642.] Electromagnetic wave scattering by small impedance particles of an arbitrary shape and applications, Challenges, 5, (2014), 35–42.
doi:10.3390/challe5010035
Open access Journal: http://www.mdpi.com/journal/challenges

[643.] Electromagnetic wave scattering by small perfectly conducting particles and applications, J. Math. Phys., 55, 083505, (2014).

[644.] Creating media with prescribed permeability using the asymptotic solution to EM wave scattering problem, Mikon-2014, Gdansk, Poland, vol. 1, (2014), pp. 356–359. (with M. Andriychuk)

[645.] A symmetry result for strictly convex domains, Analysis, 35 (1), (2015), 29–32.

[646.] Existence and uniqueness of the global solution to the Navier-Stokes equations, Applied Math. Letters, 49, (2015), 7–11.
DOI: http://dx.doi.org/10.1016/j.aml.2015.04.008
http://authors.elsevier.com/sd/article/S0893965915001445
http://authors.elsevier.com/a/1R1KV3BGwepX9b (link to download)

[647.] Representation of vector fields, Global Jour. Mathematical Analysis (GJMA), 3 (2) (2015) 73–76.
open access: www.sciencepubco.com/index.php/GJMA; http://www.sciencepubco.com/index.php/GJMA/article/view/4577 doi: 10.14419/gjma.v3i2.4577

[648.] Inverse scattering on the half-line revisited, Reports on Math. Phys. (ROMP), 76, N2, (2015), 159–169.

[649.] A fast algorithm for solving scalar wave scattering problem by billions of particles, Jour. of Algorithms and Optimization, 3, N1, (2015), 1–13. (with N. Tran)
Open access: http://www.academicpub.org/jao/Issue.aspx?Abstr=false

[650.] When are the zero-energy solutions to the Schrödinger equation bounded at infinity? Jour. Math. Sci.: Advances and Applications (JMSAA), 33, (2015), 1–4.
Open access: http://scientificadvances.co.in

[651.] Existence of the solution to electromagnetic wave scattering problem for an impedance body of an arbitrary shape, Applied Math. Lett., 41, (2015), 52–55. (with M. Schechter)

[652.] Scattering of EM waves by many small perfectly conducting or impedance bodies, J. Math. Phys. (JMP), 56, N9, 091901, (2015).

[653.] A short proof of the existence of the solution to elliptic boundary problems, Global Journ. of Mathem. Analysis (GJMA), 3, issue 3, (2015), 105–108.
Open access: http://www.sciencepubco.com/index.php/GJMA/article/view/4731 doi: 10.14419/gjma.v3i3.4731

[654.] EM wave scattering by many small impedance particles and applications to materials science, The Open Optics Journal, 9, (2015), 14–17
Open access: http://benthamopen.com/TOOPTSJ/VOLUME/9/

[655.] Creating materials in which heat propagates along a line, Boll Union. Math. Ital. (BUMI), 8, N3, (2015), 165–168. published Sep.8, (2015) online DOI 10.1007/s40574-015-0033-1

[656.] Representation of big data by dimension reduction, Fundamental Journ. of Math. and Math. Sciences, 4, N1, (2015), 23–34. (with Cong Van)

[657.] Large-time behavior of solutions to evolution equations, Handbook of Applications of Chaos Theory, Chapman and Hall/CRC, 2016, pp. 183–200 (ed. C. Skiadas). ISBN 9781466590434-CAT# K20464

[658.] Integral equations and applications, Handbook of Applications of Chaos Theory, Chapman and Hall/CRC, (ed. C. Skiadas), pp. 163–182.

[659.] Antenna synthesis by the modulus of the diagram, Journal of Advances in Applied Mathematics (JAAM), 1, N1, (2016), 1–11. open access: http://www.isaac-scientific.org/images/PaperPDF/20002_20151221151344143l3.pdf (with M. Andriychuk)

[660.] Heat transfer in a complex medium, In the book "The Foundations of Chaos Revisited:
From Poincare to Recent Advances", Springer, 2016, pp. 119–136. ISBN 978-3-319 29699-9 (print); 978-3-319 29701-9 (online)

[661.] Uniqueness of the solution to inverse obstacle scattering with non-over-determined data, Appl. Math. Lett., 58, (2016), 81–86.

[662.] A simple proof of the closed graph theorem, Global Journ. Math. Anal. (GJMA), 4, N1, (2016), 1.
open access Journal. doi: 10.14419/gjma.v4i1.5534

[663.] Inverse obstacle scattering with non-over-determined data, 2016 International Conference on Mathematical Methods in Electromagnetic Theory, pp. 85–88.

[664.] EM Wave Scattering on a Set of Small Particles and Creation of Materials with Desired Refraction Coefficient and Magnetic Permeability, 2016 International Conference on Mathematical Methods in Electromagnetic Theory, pp. 410–413. (with M. I. Andriychuk)

[665.] Creating materials in which heat propagates along a line: theory and numerical results, Pure and Applied Functional Analysis, (PAFA), 2, N4, (2017), 639–648. (with Cong Tuan Son Van)
Open access Journal

[666.] Solution to the Pompeiu problem and the related symmetry problem, Appl. Math. Lett., 63, (2017), 28–33.

[667.] Inverse problems for parabolic equations with coefficient depending on time, Engineering Science Letters (ESL), 1,(2017), 1–4.
Open access: http://esl.scik.org/a-g-ramm-inverse-problems-for-parabolic-equations-with-coefficient-depending-on-time-2017-2017-article-id-1-20-september-2017/

[668.] Global existence and estimates of the solutions to nonlinear integral equations, Global Journal of Math. Analysis, 5(1), (2017), 19–20.
Open access: http://www.sciencepubco.com/index.php/GJMA/article/view/7306

[669.] A numerical method for solving 3D inverse scattering problem with non-over-determined data, J. Pure Appl. Math., 1, N1, (2017), 1–3.
open access Journal

[670.] **Scattering by obstacles and potentials**, World Sci. Publ., Singapore, 2017.

[671.] On the denseness of the set of scattering amplitudes, International Review of Physics, 11, N4, (2017), 96–98.

[672.] Perturbation of zero surfaces, Global Journal of Math. Analysis, 5(1), (2017), 27–28. http://www.sciencepubco.com/index.php/GJMA/article/view/7474

[673.] Global existence, uniqueness and estimates of the solution to the Navier-Stokes equations, Appl. Math. Lett., 74, (2017), 154–160.

[674.] **Creating materials with a desired refraction coefficient,** IOP Concise Physics, Morgan & Claypool Publishers, San Rafael, CA, USA, 2017.

[675.] Completeness of the set $e^{ik\beta\cdot s}$, Global Journ. of Math. Analysis (GJMA), 5(2), (2017), 43–44. https://www.sciencepubco.com/index.php/GJMA/article/view/7975 doi: 10.14419/gjma.v5i2.7975 open access journal

[676.] Finding a method for producing small impedance particles with prescribed boundary impedance is important, J. Phys. Res. Appl., 1:1, (2017), 1–3.
open access Journal.

[677.] Existence of the solutions to convolution equations with distributional kernels, Global Journal of Math. Analysis, 6(1), (2018), 1–2. open access Journal: https://www.sciencepubco.com/index.php/GJMA/article/view/8632/2983rs (ESL), (2018), 2018:2, 1–5.

[678.] On the importance of producing small impedance particles with prescribed boundary impedance, Engineering Science Letters (ESL), 2018, 2018:2, 1–5.

[679.] Inverse problem of potential theory, Appl. Math. Lett., 77, (2018), 1–5.

[680.] A uniqueness theorem for inverse scattering problem with non-overdetermined data, Engineering Science Letters (ESL), 2018, 2018:3, 1–5.

[681.] Many-body wave scattering problems for small scatterers and creating materials with a desired refraction coefficient, in the book "Mathematical Analysis and Applications: Selected Topics", Wiley, Hoboken NJ, 2018, Chapter 3, pp. 57–76. (ed. M. Ruzhansky, H. Dutta, R. Agarwal)

[682.] Inverse obstacle scattering with non-over-determined data, Global Journ. of Math. Anal. (GJMA), 6 (1), (2018), 2–6. https://www.sciencepubco.com/index.php/GJMA/article/view/8887

[683.] A numerical algorithm for solving 3D inverse scattering problem with non-over-determined data, J. Appl. Math. Stat. App., 2, N1, (2018), 11–13. (with Cong Van)
open access Journal http://www.alliedacademies.org/journal-applied-mathematics-statistical-applications/inpress.php

[684.] Solution of the Navier-Stokes problem, Appl. Math. Lett., 87, (2019), 160–164.

[685.] Global existence and uniqueness of the solution to a nonlinear parabolic equation, Journ. of Advances in Math., 14, N2, (2018), pp. 1–4.

[686.] Estimates of solutions to nonlinear evolution equations, Journ of Advances in Math., 14, N2, (2018), pp. 1–6.

[687.] Symmetry problem 1, Journ. of Adavances in Math. (JAM), 15, (2018), 1–4.

[688.] Necessary and sufficient condition for a surface to be a sphere, Open J. Math. Anal. (OMA), 2, (2018), issue 2, 51–52.
Open access: https://pisrt.org/psr-press/journals/oma/

[689.] Old symmetry problem revisited, Open Journ. Math. Analysis, (OMA), 2, N2, (2018), 89–92.

[690.] Inverse scattering with non-over-determined data, Journ. of Advances in Math., 16 (2019), pp. 1–4.

[691.] **Symmetry Problems. The Navier-Stokes Problem**, Morgan & Claypool Publishers, San Rafael, CA, 2019.

[692.] On the Navier-Stokes problem, Journ of Advances in Math., 16, (2019), pp. 1–5. ISSN 2347-1921
Open access Journal.

[693.] Global existence of solutions to nonlinear Volterra integral equations, Journal of Appl. Anal. (JAA), (2019), 45–47.
open access Journal

[694.] Symmetry problems for the Helmholtz equation, Appl. Math. Lett., 96, (2019), 122–125. https://doi.org/10.1016/j.aml.2019.04.008

[695.] **Inverse obstacle scattering with non-over-determined scattering data**, Morgan & Claypool Publishers, San Rafael, CA, 2019.

[696.] Global existence of solutions to differential equations, SEMA, 76, (2019), 625–628. https://doi.org/10.1007/s40324-019-00199-6

[697.] Estimating the size of the scatterer, ROMP (Reports on Math. Phys.), 85, N3, (2020), 331–334.

[698.] On a hyper-singular equation, Open Jour. of Math. Anal., 4(1), (2020), 8–10. https://pisrt.org/psr-press/journals/oma

[699.] **Creating materials with a desired refraction coefficient,** IOP Publishers, Bristol, UK, 2020 (Second edition).

[700.] Concerning the Navier-Stokes problem, Open J. Math. Anal. (OMA), 4(2), (2020), 89–92.
Open access: https://pisrt.org/psr-press/journals/oma/

[701.] How can one create a material with a prescribed refraction coefficient? Sun Text Review of Material Science, 1:1, (2020), 102.

[702.] On hyper-singular integrals, Open Journal of Math. Analysis, (OMA), 4(2), 2020, 101–103.
Open access: https://pisrt.org/psr-press/journals/oma/

[703.] Theory of hyper-singular integrals and its application to the Navier-Stokes problem, Contrib. Math. 2, (2020), 47–54.
Open access Journal: www.shahindp.com/locate/cm; DOI: 10. 47443/cm.2020.0041

[704.] Navier-Stokes equations paradox, Reports on Math. Phys. (ROMP), 88, N1, (2021), 41–45.

[705.] Symmetry problems in harmonic analysis, SeMA, 78, N1, (2021), 155–158.

[706.] Stability of solutions to some abstract evolution equations with delay, Contributions to Math., 3, (2021), 1–10. (with N. S. Hoang.)
open access: www.shahindp.com/locate/cm DOI: 10.47443/cm.2020. 0041

[707.] **The Navier-Stokes problem**, Morgan & Claypool publishers, 2021.
isbn 978163639243

[708.] Estimate of the size of a body from its scattering amplitude, Journal of inequalities and special functions, 12, N2, (2021), 12–15. (with N. S. Hoang.)

[709.] Symmetry problems for PDE, Symmetry, 2021, 13, 240.
open access http://doi.org/102290/sym13122240

[710.] Comments on the Navier-Stokes problem, Axioms, (2021), 10 (2), 95.
open access Journal, https://doi.org/10.3390/axioms10020095

[711.] Some symmetry results for PDE, J. Math. Stat. Res., 3, (2021), N3, pp. 1–2.
open access DOI: https://doi.org/10.36266/JMSR/152

[712.] Stability of solutions to nonlinear evolution problems, Bull. of Math. Anal. and Appl., 14, N1, (2022), 28–30.

[713.] Wave scattering by many small bodies and creating materials with a desired refraction coefficient, Univ. Journ. of Laser, Optics, Photonics and Censors, 2, N1, (2022), 62–73.

[714.] Wave scattering by many small impedance particles and applications, Reports on Math. Phys., (ROMP), 90, N2, (2022), 193–202.

[715.] On hyper-singular multidimensional equations, Far East Journal of Theoretical and Applied Sciences, Volume 1, 2022, Pages 1–4.
Open access, http://www.pphmj.com/journals/fjtas.htm
http://dx.doi.org/10.17654/TAS2022001

[716.] Applications of analytic continuation to tables of integral transforms and some integral equations with hyper-singular kernels, Open Journal of Optimization, (2022), 11, 1–6.
www.scirp.org/journal/ojop

[717.] Dirichlet problem with $L^1(S)$ boundary values, Axioms, 2022, 11, 371.
Open access, https://doi.org/10.3390/axioms11080371

[718.] When does a double-layer potential equal to a single-layer one?, Axioms, 7 (10), 19287–19291.
Open access, https://doi.org/10.3390/axioms

[719.] Is creating materials with a desired refraction coefficient practically possible? Characterization and Application of Nanomaterials, (2023), vol. 6, N1, 1–5.
Open access Journal

[720.] Boundary values of analytic functions, Far East Journal of Appl. Math., 116, N3, (2023), 215–227.
http://dx.doi.org/10.17654/0972096023011

[721.] A counterexample related to the Navier-Stokes problem, Far East Journal of Appl. Math., 116, N3, (2023), 229–236.

[722.] **Analysis of the Navier-Stokes problem. Solution of a Millennium Problem**, Springer, 2023.
isbn 978-3-031-30722-5

[723.] **Wave scattering by small bodies. Creating materials with a desired refraction coefficient and other applications**, World Sci. Publishers, Singapore, 2023.

[724.] New definition of singular integral operator, Annals of Communications in Mathematics, Volume 6, N4, (2023), 220–224
ISSN: 2582-0818 © http://www.technoskypub.com

[725.] The Navier-Stokes problem. Solution of a millennium problem related to the Navier-Stokes equations, Modern Mathematical Methods, 2 (2024), No. 1, pp. 19–26.
https://modernmathmeth.com/

[726.] Distributional boundary values of analytic functions, Annals of Communications in Mathematics, Volume 7, N1, (2024), 42–46.
open access

[727.] On the Laplace transform, Journ Appl. Analysis, 2024; 30(2): 209–213.

[728.] Solution to some hypersingular integral equations, Journ. of Mathematics and Statistics, 20, N1, (2024), 45–18.
open access

[729.] Dirichlet problem with $L^1(S)$ boundary values, Annals of Communications in Mathematics, Volume 7, N2, (2024), 108–113.
open access

[730.] Absence of positive eigenvalues of the Laplacian in domains with infinite boundaries, Annals of Communications in Mathematics, Volume 7, N3, (2024), 264–266.
open access

[731.] Solution to the millennium problem related to the Navier-Stokes equations, Lobachevskii Journal of Mathematics, 2024, Vol. 45, No. 8, 3726–3735.

[732.] Inverse obstacle scattering with non-overdetermined data, Annals of Communications in Mathematics, Volume 7, N3, (2024), 252–253.

[733.] On the Riemann problem, Annals of Communications in Mathematics, Volume 7, N4, (2024), 451–454.
open access

[734.] Materials with a desired refraction coefficient, Annals of Communications in Mathematics, 8, N1, (2025), 38–42.
open access

[735.] Study of nonlinear PDE with power nonlinearities, Open J. of Math. Anal., 9, N1, (2025), 11–13.
open access

[736.] Singular Integral Equations on a Set of Distributions, Journ. Comp Pure Appl Math, 3(1):1-03. doi: https://doi.org/10.33790/cpam110 0113.
open access

[737.] Uniqueness and non-uniqueness of the Radon transform, Poincare Journal of Analysis and Applications, Vol. 12, No. 1, (2025), 1–5.
open access: DOI: 10.46753/pjaa.2025.v012i02.001

[738.] Non-uniqueness of the solution to inverse scattering problem by many small scatterers, J. all Phys. Research and Appl, 1, N2, (2025), pp. 1–2.
open access

[739.] Nonlinear elliptic problems, Australian J. Math and Anal., 22, N1, (2025), Art. 11.

[740.] On the Fourier series,
open access

[741.] Numerical solution to inverse scattering problem, Jour. Comp. Pure Appl. Math., 3, N1, (2025), 1–5.
open access

[742.] Uniqueness result for a system of PDE, Ann. Math and Phys., 8, N2, (2025), 071–072.
open access

[743.] Wave-focusing materials, All Physics Research Appl., 1, N2, (2025), 1–2.
open access

[744.] Finding mass distribution from the exterior potential, Annals of Communications in Mathematics, 8, N2, (2025), 184–187.
open access

[745.] A uniqueness result for a coupled system of elliptic PDE, Ann. Math. and Phys., 8, N2, (2025), 071-972.
open access https://dx.doi.org./10.17352/amp.000148

Brief Description of the Work of Alexander G. Ramm

The work of Alexander G. Ramm can be divided into several areas:

(1) PDE, ODE and integral equations,
(2) spectral and scattering theory for differential operators, especially for the Schrödinger operators,
(3) static problems and wave scattering by small bodies of arbitrary shapes,
(4) random fields estimation theory,
(5) nonlinear passive systems,
(6) inverse scattering problems,
(7) theoretical numerical analysis and ill-posed problems,
(8) non-selfadjoint operators and their applications in scattering theory,
(9) signal and image processing,
(10) local tomography,
(11) mathematical geophysics,
(12) electromagnetic theory and mathematical physics,
(13) creating materials with a desired refraction coefficient,
(14) symmetry problems for PDE,
(15) the Navier-Stokes problem in R^3,
(16) integral equations with hyper-singular kernels.

The breadth and volume of the work do not allow one to describe the work in detail. Therefore only the highlights will be mentioned.

(1) In a long series of papers starting with papers 3–7, 11, 118, 144, 190, 670, 730, (the numbers cited are from Alexander G. Ramm (AR) list of publications), a thorough study of the spectral properties and eigenfunction expansions is given for the first time for Schrödinger

operators in domains with infinite boundaries; a sufficient condition is given on the infinite boundary for the Schrödinger operators to have no positive eigenvalues on the continuous spectrum.

(2) Iterative methods are developed for solving interior and exterior boundary value problems for Laplace's equation, analytic formulas for the S-matrix for acoustic and electromagnetic wave scattering by small bodies of arbitrary shapes are derived and applied successfully to numerical and physical problems (see monograph 144, 612);

(3) analytic theory of random fields estimation is developed (monographs 246, 486), which is an original detailed study of a new class of multidimensional integral equations basic in estimation theory. No results of this type have been known. Many results known for one-dimensional estimation theory are very particular cases of the general theory developed in the monograph 486. The theory has many application in signal processing, and in geophysics in particular. Monograph 246 was translated into Russian by MIR publishing house in 1996.

(4) In the pioneering papers 72 and 80 (also 125, 128, 137, 142) the mathematical foundations of the EEM and SEM methods are given. These methods are now very popular in electrical engineering sciences.

This research was supported by AFOSR from 1979 till 1983;

(5) A thorough study of existence, global stability and calculation of the stationary regimes in passive nonlinear systems is given in paper 129. The results are optimal as shown by examples.

(6) A deep study of inverse scattering problems is given in a long series of papers(see monographs 190, 313, 470, 670 and papers 252, 425, 460, where a summary of some of the author's results is given. In papers 584, 589, 603, and in monograph 670 the problem which has been open for many decades is solved: uniqueness of the solution to the basic non-over-determined inverse scattering problems is proved by the author.

Exact inversion of the low-frequency scattering data is given in the monographs 190, 670.

A powerful method, Property C method, based on the notion of completeness of the set of products of solutions of PDE is developed and applied to many important inverse problems, 407, 470, 670. In these works several problems are solved which have been open for decades. For example, the first global uniqueness theorems in geophysics and potential scattering with fixed-energy data are obtained, the first mathematically justified method for solving the 3D inverse scattering problem with noisy fixed-energy data is given. and for the first time stability estimates for the solution to the inverse scattering problem with noisy fixed-energy data are obtained, 313, 470, 670.

The first variational principle for solving inverse scattering problems which is equivalent to the inverse problems was found; this work is published as a monograph 313, which is an expanded version of monograph 278, translated into Russian in 1994. In paper 393 a fundamentally new uniqueness theorem is obtained: it says that a compactly supported real-valued square-integrable spherically symmetric potential is uniquely defined by any part of the fixed-energy phase shifts with the angular momenta j running through an arbitrary set J of non-negative integers such that $\sum_{j\in J, j\neq 0} \frac{1}{j} = \infty$.

Property C is defined and proved for ordinary differential equations (ODE) and its many new applications are demonstrated. Most of the known results for one-dimensional inverse problems are obtained by using this property, and many new results (387, 402, 470, 670). Among the classical results which are obtained by using property C for ODE are Marchenko and Borg's uniqueness theorems concerning recovery of the potential from two spectra (Borg) and from scattering data or spectral function (Marchenko).

Inverse problems for an inhomogeneous Schrödinger equation are studied for the first time (391, 413), a non-over-determined three-dimensional inverse problem of recovery of a potential from the diagonal values of the spectral function known on the boundary of a bounded domain and all real values of the spectral parameter is considered and a uniqueness theorem is proved for this problem (412).

A new approximate method for solving the inverse scattering problem with fixed energy data is given for a spherically symmetric potentials which are known for $r > a$ but unknown for $r < a$, where $a > 0$ is an arbitrary large fixed number (394). Numerical results are obtained by this method (400).

Krein's method in inverse scattering is justified and its consistency is proved for the first time, (405).

Analytical theory is given for inversion of the surface scattering data in the ground-penetrating radar problem for two functions: permittivity and conductivity of the ground, under the assumption that these functions depend on the vertical coordinate only (367, 380, 398).

A method for recovery of a quarkonium system from experimental data is developed (375).

Inverse problem of finding point scatterers from the surface scattering data is posed and solved (395, 411).

For the first time uniqueness theorems are proved for three-dimensional scattering problems with non-overdetermined data (papers 562, 584, 589, 603).

Stability of the Pompeiu property is established (363) and further results are obtined (382, 535). A complete solution to the Pompeiu problem, a proof of the Schiffer's conjecture and a general result on the symmetry problem for the Helmholtz equation are obtained in the monograph (691) and numerous papers of the author on this topic. First results on symmetry properties for harmonic analysis are in paper 705. In paper 709 many of the author's symmetry results for PDE are derived.

In a series of papers, starting with (506 and cited in the monographs 635 and 699 a method for constructing "smart materials" is given. It is proved that one can distribute small particles in a bounded domain so that the resulting material has a desired refraction coefficient or the a priori chosen radiation pattern (wave-focusing property).

In paper 632 theory of scalar wave scattering by one and many small bodies of an arbitrary shape is developed for various boundary conditions (Dirichlet, Neumann, impedance, transmission). In paper 628 theory of EM (electromagnetic) wave scattering by one and many small impedance bodies of an arbitrary shape is developed. Methods for creating materials with a desired refraction coefficient are given on the basis of the above theory. These results and their generalizations are presented in monographs 635, 674, 699.

(7) Mathematical justification of the widely used T-matrix approach in scattering theory is given (monograph 190).

In a series of papers (starting with 506 and cited in the monographs 635 and 699) several ill-posed problems are investigated. In particular, the now widely used stable differentiation procedure based on the regularization by the choice of the step size in the divided difference formula has been introduced for the first time in paper 32.

The important feature of this and my other works on ill-posed problems is the error estimates with explicitly written estimation constants.

A theory for the stable solution of a class of the Fredholm equations at a characteristic value is constructed in several papers and presented systematically in the monograph 144. This theory was a basis for the theory of wave scattering by small bodies of arbitrary shapes in this monograph.

Numerical methods were given for solving integral equations of estimation theory in distributions. This theory is summarized in the monograph 246. The basis of it is a theory, developed by the author, of a class of multidimensional integral equations whose kernels are kernels of positive rational functions of arbitrary self-adjoint elliptic operators.

In a series of papers (452, 454, 456, 457, 469, 485, 491, 500–502, 522, 525, 539, 542–545, 549–550, 554, 567, 574, 575, 579, 581), some of which are joint with Ramm's Ph.D students, and in monographs 499, 612, a general method, Dynamical Systems Method, (DSM), for treating linear and, especially, nonlinear ill-posed problems by solving a suitable Cauchy problem in a Hilbert space was developed. Convergence theorems are proved. Discretization of the Cauchy problem leads to a variety of iterative methods for solving ill-posed nonlinear problems and convergence theorems for these methods are obtained. In monograph 612 these results are illustrated by numerical examples. A novel approach to solving exterior and interior boundary value problems and scattering problems, based on the theorem, proved by A.G. Ramm and called by him Modified Rayleigh Conjecture (MRC), has been developed and tested numerically (papers 430, 461, 475, 481, 493).

(8) The theory of weakly non-selfadjoint operators was applied to scattering theory (50, 72, 80). For the first time completeness of the set of root vectors of some non-selfadjoint integral operators arising in diffraction and scattering theory was proved. This gave a mathematical justification of the EEM (eigenmode expansion method), a popular method in electrical engineering;

(9) A.G. Ramm (jointly with his Ph.D student A. Katsevich) developed new methods in signal and image processing, edge detection, local tomography; a very general test of randomness against fairly broad alternatives is found and justified mathematically (348).

New methods were developed for finding jumps of functions from local tomographic data. These methods turned to be practically important.

These results were tested numerically and practically and demonstrated their effectiveness. Monograph (348) contained these results.

Two patents (5,539,800 of July 23, 1996 and 5,550,892 of Aug. 27, 1996) have been issued by the US Patent Office to A.G. Ramm and A.I. Katsevich "Enhanced local tomography" and "Pseudolocal tomography".

(10) A systematical study of the singularities of the Radon transform is given, a complete description of the asymptotics of the Radon transform near a point of its singular support is obtained and applied to the important problem of tomography: finding singularities of a function from its tomographis data; these results are published in a series of papers and appeared in the monograph 348, see also paper 414.

(11) The basic uniqueness theorems for model inverse problems of geophysics have been proved, examples of non-uniqueness were constructed, the theory of inversion of low-frequency data has been developed (monographs 190, 278 and 670).

(12) Theoretical investigation of a number of antenna synthesis problems, including a non-linear synthesis problems have been investigated. Degree of non-uniqueness of the solution to the general synthesis problem has been described (monograph 118, 167). There are many other results of various nature and in different branches of mathematics (general relativity, asymptotics of the spectra of linear operators and quadratic forms, approximation theory, variational estimates of capacitances and polarizabilities, methods for calculation of resonances in open systems and quantum mechanics, perturbation theory for resonances, impedance tomography, singular perturbation of integral equations, quantum chaos, etc. The characteristic features of the works is a systematic usage of functional analysis and classical analysis,numerical methods, PDE, physics and theoretical engineering and their combinations. Broad interests made it possible to interact with mathematicians and engineers with quite diverse interests.

(13) In 2007–2025 A.G. Ramm has published a series of papers (506–511, 513–516, 518–520, 523, 533, 536, 537, 540, 552, 553, 564, 590, 595, 597, 622, 632, 634, 649, 652, 654, 655, 660,719, 734, 743, and in monographs 635, 674, 699, 723) in which he has developed a method for creating materials with a desired refraction coefficient. This method is based on Ramm's solution to many-body scattering problem by many small particles embedded in an inhomogeneous medium. The refraction coefficient can be created so that the new material has a desired wave-focusing property, or it may have a negative refraction property, which

means that the group velocity in this material is directed opposite to the phase velocity. These results are presented in monographs 635 and 699. The important new mathematical problem (on which this theory is based) is the many-body wave scattering problem for small bodies. AGR solved this problem asymptotically, as $a \to 0$ under the assumption $a \ll d \ll \lambda$. Here a is the characteristic size of a small body, d is the minimal distance between neighboring bodies, and λ is the wavelength. The multiple scattering is essential under these assumptions. The a can be as small as $20nm$. These results will be immediately applicable practically if small impedance particles with a desired refraction coefficient can be produced in practice. **Using this theory AGR gave a recipe for creating materials with a desired refraction coefficient.**

(14) In 2017–2019 A.G. Ramm was working on symmetry problems for PDE. His new results, including the proof of the Schiffer's conjecture and a solution to the Pompeiu problem are presented in the monograph 691, papers 694, 705, 709, and the author's papers cited there.

(15) A.G. Ramm has solved **the millennium Navier-Stokes problem in R^3**. His solution is published in papers 704, 725, 731, and in monographs 707 and 722. He proved that the Navier-Stokes problem (NSP) is contradictory and has no solution.

This follows from **the NSP paradox**, proved by A.G. Ramm in paper 704 and in the monographs 707, 722).

NSP paradox: *if the initial data $v(x,0) \not\equiv 0$, the solution to the NSP exists for all $t > 0$ and the exterior force $f = 0$, then $v(x,0) = 0$.*

(16) In 2017–2019 A.G. Ramm has proved for the first time uniqueness of the solution to the inverse scattering problem for compactly supported potentials and *non-over-determined* scattering data. These results are published in monograph 670 and in the author's papers cited there, in particular, in 584, 589, 603. His theory includes a proof of uniqueness of the solution to inverse obstacle scattering problem with *non-over-determined* data. These results are presented in papers 682, 690 and in monograph 670, 695.

(17) In 2018–2022 A.G. Ramm has developed a theory for solving convolution integral equations with hyper-singular kernels. These results are presented in papers 698, 715, 716, 728, 736, and in the monographs 707 and 722.

Prof. Ramm is elected a member of Electromagnetic Academy, MIT, (June 1990), a member of New York Academy of Science, he is an associated editor of many professional Journals.

He received many other honors.

Ramm has directed 11 Ph.D students, some of them are now professors at various Universities.

His research has been supported by AFOSR, NSF, ONR, NATO, USIEF, Fulbright committee, SERC of Canada and United Kingdom, DFG, Research Councils of Italy and many Universities in Europe, and Asia.

He was a guest of Academia Sinica in Taipei, Beijing, Shanghai and Hefei, of Indian Institute of Science in Bangalore, of Royal Institute of Technology (Sweden), of the Universities of Bonn, Heidelberg, Stuttgart, London, Manchester, Leicester, Paris, Florence, Rome, Madrid, Göteborg, Uppsala, Marcel, Novosibirsk, Milan, Cagliari, University of Mexico UNAM, University of Grenoble, Technion in Haifa, Tokyo Metropolitan University, Kyoto University, and gave lectures at many other Universities throughout the world.

A.G. Ramm worked as a research consultant for Los Alamos National Laboratory, oil industries and electronics industries. In 1997 he was awarded a Commerce Bank distinguished graduate faculty member research award.

Professor A.G. Ramm was an invited Distinguished Foreigh Professor of the Academy of Science of Mexico in October 1997 and gave lectures at UAM and UNAM in Mexico City. He was distinguished foreign professor at the University of Cairo in 2004 and 2006, Mercator Professor at TU Darmstadt in 2007, Invited Plenary Speaker at 7-th PACOM in 2009, Distinguished Visiting Professor invited by Royal Acad. of Engineering UK in 2009, Visiting Professor at IMPAN, 2010, MPI (Max Planck Institute) in 2011, Beijing Institute of Technology (BIT) in 2013. Fulbright Research Professor in Israel (Technion) in 1991–1992, in Ukraine (University of Lviv) in 2015.

Professor A.G. Ramm has taught a wide variety of various courses at all levels. He worked with MS and Ph.D. students and has experience in designing curriculum for courses in ODE, PDE, Applied functional analysis and theoretical numerical analysis, Integral transforms and applications, Tomography and the Radon transform, as well as the standard calculus sequence. His lectures are well organized, clear, and adjusted to the level of the audience.

Professor A.G. Ramm had worked much with engineers both from academia and industries and because of his broad knowledge of the basic mathematical and physical sciences and theoretical electrical engineering, he is able to communicate easily with engineers. He has always supported close connections between the mathematics department and engineering school, and was interested in developing the relations with industries.

Ramm has received Distiguished Graduate faculty award (1996).

He has received Khwarizmi Intenational Prize for mathematical research (2004).

Ramm was a Distinguished Visiting professor supported by the UK Royal Academy of Engineering in Sep.–Oct. 2009. He was a Mercator Professor in 2007, Distinguished HKSTAM speaker (2005), London Math. Society speaker (2005), research CNRS professor in France (2003), Distinguished Visiting Professor at the University of Cairo (2004), (2006), CNRS Professor (2003), Distinguished Foreign Professor at the Academy of Science of Mexico (1997).

Index

www.ingramcontent.com/pod-product-compliance
Ingram Content Group UK Ltd.
Pitfield, Milton Keynes, MK11 3LW, UK
UKHW022109170926
8624IPUK00024B/313

9 789819 823482